Micro and Small Enterpris[es in] Developing Coun[tries]

A Challenge for Sustainability in Colombia

ACADEMISCH PROEFSCHRIFT

ter verkrijging van de graad van doctor
aan de Universiteit van Amsterdam
op gezag van de Rector Magnificus
prof. dr. D.C. van den Boom
ten overstaan van een door het College voor Promoties ingestelde
commissie, in het openbaar te verdedigen in de Agnietenkapel
op woensdag 21 januari 2015 te 12:00 uur

door

Gloria Ana Maria Monica Sanz Galindo

geboren te Bogotá, Colombia

The doctoral research was supported by a research grant from Colfuturo Institution, financing postgraduate studies abroad, at earlier stages in France and carried out within the framework of the European research project SWITCH (Sustainable Urban Water Management Improves Tomorrow's City's Health). SWITCH is supported by the European Commission under the 6th Framework Programme and contributes to the thematic priority area of "Global Change and Ecosystems" [1.1.6.3] Contract n° 018530-2. It run from years 2006-2011 and was lead by UNESCO-IHE and by the Universidad Nacional, IDEA institute in Bogotá, Colombia.

CRC Press/Balkema is an imprint of the Taylor & Francis Group, an informa business

Published by:

CRC Press/Balkema

PO Box 11320, 2301 EH Leiden, The Netherlands

e-mail: Pub.NL@taylorandfrancis.com

www.crcpress.com – www.taylorandfrancis.com

ISBN 978-1-138-02769-5 (Taylor & Francis Group)

Promotiecommissie:

Promotor:	Prof. dr. J. Gupta
Copromotores:	Dr. R. Ahlers
	Dr. M.A. Siebel
Overige Leden:	Prof. dr. I.S.A. Baud
	Prof. dr. J.M. Baud
	Dr. M.A.F. Ros-Tonen
	Prof. dr. ir. P. van der Zaag
	Prof. dr. M.Z. Zwarteveen
Faculteit:	Faculteit der Maatschappij- en Gedragswetenschappen

Acknowledgements

Working on this research thesis was always an insightful and challenging endeavour. Complexity was its characteristic at all levels.

Using this space to mention the people and institutions that made this work possible is only a small acknowledgement of the overwhelming support that they have provided.

First and foremost, I wish to thank my supervisory team- Prof. Dr Joyeeta Gupta, Dr Maarten Siebel and Dr. Rhodante Ahlers. I am grateful that they were supportive of my research idea that entailed huge difficulties at a time when Action Research was seen with suspicion by many academics. Prof Joyeeta, thank you for your valuable teachings, guidance and especially for your patience through this long journey; for believing that the poorest of the poor can be empowered. Maarten, I express my immense gratitude because even though this thesis took distance from an exclusive technical approach to CP implementation, you always believed in the idea of building solutions for the smallest scale industries. You allowed me to get financial support through the EU project SWITCH. You made this dream possible. Rhodante, when you became part of this enterprise, the whole work made sense. Your open-mindness and methodological knowledge built bridges between the social and the technical academic worlds. You supported my soul during the hardships without renouncing to the search for academic excellence. I feel privileged for being one of your students. My everlasting gratitude because I found, in all of you, strong commitment to high values of society.

I am grateful to the University of Amsterdam (UvA) and to the Institute for Water Education (UNESCO-IHE) for hosting me, and to all the colleagues and friends at UNESCO-IHE, who were extremely cooperative, friendly and helped me feel at home when I was at the institute. Special thanks to Jolanda Boots, Maria Laura Sorrentino, Guy Beaujot and Dennis Thijm. I am also grateful to Chris Holtslag for his kindness and enormous support with all the tables and figures.

Funding was critical during the first steps in The Netherlands. I had started my research work in France with financial assistance from Colfuturo and two years passed-by before some financing was obtained with the work on conflict resolution from the Chamber of Commerce of Bogotá. My gratitude to Jerónimo Castro, Chief Executive of Colfuturo who has been for more than thirty years like my big brother. I would like to acknowledge the essential aid received from the SWITCH project. Working on this project was a pleasurable journey, full of wonderful experiences and substantial learning. I would like to especially thank Carol Howe and John Butterworth for their assistance throughout it. They were especially warm and supportive to the tanners during their visits to Villapinzón.

Back home, my profound gratitude to the "father" of the environmentalism in Colombia, Prof. Julio Carrizosa. With his understanding of the Colombian complexity, he opened for me the doors of the National University in order to be able to implement the SWITCH project. In this university I found dedicated colleagues who adopted as their own the idea of working with the tanners. My sincere thanks to Laura Osorio and Tomás León and to the entire interdisciplinary team of wonderful professionals.

The fieldwork was without doubt the most interesting part of the research. I am particularly grateful to the people that backed-me up with the idea of empowering a community that was considered without having the will to change. Ana Dorly Jaramillo with her group of tanners from Cerrito Valle was my first support. Her dedication and legitimate commitment to those marginalized inspired me and gave me strength to pursue my goal. Like Ana Dorly, other "champions" took the lead in their own contexts: Nancy Patricia Gutiérrez at the Senate, Camilo Ospina at the Presidency, Nelly Yolanda Villamizar at the Judicial Court, Diana Moreno at the Ministry of the Environment, Carlos Manrique who passed away, together with Paula Samper and Esteban Restrepo as committed lawyers, and my long time friend Carlos Enrique Cavelier as a conscious business leader.

My admiration and esteem goes to the community of tanners who believed they could change their fate. Their strong spiritual beliefs and endurance inspired me. Despite the fact that they are aware that their hardships have not come to an end, they are convinced they are building a better future for their descendants. Evidalia, Don Manuel, Don Paulino, Nelson, Don Arturo…you are the best citizens this country could have.

Finally, my beloved family. Without their constant motivation and unwavering support I would have not been able to bring this endeavour to completion. I am grateful to my parents who made me aware of my responsibilities as a privileged member of our society. Their example has always inspired me. I especially think about my grandmother Maria Luisa. She inspired me with her passion for nature and for social concerns. She was a real environmentalist ahead of her time. I am fulfilling her dreams through this thesis and keeping my promise to her to struggle for a better future. I thank my in-laws for giving me a wonderful husband. My father-in-law, who is no longer with us, would have been very proud to see this thesis. He always supported social justice and academic excellence.

My deepest gratitude to my daughter Maria Luisa and my two sons, Alejandro and Emilio. They have seen a mother always studying and struggling. They have lived through difficult times and have always supported me. It is time that their mother is not any more a computer biologist!!

Last but not the least, I thank my husband Ricardo immensely. You and I have walked together through happiness, sadness, crises, sickness, prosperity and hardships. You did not allow me to abandon this dream that started in Canada at the time we decided to come back to Colombia to work for a better future in search for equality. This thesis is ours, not mine.

Finally, I would like to dedicate this thesis to Evidalia Fernández. Despite the lack of opportunities in her up-bringing, her strong values, endurance, perseverance, and resilience have made her a real leader in Colombia. I thank you for all your teachings. Brave hearts like you are needed in our country.

Thank you all

Mónica Sanz

ABBREVIATIONS AND ACRONYMS

AAUs	Urban Environmental Authorities
ACOPI	Colombian Association for the Small and Medium Size Enterprise
ADR	Alternative Dispute Resolution
AI	Appreciative Inquiry
AoI	Agents of Inclusion
APP	Public Private Alliances
AR	Action Research
ARP	Action Research Partnership
ASEAN	Association for the South East Asian Nations
BAT	Best available technology
BATNA	Best Alternative to a Non-Negotiation Agreement
BOD	Biochemical Oxygen Demand
CAR	Autonomous Regional Corporation of Cundinamarca
CARs	Autonomous Regional Corporations
CECODES	Colombian Enterprise Council for the Development
CNPML	National Centre of Cleaner Production
COD	Chemical Oxygen Demand
CP	Cleaner Production
CR	Conflict Resolution
CRPML	Regional Centre for Cleaner Production of Valle del Cauca
CVC	Autonomous Regional Corporation of Valle del Cauca
DAMA	Environmental Administrative Department of the Bogotá District
DANE	National Department of Statistics
DNP	National Department of Planning
EAAB	Enterprise of Water and Sanitation of Bogotá
EIP	Eco-Industrial Park
GDP	Gross Domestic Product
GINI	Income distribution of a country's residents
GNP	Gross National Product
GOP	Good operational practices
GWP	Global Water Partnership
IDEAM	National Institute of Hydrology, Meteorology and Environmental Issues
IIFT	Indian Institution of Foreign Trade
INDERENA	National Institute of Natural Resources Development
IPCC	Intergovernmental Panel on Climate Change
IWRM	Integrated Water Resources Management
MADS	Ministry of Environment and Sustainable Development
MASL	Meters above sea level
MCIT	Ministry of Commerce Industry and Tourism

McSE	Micro and Small Enterprise
Micro-case	Continuous data collection process to strategic actors
MLMW	Minimum Legal Monthly Wages
N	Negotiation
NBI	Unsatisfied Basic Needs
NCPC	National Cleaner Production Centre
OECD	Organization for Economic Cooperation and Development
OST	Open Space Technology
PAR	Participatory Action Research
PAT	Three-Year Action Plan
PGAR	Ten-Year Regional Environmental Management Plan
PI	Policy Instruments
POIAs	Annual Investment Operating Plans
POMCAs	Watershed Administration and Management Plans
POT	Territorial Development Program
SASI	Systematic Approach for Social Inclusion
SBA	Small Business Administration
SD	Sustainable Development
SDA	District Environmental Secretariat
SINA	National Environmental System of Governance
SME	Small and Medium Enterprise
SP	Stakeholder Participation
SPA	Social Process Approach
SWITCH	Sustainable Water in Tomorrow's Cities Health
TSS	Total Suspended Solids
UNAL	National University
UNCTAD	United Nations Conference on Trade and Development
UNDP	United Nations Development Program
UNEP	United Nations Environmental Program
UNIANDES	University of Los Andes
UNIDO	United Nations Industrial Development Organization
USEPA	United States Environmental Protection Agency
USITC	United States International Trade Commission
WB	World Bank
WHO	World Health Organization
WRM	Water Resources Management
WWAP	World Water Assessment Program
ZOPA	Zone of Possible Agreement

Executive Summary

The problem: The relationship between environmental agencies and polluters is especially complex in the context of micro and small sized enterprises (McSEs) in developing countries. McSEs employ 1-49 employees and are a subset of small and medium sized enterprises (SMEs) (1-249 employees). They share vulnerability with respect to access to technology, markets, credit, and level of education. They have limited monitoring, a low level of capacity to work through collective strategies and are above all mostly informal (Alvarez and Durán, 2009). McSEs face the paradoxical situation of having the potential to be the driving force behind equitable economic and industrial development while being extremely vulnerable because of their limited access to educational, legal, technological and financial opportunities (Blackman, 2011; UNCTAD, 2008; UNIDO, 2009; 2005; Ashton *et al.*, 2002). They also cause substantial collective environmental harm (Blackman, 2010; Baas, 2007; Van Berkel, 2007; Blackman, 2006; Siaminwe *et al.*, 2005; Frondel *et al.*, 2005). In fact, small-scale industries usually employ environmentally unfriendly obsolete processes. They may collectively be responsible for 70% of all local and global industrial pollution (Le Van Khoa, 2006; Soni, 2006; Hillary, 1997). Environmental agencies in developing countries (especially in South America) pay little attention to pollution prevention policies, barely cope with control and pollution remediation and focus mainly on end-of-pipe solutions that entail investments that are usually unaffordable for McSEs (Thabrew *et al.*, 2009; Van Berkel, 2007; Montalvo, 2003). The literature points to three issues:

(a) The use of regulatory, market and persuasive policies bypass McSEs because of their specific characteristics, and the only possibilities left to deal with McSEs are legal actions (Blackman, 2010; Sanchez-Triana *et al.*, 2007; Delli Priscoli, 2003). This leads to lawsuits against the polluters as well as the defaulting environmental authorities.

(b) The implementation of Cleaner Production (CP) in McSEs in developing countries has, in general, been ineffective (Blackman, 2011; 2010; 2006; 2008; 2000; Montalvo and Kemp, 2008; Baas, 2007; Ashton *et al.*, 2002; Blackman *et al.*, 2012; 2009; 2007a; 2007b; 2006) even though CP has high potential benefits for industries. Cleaner Production strategies reduce emissions and waste, and promote cost savings due to reduced resource use and reduced waste generation (Montalvo and Kemp, 2008). The mainstream approach to CP, based on quantitative approaches that correlate data but not context, can ignore crucial aspects related to the complexities of McSEs (Del-Río-González, 2009; Howgrave-Graham and Van Berkel, 2007). This approach has not focused on a profound understanding of their specific contexts and has not recognized specific needs (Van Berkel, 2011; 2010; 2007; Baas, 2007; Sánchez-Triana *et al.*, 2007; Mitchell, 2006; Ashton *et al.*, 2002; Baskerville, 1997), nor focused on how mechanisms support CP implementation and, hence, foster change; to build consensus; and how the choice of technology is being implemented (Montalvo and Kemp, 2008; Baas, 2007; Montalvo, 2003).

(c) Although stakeholder participation may support the management of environmental problems, it does not guarantee successful problem solving and lacks effective tools for consensus building and decision-making (Estacio and Marks, 2010; Hirsch *et al.*, 2010; Jacobs, 2010; Reed *et al.*, 2009; Terry and Khatri, 2009; Wiber et al., 2009; Ataov, 2007; Hayward *et al.*, 2004; Van de Kerkhof, 2004; Karl, 2000; Berk *et al*, 1999; FAO, 1995; Grimble *et al*, 1995; Fiorino 1990), or in the specific case of vulnerable communities, that their interests will be respected (Delli Priscoli, 2003).

The situation of McSEs is complex and can vary from country to country. In *practice* the McSEs' interests in Latin America have traditionally been overlooked in comparison to those of other groups

(Blackman, 2010; Blackman *et al.*, 2009; Caro and Pinto, 2007; Tokman, 2007). McSEs in the water sector are generally not included in the policy incentives targeted at the private sector and mainly end-of-pipe measures are often unaffordable leading to a cycle of exclusion, vulnerability and social unrest. Not surprisingly, McSEs come into frequent conflict with authorities.

In *theory*, the above problematic issues suggest the existence of gaps in knowledge on the theoretical fields of policy instruments, cleaner production and stakeholder participation. An improved theoretical framework is needed regarding the concern of McSEs in developing countries.

Research questions: This thesis addresses the question: How can Micro and Small Industries be effectively engaged to achieve national water and environmental policy goals and approaches?

Against this background, this dissertation focuses on Colombia as a single, but layered case study. It studies the national and three sub-national situations of micro-tanneries impacting water bodies.

The three sub-research questions are:

1. What is the current institutional framework in Colombia for integrating and supporting Micro and Small Industries in their adoption of Cleaner Production?

2. Which perceived mechanisms support Cleaner Production and how do these influence the adoption of Cleaner Production by Micro and Small Industries in Colombia?

3. How can action research methodology be developed and tested to help McSEs implement Cleaner Production?

Choice of Case Study: As mentioned above, this thesis is a single, layered case study of Colombia. Within Colombia, it focuses on the water sector, which provides a microcosm of the social, economic and environmental issues that face McSEs. These enterprises constitute a very important group employing 67.9% of the work force, representing 99.4% of the total number of enterprises in 2007 and linked to the informal sector of the economy (DNP, 2007; Portafolio, 2005). Within the water sector, this thesis deals with the highly polluting micro-tannery industry.

Three sub-case study areas were selected. San Benito is the largest tannery community in Colombia with around 365 tanneries of which 90% are micro tanneries. Cerrito is a fairly small community of basically micro-tanneries (21) with only a medium sized industry in the area. Both communities are located on riverbanks which should be an area set aside for the protection of the river. The third sub-case study was in Villapinzón, a village with 120 micro tanners located 10 km from the source of the Bogota River, which is used for the potable water of the north of the Bogota region, and for the crop region for the 8 million inhabitants of Colombia's capital.

The theoretical framework: The theoretical framework builds on a number of existing theories. It draws on the insights from the literature on Policy Instruments, Cleaner Production, Stakeholder Participation, Complex Negotiation (including Conflict Resolution) and Action Research.

Methods: Seven methodological steps have been undertaken:

First, the overall research question and the sub-questions were initially dealt with through a thorough literature survey.

Second, a methodology was developed for addressing the research questions. Two methods were developed: the case study method from Yin (2009) combined with the extended case method (Burawoy, 1998), and an action research approach –created during the research- called SASI

(Systematic Approach for Social Inclusion). The combined case study method aimed at (a) acquiring a deep understanding of the multiple forces operating at McSEs implementing CP, and (b) getting acquainted with the chain of events (like behavioural change on CP implementation) during a long period of time. SASI is based on five principles, includes six spiral steps, and derives from methods and techniques from complex negotiation (including conflict resolution) (Holman, 2010; Holman *et al.*, 2007; Fisher *et al.* 1991; Lax and Sebenius, 1991) and action research (Burnes, 2004; Schein, 1996; Lewin, 1947). All of these are characterized by the highest level of stakeholder participation in policy making and scientific practice (Van de Kerkhof, 2004) or by transformative approaches, and were inspired by approaches on sustainability from cleaner production. Developing action research provided an ideal opportunity because it allowed 1) to respond to an urgency that existed because of a crisis situation, 2) to deal with complexity and, maybe, 3) to innovate.

Third, the *first 'what' sub-research question* was addressed through content analysis of policy and legal documents, analysis of newspaper clippings and in-depth interviews with decision-makers. Three out of the 18 interviews became micro-cases providing continuous contacts and information.

Fourth, the *second 'which' and 'how' sub-research questions* were answered through the combined case study methodology at the local level for which two case studies were chosen (Yin 2009; Burawoy, 1998). In the two tannery cases: San Benito and Cerrito in which the involvement of the researcher (i.e. myself) was limited, the research analysed over a period of six years the contextual conditions relevant for CP implementation.

Fifth, the *third sub-research question* was addressed by testing the SASI approach with the micro-tannery community of Villapinzón, which represented an extreme case of pollution and social exclusion of McSEs in the tanning sector. Once the first SASI steps were met, the CP implementation could start through different quantitative and qualitative methods such as doing material balance analysis, establishing a monitoring system that starts with indicators based on conflict assessment, defining the CP options, or doing composting from the dehairing and grease residues. This approach was developed over a period of six years.

Sixth, the data collected through the layered case study was analysed, compared and integrated in order to draw conclusions for both theory and practice.

Seven, contributions, recommendations, short comings and perspectives were elaborated.

Results: With respect to the *first sub-research question* on content analysis, the results are:

(a) There has been progressive improvement in including micro, small and medium sized enterprises in the policy process in Colombia. Even though bodies dealing with them as a whole existed since the 1950s, specific laws addressing them were only developed in 1988. Since then, the institutional FADD has been evolving from being repressive to being more open especially for the small and medium sized enterprises but less so for the informal and traditionally marginalized micro enterprises. Only recently have the micro enterprises been recognized by the Ministry of Commerce, Industry and Tourism (MCIT) through (1) Law 905 (2004) which differentiates between micro, and small and medium sized enterprises, and (2) a superior council for the micro-enterprise was created in 2005 to give advice to the vice ministry of Enterprise Development.

(b) In the environmental (including water and CP) field, McSEs are even less targeted. The two top councils that advise the Ministry of Environment on decrees that establish regulations do not include them. The water policy prioritizes water for human consumption but does not (i) include a *priority of use* clause, (ii) determine the economic value of water, or (iii) establish simple and clear procedures including administrative or legal appeal that can be used by McSEs (Sánchez-Triana *et al.*, 2007). The CP policy (i) is rarely linked to command-and-control regulations; (ii) includes technological options that do not match the needs and interests relevant to McSEs; (iii) uses exclusive top-down consultancy approaches; (iv) has a limited budget (representing only 2.8% of the regional environmental agencies' budgets and (v) focuses on projects, not on processes. However, the CP policy states that it is compulsory that 20% of a credit line from the Inter American Development Bank be used to educate the smallest industries on CP.

(c) Participation: Although participatory processes were introduced in the Colombian Constitution (1991), they do not specify precisely how and when to implement these mechanisms.

Furthermore, the content analysis reveals lack of coherence between the national, regional and local levels, expert interviews show that (a) there is lack of political will; (b) there is lack of bureaucratic commitment: although the government has tried to focus on polluting industries through the creation of a joint agenda between the Ministry of Industry and Commerce and the Ministry of the Environment in 2006, in practice they almost never meet (only one meeting in 2009) (mc1); (c) there is lack of awareness at the employee level in public entities; and (d) since micro industries are part of the private sector, they cannot receive support from the government except for the recently created APP (public private alliances). However, as informal enterprises they have limited access to banks, are excluded from formal capacity building programs or simply cannot afford to pay the costs of such CP programs, which are set for the biggest industry.

The *second sub-research-question* was answered through the local case studies of micro-tanneries in San Benito and Cerrito in relation to the mechanisms supporting CP implementation. In San Benito there was some success with CP implementation based on private consultancy projects for small and medium sized enterprises but not for the micro firms. Since 2003, in Cerrito, the regional environmental authority has supported CP implementation through participative approaches and has improved the quality of the river water. The comparative assessment suggests that successful CP implementation programs for McSEs:

- Focus on attitudinal and organizational mechanisms such as having long-term support, high degree of participation, multidisciplinary capacity building programs based on learning through trial-and-error, on facing first the social and economic issues (specific contexts), on 'policy entrepreneurs' who take on a leadership role, and which do not rely on quick short term savings;

- Run parallel to appropriate command-and-control policies; and are not slowed down by interrelated jurisdiction of spatial planning and environmental policies that create uncertainty and risk for the micro-entrepreneurs.[1]

---

[1] For example, a 1976 policy imposes a 30 m zone on riverbanks where no industrial activities can be deployed. However, it does not recognize the rights of the industries that existed previously, nor does it offer relocation support.

The *third sub-research question* was answered through action research. It took place in Villapinzón, where since the 1990s the regional authority has considered 66 proposals using end-of-pipe solutions to deal with water discharges from the tanneries. Until 2004, no effective solutions had been implemented; all the tanners had been sued in court and excluded from credits and support from authorities. Forced closures had been carried out. More than eighty tanners were closed down including fifty on the riverbank that were to be relocated when the research was starting.

In order to analyse this problem, the SASI method was created in April 2004 (see method). The method aims at reducing the level of conflict between the tanners and the environmental authority, and at developing integral approaches to their complex problems focusing at implementing CP alternatives through participative strategies and technical support from academia (including my colleagues). The method aims at (a) reducing social exclusion through empowerment and by raising awareness on the characteristics of McSEs from authorities; and (b) replacing exclusive end-of-pipe approaches by stimulating CP implementation.

My action research helped to address the social exclusion and:

(1) The tanners became united and chose a positive representative leader.[2] The tanners changed their polluting practices. Their negotiating power increased as they received support from the President's legal office, from the magistrate ruling on the court order[3] on the Bogotá River, the Governor, the Bogota Chamber of Commerce, and the National University. This led to press coverage and the Environmental Authority started to listen and negotiate with them. They gained political and organizational influence: A new mayor was elected because of their support in 2010 and represents their interests for the first time. The tanners took the lead on spatial planning solutions in order to formalize the industrial area at the village council, which were delaying the technical solutions on CP implementation. They acquired a multi-level presence: The tanners are participating in the national tannery committee, which gathers formal representatives from all the tannery communities.

(2) The environmental authority CAR (Autonomous Regional Agency of Cundinamarca) co-financed the CP implementation project and recognized that their past punitive approaches were inappropriate as (a) the pollution fines and lawyers fees were unaffordable; (b) there were few positive incentives to encourage the tanners to participate and implement policies; (c) forced closures had not differentiated between the industries willing to change provided they were offered support, from the ones that were not. CAR realized that their punitive approach stimulated clandestine industrial activities and corruption. Their participation in SASI helped them understand that since their relationship with the tanners was of a long term and interdependent nature, they needed to help build solutions. They learnt that support works better than just punishment.

In terms of replacing exclusive end-of-pipe approaches, this action research led to the adoption of technical solutions based on CP and financed partially (15%) by the tanners, once impending social issues were handled, mutual trust was built among all (including the researcher), and tanners and authorities were on speaking terms. The tanners were thus able to reduce their levels of pollution.

---

[2] This leader was appointed as one of the 35 best leaders of Colombia in August 2011 by the NGO Liderazgo y Democracia among 300 nominees (Semana, 2011).

[3] The magistrate ruled that the tanners were required to implement CP and that the authority was expected to support that initiative. The researcher was appointed as a supervisor in the ruling, once enforced (Court order, 2004). (As of 2011, the court order has not been officially enforced, because of complex appeal processes).

Between 2004-2009 there were reductions in pollution loads of 32-68% in Chromium, 60-72% in BOD_5, and savings in water use between 24-68% on the liquid discharges to the river. The tanners are composting the dehairing residues and have carried out innovations in their own production processes. I played an active role in the process. Initially, I played the role of helper and motivator. As time passed, my influence diminished. My role was constantly and spontaneously backed up by 'identified champions' (positive leading persons supporting commitment and change) within the different stakeholder groups.

Since there is a need to further develop the research field of McSEs, this dissertation has elaborated on the virtues of how to choose methods for complex environmental inquiry such as the McSEs based on the kind of research question suggesting (or not) close interaction with actors, intervention for change processes, or for leading negotiations. The results on this matter focus at deploying transformative modes of participation.

Conclusions:

It is concluded that:

First, the characteristics of McSEs exclude them in general from (a) voluntarily adopting cleaner technologies; (b) having access to public participation and decision making processes and decision makers associated with their own issues; (c) enhancing the capacity to work cooperatively (associatively); (d) having access to new market opportunities; (e) improving their products' quality; and (f) improving the quality of life of 67.9% of the work force in Colombia. McSEs are also more vulnerable to exogenous forces than the medium sized enterprises where national challenges such as corruption and unclear property rights can create barriers to their viability.

Second, water and environmental policies in Colombia do not take McSEs' specific characteristics and specific contexts into account; the disincentives are perverse and incentives are out of reach. In the environmental field, the informal sector is often associated with illegality. Since there is no recognition of the McSEs' characteristics, the medium sized enterprises capture the limited financial resources. The standard approach stimulates exclusive end-of-pipe approaches based on centralised treatment options that can be unaffordable for the majority of the population. Phased approaches are hardly considered for law enforcement and are left to the officials' will at the local level.

Third, the case studies show that the traditional top-down command-and-control methods (San Benito) are less successful than participatory, learning based on trial-and-error methods (e.g. in Cerrito) and that action research (e.g. in Villapinzón) which helps to provide leadership in the policy development and implementation process fosters positive change, responds to urgency to act in complex situations and provides effective sustainable results. The case of Cerrito illustrates the power of having political will and leadership from authorities prioritizing the McSEs' agenda, approaches based on prevention such as CP implementation, and a CP policy working in parallel with command-and-control, and stakeholder participation towards social inclusion. In particular, the application of the SASI method empowered vulnerable McSEs by (a) making them experts in their own field, (b) helping them pool their resources, (c) helping them out of their exclusion from the policy process, and (d) proactively participating in a change process adopting approaches towards prevention. The SASI method also fostered the transition (a) to prevention oriented policies and approaches from the institutions, (b) towards raising awareness on the McSEs' needs, and (c) towards supporting real participatory approaches.

Fourth, the long-term combined case study research goes beyond the traditional case study methodology (Yin, 2009) because it allows greater interaction with the actors and to get acquainted with the change process. The layered case study approach allowed for vertical comparison.

Fifth, in relation to action research, the SASI process goes beyond analysing a contemporary situation because it allows building understanding while responding to the urgent need to act and to foster change in the mid-term. It contributes to existing action research methods by creatively combining insights from the theories of complex negotiation (integrative negotiation and conflict resolution) in order to (a) follow a systematic process with specific targets per step inspired by integrative negotiation and leading to decision-making, (b) offer strategies to overcome social exclusion and conflict from methods such as Open Space Technologies (OST) and Appreciative Inquiry (AI) as visioning, and (c) boost the negotiation power of the marginalized groups through internal strengthening and strategic alliances by (i) respecting stakeholders' own decisions while supporting the common good, (ii) understanding the situation of the McSEs and privileging contextual understanding over generalizations (Baskerville, 1997); (iii) undertaking top-down and bottom-up analysis; (d) promoting the culture of participation based on positive[4] and not on claim oriented approaches; (iv) promoting the dynamic creation of knowledge to deal with changing situations; (v) ensuring medium-term trust and commitment; (vi) promoting behavioural change (e.g. the tanners were willing to clean the river once they felt that their interests were taken into consideration, and win-win situations were possible through CP implementation); (vii) stimulating third party involvement and commitment that goes beyond mediation; (viii) using Appreciative Inquiry which led other 'manipulative actors' in the field to lose their influence and chances for corruption; (ix) innovating on technical processes once people get out of the vicious cycles, (x) supporting the transition from informality to formality and lead to effective law enforcement, and (xi) developing a comprehensive holistic, self-enforcing, solution that turned out to be supported by academia. However, SASI is a more expensive and time-consuming process because it requires building consensus and change; entails longer periods for reporting, writing, and publishing scientific papers (Bodorkós and Pataki, 2009; personal communication SWITCH project, 2011); requires skills for organizing and making multidisciplinary teams work efficiently; requires commitment from the participants; needs constant feedback and reflection, and needs to assess the risk of intervening in the lives of others.

Sixth, this research shows that committed action researchers can be Agents of Inclusion (AoI)s provided they can switch between the role of researcher, helper, facilitator or negotiator as needed (Bodorkós and Pataki, 2009; Schein 1996; Lewin 1946), can maintain independence, are open to discuss all issues, are willing to take a holistic approach, assume self-reflexivity, commit on a long-term basis, and can play a leadership role in empowering people to find appropriate solutions. Clearly as the target group becomes empowered, the role of the action researcher is minimized and the process becomes sustainable; while the researcher gains in knowledge.

Seventh, both methods used, the combined case study methodology and SASI, go beyond the methodology that inspired them, which are the case study from Yin (2009) and managed learning from Lewin (1946) and Schein (1996).

---

[4] The theory of conflict resolution states that a way to start building solutions among stakeholders is to look for any positive issue in common among stakeholders, which stimulate positive discussions and not concentrating only on the negative aspects of a conflict.

Eight, CP programs in developing countries, especially in Latin American countries like Colombia, Ecuador, Peru, Chile, and Mexico focus on top-down approaches and are not targeted at the McSEs. CP implementation is not seen as a process that needs to be monitored and agreed upon in terms of the industries' particular contexts. The authorities are reluctant to implement CP because they lack the appropriate control instruments (CP policy does not run in parallel with command-and-control). Besides, there is the belief that CP is for formal enterprises that are already clean and want to become cleaner. CP is supported by private initiatives that are too expensive for McSEs and that do not focus on holistic approaches but only on the technical issues. CP implementation in the context of McSEs face difficulties with respect to exclusive quantitative monitoring and evaluation because they do not have the right tools. Qualitative innovative approaches can balance these weaknesses. SASI was successful in supporting CP implementation because it worked towards building common grounds between stakeholders, understanding each other's interests and focusing on integrated solutions. Cleaner Production did not threaten the McSEs' own identities and their right to exist because the technical options could be adapted to their specific situations. Successful CP implementation focused, hence, on the attitudinal and organizational (systemic) mechanisms in addition to classical ones. Tanners, researchers, and tannery experts made the choice of technology (the authority remained as observer). The spatial planning problems slowed down the CP implementation processes. In Villapinzón, solving them took three times as much time (compared with the technical discussions based on prevention) to be resolved by all the stakeholders. In San Benito, the spatial planning is still hampering the CP implementation from being effectively diffused.

Ninth, approaches based on action research like SASI and the respective researchers' roles are best suited for handling complexity at the local level and linking teaching, research and real world engagement Denzin and Lincoln (2011).

Although the research was conducted in Colombia, the research results are likely to resonate with other countries facing environmental pollution from marginalized communities. Further research needs to confirm this assumption.

This thesis faced complexity as a backdrop. At the overall and global level, this thesis contributed to bringing the environmental debate at the level of 'acting', of designing and implementing solutions, of prioritizing behavioural change at the level of institutions, practitioners, community members, academicians, and decision-makers; and not just criticizing (Delli Priscoli and Wolf, 2009). It supported a more reflexive attitude towards water. At the local and national levels, it focused the debate towards values such as equity, high degree of participation, and conflict resolution with respect to the micro and small enterprises in developing countries. It adopted a proactive attitude towards their complex problems by deploying dialogue and by targeting the McSEs' legitimate concerns such as their needs and interests. It was successful in changing attitudes by means of (socializing) technology, adapting it to the local needs through action research on CP implementation.

This research aimed to understand how small and micro enterprises could be effectively engaged to achieve national water and environmental policy goals and approaches. This research shows that the very characteristics of these enterprises put them often beyond the reach of formal policymaking processes. While participatory approaches, if effectively implemented may make a partial contribution towards incorporating the concerns of the marginalized communities, action research through, for example, the implementation of CP via the SASI method can lead to long-term sustainable solutions through empowering these communities, to understand the significance of sustainable development and to stand up for themselves. Changed tanners were the key factors towards sustainable and long lasting results even if the institutions were not strongly committed. Only this can help end the vicious cycle of poverty and exclusion for these communities.

Table of Contents

Table of Figures

Table of Tables

1 Introducing Micro and Small Sized Enterprises

1.1 Introduction

The overall objective of this dissertation is to contribute to the integration of micro and small sized enterprises impacting the environment into environmental policy and approaches.

This research deals with the complex problem of micro and small sized enterprises (McSEs) impacting on the environment (especially water bodies) in the context of developing countries. In fact, it is acknowledged today that micro and small sized enterprises (McSEs)[5] belonging to the industrial sector, heavily impact on the environment. Some authors consider them responsible for 70% of the industrial pollution in developing countries (Le Van Khoa, 2006; Soni, 2006; Hillary, 1997) and consequently stimulating an impending water crisis (Rosegrant *et al.*, 2002).

Focusing on these industries implies looking at the local and extra-local forces at work such as those springing from social, economic and environmental issues, from informal and formal sectors, from rural and urban settings, and from the way these interact amongst each other in the context of developing countries.

This chapter elaborates on the problem definition (1.2), the research questions and objectives (1.3), the steps in the thesis (1.4), the focus, strengths, risks and limitations of the thesis (1.5), the structure of the thesis (1.6), and the implications for the research (1.7).

1.2 Problem definition

1.2.1 Introduction

In general, there is still reluctance towards the integration of the principles of sustainability, which entails meeting the present needs of people without compromising on the ability of future generations to meet their own needs (World Commission on Environment and Development, 1987), into country policies and programs. This situation gives rise to two major challenges (Thabrew *et al.*, 2009):

1. To focus the environmental concerns through technical solutions based on prevention;
2. To integrate the environmental concerns in cross-sectoral planning and implementation of a development strategy that by itself asks for improved methods of stakeholder participation.

Some authors like Stiglitz (2012) and Gagnon *et al.* (2007) suggest that besides the above mentioned challenges, the concept of distributive justice or the fight against social and economic inequities ought to be emphasized in today's world in order to reach sustainability. Quental *et al.* (2011) propose that the sustainability concerns should move from emphasizing pollution control and the health of national resources to a position where human development is at the heart of the discussions.

These major challenges are especially meaningful at the level of micro and small sized enterprises that belong to the group of McSEs in the water sector in developing countries because they entail multiple

---

[5] McSEs include micro enterprises (1-9) and small enterprises (10-49) employees. The term McSEs is considered comprehensible and less prone to confusion with the commonly used term SMEs that refers to micro, small and medium sized enterprises.

problems and urgent intervention (Fandl, 2010; Mondragón-Velez *et al.*, 2010; Revell *et al.*, 2010; Mel *et al.*, 2008; Corredor C., 2007; Caro and Pinto, 2007; Blackman, 2006; Herrero and Henderson, 2003; Ashton *et al.*, 2002; Visvanathan and Kumar, 1999). These problems include the characteristics of the McSEs in developing countries (1.2.2), the McSEs' relationships with environmental agencies (1.2.3), and the existing gaps in the theoretical fields of policy instruments, cleaner production and stakeholder participation in the context of McSEs (1.2.4). The effort in this thesis is framed by the need to develop peace building strategies in a complex world (UNEP, 2010; WWAP, 2009) because the depletion of water resources has fuelled confrontations (WWAP, 2006).

1.2.2   Characteristics of McSEs in developing countries

Definition of SME

Most of the literature refers to micro, small and medium sized enterprises as being part of the same group under the abbreviation SME. These firms are nevertheless not uniform worldwide. The criteria for defining them depend usually on economic development and the policy purposes. The World Bank has reported over 60 definitions for SMEs. The most commonly used criteria relate to the following, which address enterprises on the commercial, service and industrial sectors (IIFT, 2012; Ashton *et al.*, 2002):

- Number of employees
- Annual turnover
- Annual balance sheet

The European Commission (2003) for instance adopted the latter in order to define the categories for micro, small and medium sized enterprises (Table 1-1).

Table 1-1 Definition of SMEs in Europe

Enterprises	Head Count	Annual turnover	or	Annual Balance sheet
Micro sized	<10	< € 2 million		< € 2 million
Small sized	<50	< € 10 million		< €10 million
Medium sized	<250	< € 50 million		< € 43 million

Source: European Commission (2003)

In Europe, such firms represent 99% of an estimated 23 million enterprises and are responsible for two-thirds of all employment (meaning 75 million jobs). For these enterprises, most benchmarks or social marketing metrics are too cumbersome to implement (EU, 2005).

In the ASEAN (Association for the South East Asian Nations) countries, the general definition is limited to the manufacturing sector. In the USA, an SME entails an organization of up to 1500 employees and is qualified in terms of the nature of the productive activity (USTIC, 2010). The US International Trade Commission reports in 2010 that firms with less than 500 people are an important source of innovation, are more efficient than larger firms, and constitute an opportunity for acquiring entrepreneurial skills. They represented half of the non-agricultural GDP from 1998-2004. The Small Business Administration (SBA) states that despite their importance, these employees earn less than those in larger firms and are more prone to on-the-job injuries and fatalities (USITC, 2010). In India,

SMEs are classified by the investment in plant and machinery, and the kind of productive activity (Development Act, 2006)[6].

Since this thesis focuses on developing countries in Latin America like Colombia, the SMEs' definition will address this region.

Characteristics of SMEs in Latin America - Adopting the abbreviation McSEs

In contrast to the European Commission, Latin America has not adopted a common definition for SMEs (Alvarez and Durán, 2009). In Latin America they also hire a great bulk of the population. Countries like Chile characterizes them by their annual sales and number of employees; Costa Rica by the number of employees (medium sized businesses go up to 100 employees; and Guatemala by the number of employees (medium sized only up to 50), annual sales and total assets. Venezuela defines SMEs only for the enterprises in the manufacturing sector. The case of Peru is especially interesting: aiming at promoting competitiveness and formality, its government has eliminated the group of medium sized businesses and is focusing the policies on the micro and small enterprises (up to 100 in terms of the number of employees). Colombia characterizes SMEs by the number of employees and the total assets in terms of the number of minimum legal monthly wages (MLMW) according to Law 905 of 2004. The latter has similarities with the EU definition and will be used in this thesis (Table 1-2).

Table 1-2 Definition of SMEs in Colombia

Enterprise	# of employees	Total Assets (MLMW)
Micro	<10	500
Small	<50	501-5000
Medium	<200	5001-30000

Source: Colombian Law 905 of 2004

Overall, in Latin America, micro and small sized enterprises (

Table 1-3) share vulnerability with respect to access to technology, markets, credit, and level of education. They have limited monitoring and are above all mostly informal. Hence, micro and small sized enterprises have more in common than the small and medium sized ones because the medium sized enterprises are already formal and have access to the limited national financial support (Alvarez and Durán, 2009).

Table 1-3 Characteristics of SMEs

Micro	Small	Medium
Mainly informal	Partially informal	Formal

---

[6] msme.gov.in/MSME_Development_Gazette.htm in 03/03/12

Low education	Low education	Higher education
No access technology	Limited access technology	Access to technology
No formal financing/costly	Limited access to financing	Access to financing
Limited associations	Associations	Associations
Very narrow markets	Narrow markets	Full markets
No social security	Limited social security	Social security
No monitoring	Limited monitoring	Monitoring

Until very recently the literature has acknowledged that the micro and small sized enterprises are the ones involving the biggest challenges because it is very difficult to handle them (Fandl, 2010; Mondragón-Velez et al., 2010; Mel et al., 2008; Corredor C., 2007; Caro and Pinto, 2007; Herrero and Henderson, 2003; Ashton et al., 2002).

Their interests have traditionally been overlooked in comparison to those of other groups (Caro and Pinto, 2007; Tokman, 2007; Herrero and Henderson, 2003).

In India, micro and small sized enterprises can be known and grouped together as MSEs but since it is easy to get confused between SMEs/MSEs, this thesis proposes to name them McSEs. Hence, McSEs will stand for Micro and Small sized enterprises throughout this dissertation in the industrial sector.

Analysis of the McSEs' situation in developing countries

In most open economies, about 10 - 15% of the Gross Domestic Product (GDP) comes from the government and 85-90% from the private sector (UNIDO, 2009; 2005). Raising standards in the developing world through the performance of the private sector has been considered to be a major target. The situation of the McSEs is nevertheless complex in these countries. In fact, even though McSEs in developing countries are viewed as having the potential to adapt and to innovate in today's world because of their size, they are also vulnerable in terms of having limited access to financial resources, information, technology, and limited opportunities for building their own capabilities (Blackman, 2011; UNCTAD, 2008; UNIDO, 2009; 2005; Ashton et al., 2002). Authors like Beck and Demirgue-Kunt (2006) and Amonoo et al. (2003) point at the fact that McSEs are especially vulnerable to the interest rates for credit demand and loan repayment whenever they have access to formal financing. Since McSEs in the industrial sector cause substantial environmental impact, appropriate solutions are not easily   implemented, as they cannot afford to eliminate the harm they cause to the environment through technology and innovation (Moors et al., 2005).

Designing the appropriate strategies for McSEs entails looking at the multiple challenges they face at the local, regional, national and even global levels. McSEs are indeed confronted on the one hand with a globalized world of (a) competitiveness, (b) growing environmental concerns and may not be able to cope (Soni, 2006); but on the other hand with (c) growing awareness of their role in the fight against poverty.

In order to face both, (a) competitiveness and (b) environmental concerns, collective solutions have been found to work effectively in some contexts. Small and micro sized enterprises are grouping together in clusters or in the so-called industrial parks (Shi et al., 2010). McSEs are learning to work in a cooperative manner not only for marketing purposes but to find technological solutions to their discharges. The results have not been always positive because as authors like Blackman (2005; 2006) state, what lies at the heart is behavioural and cultural change, and adaptation to technology and innovation. In this sense, Blackman and Kildegaard (2010; 2003) report that McSEs' trade associations are becoming more important as sources of technical information in developing countries.

McSEs' clusters or sectored agglomerations seem to be a common practice in developing countries (Shi *et al.*, 2010). By being close to each other they develop a competitive advantage from the:

- Proximity to the source of raw inputs,

- Presence of skilled labour force,

- Availability of tailored services, and

- Availability of clients.

These clusters open possibilities on the one hand for association, vertical specialization and horizontal cooperation (Soni, 2006; UNIDO, 2005). For control purposes, agencies may more easily monitor their industrial activities and may also design incentives that stimulate good practices. Working with clusters entails building a process based on big groups undergoing technological changes and it has been observed that for change to last, it must be congruent with local cultural and social contexts (Sakr *et al.*, 2011; Barreto, 2001; Schein, 1996). On the other hand, promoting such small-scale technologies is not easy because of the scale of these units, the educational level of employees, the costs of these technologies and general awareness levels. In general, the industrial parks are called eco-industrial parks (EIP) and a significant number of them are failing (Sakr *et al.*, 2011; Shi *et al.*, 2010). Limiting factors are related to: *'creation of symbiotic relationships, information sharing and awareness, financial benefits, organizational structure, and legal and regulatory frameworks'* (Sakr *et al.*, 2011: 1168).

With respect to (c) growing awareness of the role of McSEs in the fight against poverty, the United Nations considers that focusing at improving the McSEs' performance constitutes an opportunity for achieving the Millenium Development Goals. The target is to support self-employed workers that have not had the opportunity to get out of vicious cycles because of the weak operating environment for businesses in most developing countries (UNIDO, 2005: 4). Setting supportive strategies entails (Stiglitz, 2003; Ocampo, 2002) designing a new development agenda based on inclusion (Table 1-4). There is consensus over the fact that inclusion and development are strongly tied to each other and that as such, inclusive processes lead to better development (Corredor C., 2007; Rocha, 2007; UNIDO.OECD, 2004). Since McSEs are overlooked by policies and approaches (Blackman, 2010), (Blackman *et al.*, 2009) social exclusion can be considered a common factor for the micro sized enterprises (Mondragón-Vélez *et al.*, 2010; Rocha, 2007; Herrero and Henderson, 2003). Walker and Preuss (2008) suggest for example to create networks with the public sector and the smallest firms with respect to sourcing for green markets or eco-labelling.

Box 1-1 The new development agenda

The new development agenda opposes the policies set in the nineties by the Washington Consensus[7] aimed at sustaining macroeconomic stability in order to reach a more competitive market economy through open and unregulated markets, and privatising public enterprises (Perry, 2003). Supporters of the new development agenda consider that as a developmental agenda, the Washington Consensus' approach was incomplete and deepened inequities (Fundación Agenda Colombia, 2007; Stiglitz, 2012-2002-2003; Perry, 2003). Two mistakes were committed: 1- Not setting limits to the markets; 2- Turning means into ends. As a result, through the lenses of this orthodoxy the State remained aloof and economic planning was eliminated, conceived as a conscientious exercise of governments to assign resources to address the most important economic and social issues facing the majority of the population (Corredor A., 2007). Social concerns came to be managed by trust funds; public services such as health and education were handled through the lens of private efficiency, and industrial and rural matters were weakened (Stiglitz, 2003). It has been even reported that since those developing economies did not pay enough attention to the small industries' sector, stagnation resulted (Caro and Pinto, 2007; UNIDO, 2005). With respect to economic indicators such as inflation, interest and exchange rates, they became ends by themselves and essential variables such as employment, quality of life and growth lost their importance as main goals (Fundación Agenda Colombia, 2007; Stiglitz, 2012-2003).

Two key concepts deserve an explanation because they are tied to the problem definition of McSEs in the developing countries: Social exclusion and informality.

Social exclusion

A human being's life is interwoven into the social fabric of the society and community he/she belongs to, which essentially determines his/her potential as a member of that social order. The level of inclusion or participation in the activities of said society, or on the contrary being excluded from these, can either foster or hinder the vital development of a human being (Rawls, 1971).

The scope of social exclusion takes into account the social, economic, political, cultural and recreational spheres of a given society and is therefore a multidimensional and multifactor phenomenon. Economic exclusion refers to the incapacity to earn a living and meet basic needs; social capital exclusion is understood to refer to the limits imposed by a society or community on the full social participation of its members in a wide range of activities spanning the social, economic, political, cultural and recreational realms; human capital exclusion alludes to the lack of access to education and health services as well as to the suitable or inadequate quality of such services; and political exclusion refers to the impediments that exclude a person from making full use of the rights he/she has as a citizen bringing about the violation of his/her political and civil rights (Garay, 2002).

The pervading exclusion of minority groups in a society and the resulting increase in the economic gap between minorities and the dominant segments of society has further distanced social classes. The failure to resolve this increasing gap affecting minorities globally has enlarged the population of the disadvantaged members of society who's social, political and economic rights are easily infringed upon by those wielding a greater political and economic power and has deepened inequality (Fandl, 2010; Mondragón-Vélez et al., 2010; Rocha, 2007; Corredor A., 2007). Latin America has historically been challenged by such phenomenon, as social exclusion is a major factor of the great social

---

[7] The Washington consensus is a group of policies originally set for Latin American countries in the 90's supported by the International Financial Organizations such as the International Monetary Fund, IMF and the World Bank, WB and economic entities established in Washington, D.C.

differences. The last report by the program on development by the United Nations (UNDP), called Acting on the Future: Breaking Intergenerational Transmission of Inequality, states that the Latin American and the Caribbean regions host 10 of the 15 most unequal countries of the world with Bolivia and Haiti being at the top (UNDP, 2010).

Inequality is a major issue being observed from a different perspective over the last few years by economists. In fact, the tendency today is not to target primarily on growth but on how the resources are distributed (Stiglitz, 2012; DNP, 2012; 2011; DANE, 2011; WB, 2011b; PND, 2010; The Economist, 2011; UNDP, 2010). In Latin America, the Gini co-efficients[8] are the highest. In some countries like Brazil and Argentina the Gini co-efficient has been nevertheless dropping over the last decade (from 57.6 to 54 and from 52.5 to 45.8 respectively (WB, 2011b): In Colombia it is 56. This figure has remained stable over the past six years, maintaining one of the highest income inequality rates in Latin America (Echeverry *et al.*, 2011; DANE, 2009), and is even expected to increase (Semana, 2011). The latter can easily be reflected especially on the McSEs, which as a result of their social and economic backgrounds have been experiencing social exclusion reinforced by the fact that the ones in the industrial sector are polluters.

Informality

In most developing countries, the micro and small sized enterprises belong to the informal sector (UNIDO, 2009; 2005; Tokman, 2007; ILO, 2002).

A multiplicity of policies facing informality shows that there is a lack of a shared approach and hence, an absence of a common definition for this phenomenon. Defining informality only through illegality and labour precariousness has been misleading (Cruces and Ham, 2010; Mel *et al.*, 2008; Caro and Pinto, 2007; Perry *et al.*, 2007; Tokman, 2007). For Caro and Pinto (2007) informality is a multidimensional phenomenon that could be presented as being characterized by degrees of specific contexts ranging from complete illegality to subtle informality, referring to economic activities that are developed within inexisting legal frameworks or others that are partly formal, and partly informal. Perry *et al.* (2007) offered a conceptual framework of informality. They stated that informality is a function of both exclusion and exit- considered as the voluntary act to be out of the formal system for benefit-costs ratios. Mel *et al.*, (2008) have based their analysis on the nature of the owners of the McSEs because own account workers represent more than half the informal sector in low-income countries. They have tried to reconcile De Soto's (1989) and Tokman's (from the International Trade Organization - ILO) thesis. While De Soto (1989) has considered micro-enterprise owners as entrepreneurs[9] that need lower costs and simplified registration schemes in order to improve their situation, for Tokman (2007) own account workers are marginalized people that would be satisfied with improved wage work opportunities. Mel *et al.* (2008) found that the majority of own account workers are not entrepreneurs willing to grow and that the latter could explain why most of McSEs do not really grow despite institutional efforts. The present work considers the above analysis based on the nature of the owners of the smallest enterprises problematic because it could lead to dangerous generalizations. In any case, this thesis considers that the contributions of Mel *et al.* (2008) and Perry *et al.* (2007) highlight the complexity involved in informality and the relevance of considering existing different concepts in designing policies.

---

[8] an indicator of income inequality where 0.0 is the minimal inequality and 1 is the maximum inequality

[9] which are risk takers and innovators, which have a special ability and motivation for personal achievement and are willing to give up some control over their personal situations (Mel *et al.*, 2008).

For decades, the formally established manufacturers had perceived McSEs in general as illegitimate competitors to be eliminated (Fandl, 2010; Caro and Pinto, 2007; Tokman, 2007; Sverisson and Van Dijk, 2000). The latter has explained that in many developing countries, a number of regulations still currently in force directly work against the interests of small owners (Caro and Pinto, 2007; Sverisson and Van Dijk, 2000). Contrary to expectations, by 2003, the informal sector was providing half of the urban employment and its share on the labour market has been growing steadily: Since 1980, 60 new jobs out of 100 came from the informal sector. It has been observed that microenterprises represent businesses with the highest rate of labour market growth (Tokman, 2007). With respect to measuring informality, three aspects are traditionally considered, the lack of commercial registration, the lack of tax payment, and the non-payment to social security (WB, 2010).

Independent of the interpretation adopted, the characteristics of the informal sector are similar: small, unsophisticated technologies, low capital requirements per worker, and a distinction between micro- and large-sized enterprises in terms of capital requirements. Additional features include limited sharing of the property of the means of production, and a majority of waged workers labouring without contracts and protection (Tokman, 2007: 2). Informality constitutes a challenge to any developing country because it weakens the welfare of workers by increasing the vulnerability and exposure to risks, lowers a firm's productivity as access to credit is difficult, and challenges the fiscal viability of the State to provide goods and services for all (WB, 2010).

Although the phenomenon of informality is still far from offering job stability and labour and social protection, a closer look at the reasons for this informality is considered important in order to carry out a consistent and effective approach towards regulating the informal sector. The logic of survival is considered the major factor in the development of informality. Population growth, the pressure of a labour surplus for jobs are the well-known factors for survival. New activities generated by the logic of decentralization, in a context of rapid economic growth and the extra regulatory behaviour, along with the increase of labour reforms are also major items in the discussion. In fact, modern enterprises adapt to new environments by cutting production costs at the expense of employment and labour, and social protection remains absent (Tokman, 2007). For Fandl (2010), informal enterprises may reject the rule of law simply because the State has historically failed its social contract and as such, they are not willing to cede their freedom for the state protection they barely know. As a result, the work conditions deteriorate.

In all, Fandl (2010), Azuma and Grossman (2008), Tokman (2007), Caro and Pinto (2007), Perry *et al.* (2007), and McMillan and Woodruff (2002) agree on the fact that a different perspective should tackle the process of formalization. For these authors, the focus is on *easing* the integration of informal activities in the modernization process, transitioning from a negative perspective based on punishment to a more positive one. Tokman (2007) states that ad-hoc regulatory systems must be designed. He considers that structural features are more important determinants of the existence of informal activities than just access costs to legality. Herrero and Henderson (2003) focus for example on the McSEs' vulnerability to weak judicial systems which are hardly available for them in Peru and the need to develop alternative conflict resolution paths that suit their needs. UNIDO (2009; 2005) proposes to tackle healthy environments instead of basing the strategy on subsidies. Other authors recall that there is consensus that interventions aimed at individual firms are not likely to be effective unless their local environments are affected simultaneously (Fandl, 2010; Blackman, 2006; Van Hoof, 2005). Sverisson and Van Dijk (2000), citing Humprey and Schmitz (1996), have suggested that policies to support McSEs need to be customer-oriented (focusing on niches, styles and designs), collective (importance of clusters, associations), and cumulative (long-term consistency of support).

Authors on McSEs recommend bringing the research towards multidimensional and context-oriented analysis that could build insight on universal mechanisms for formalization (Fandl, 2010; Mel *et al.*, 2008). They also point out the importance of studying the cluster or the EIP concepts more closely, which did not seem to be successfully targeted in the past (Sakr *et al.*, 2011; Shi *et al.*, 2010; Van Hoof, 2005; Schaper, 2003; Sverisson and Van Dijk, 2000). Overall, equality must be sought through reformulation of policies favouring the poor because growth does not necessarily mean poverty reduction (Stiglitz, 2012; Ocampo, 2003).

Researchers working in the field of informality or on McSEs in general agree that, in the long run, growth depends on recognizing that the technological change is what really lies at the heart of the problem. McSEs operate in flexible ways but they need to be culturally prepared to produce for demanding markets (Tokman, 2007; Sverisson and Van Dijk, 2000) whereas positive assistance from the State is needed through market infrastructure (McMillan and Woodruff, 2002).

Some industrial fields are more challenging than others in terms of the level of informality and environmental impacts. The tannery industry is one of them.

The tannery industry

Converting animal rawhide and skin, primarily cattle hide into leather through tanning, an activity born out of necessity for early man, has become a thriving industry thanks to techniques developed in the 20th century. As the world's most important industry based on a by-product, the leather, industry has historically been associated with foul odours and severe environmental pollution and as a result the people involved in tanning have not particularly benefitted from a high social standing stemming from their industrial activity (UNEP, 1991).

The most polluting steps of the tanning process (dehairing, tanning) use toxic chemicals like sodium sulphate and chromium sulphate which form compounds that are highly resistant to biological breakdown, are associated with high oxygen depletion and contain potentially toxic metal and salt residues (Sivakumar *et al.*, 2009; Kanagaraj *et al.*, 2008; Kindlein *et al.*, 2008). Wastewater from the various processes in leather making has traditionally been mixed making recovery or reuse of components in the effluent exceedingly difficult and expensive (UNEP, 1991).

Rivela *et al.* (2004) reported that during the last thirty years the global output of leather products has risen by 55% due to increasing demands. Due to the low priority assigned to industrial waste and effluents in developing countries, these tanning steps were relocated mainly to developing countries from the 1960's to the 1980's as OECD countries underwent increasing pressure from strict environmental policies and abatement pollution costs (Rivela *et al.*, 2004; Biller and Quintero, 1995).

A low technological level, high environmental impacts, high employment generation, high informality, heterogeneous leather quality, globalized markets, and changing fashions that dictate consumption trends characterize the tanning industry in developing countries (Rivela *et al.*, 2004; Kennedy, 1999; UNEP, 1991). This productive activity faces pressure to lower prices and to consequently reduce profits (MAVDT, 2006).

Since 1991 UNEP has considered that there have been significant advances in the tannery sector and that there are tannery effluent treatment options that cause less environmental impact. Rivela *et al.*, (2004) reported that primary treatment was showing the largest reductions of environmental impacts. Solid wastes originating from tanning and that are recovered may be classified as by-products which, in turn, may be sold as raw material to other industry sectors. Nevertheless, these options entail collaborative approaches and extensive internal coordination from all stakeholders involved in implementing pollution abatement measures (Kennedy, 1999). Comprehensive approaches based on cleaner production processes can provide both economic and social benefits and are more effective than end-of-pipe approaches (Van Hoof and Herrera, 2007; UNEP, 1991).

Cleaner production (Chapter 2) has been envisioned as providing the key production and control practices in order to reduce the environmental impact as well as lowering production costs. Specific studies entailing recycling, for example, of the de-hairing processes from Nazer *et al.* (2006) in Palestine resulted in 58% reduction of the wastewater treatment costs, and savings in water use up to 58% and on chemicals up to 28%. Guterres et al. (2010) showed that minimization of water consumption was possible at industrial scale by reusing without damaging the quality of the leather. The implementation of cleaner technologies entails modifying the process of the industry as shown in Figure 1-1.

Figure 1-1 Conventional Tanning *vs.* CP tanning (Adapted from: PROPEL 1995)

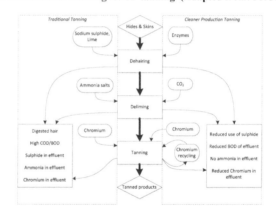

The implementation of CP has not been unproblematic:

(a) The efforts from the industry entail determining the best available options based on the circumstances of the unit operations in the tannery sector; there is no single method for producing leather. Counselling is not always available on this regard.

(b) Incentives for employees for adopting CP implementation have not been established.

(c) Regular monitoring and reporting of performance have not been set.

(d) Follow-up from authorities does not aim at pollution prevention and waste reduction.

(e) Environmental standards do not take into account local conditions and whenever compliance is difficult, a long-term plan for upgrading the operations is not available (Van Berkel, 2007; Baas, 2007; UNEP, 1991).

It is believed that in order for micro tanneries to keep up with market demands and to be able to comply with environmental legislation, these must develop a strategy of integration in the leather value chain, and to learn to work collectively through associations (Van Hoof and Herrera, 2007; Van Hoof, 2005; Rivela *et al.*, 2004; Kennedy, 1999).

The economy of a country can also be affected by the tanneries' negative impacts. Food production can be adversely impacted by the quality of the water involved in crop irrigation when wastewaters from tannery effluents are discharged directly onto adjacent water bodies or land. The costs entailed in the recovery of the water bodies and the soils are considered high opportunity costs (UNEP, 1991). In order to avoid the fact that trade liberalization may create pollution havens in Latin America, it has been reported that increased stringency of environmental enforcement is necessary (Jenkins, 2003). UNEP (2006) proposed multilateral environmental agreements on particular environmental issues.

1.2.3   The relationship between environmental agencies and McSEs

The relationship between environmental agencies and polluters is, hence, especially difficult in the context of micro and small sized enterprises in developing countries.

As shown in the previous section, on the one hand, McSEs face the paradoxical situation of having the potential to be a driving force behind equitable economic and industrial development while being extremely vulnerable because of their limited access to educational, legal, technological and financial opportunities (UNIDO, 2005). They also cause substantial collective environmental harm to water bodies (Blackman, 2010; Baas, 2007; Van Berkel, 2007; Blackman, 2006; Siaminwe *et al.*, 2005; Frondel *et al.*, 2005). In fact, small-scale enterprises usually employ environmentally unfriendly obsolete processes. It has been identified that they are collectively responsible for 70% of all local and global industrial pollution (Le Van Khoa, 2006; Soni, 2006; Hillary, 1997).

On the other hand, environmental agencies in developing countries like in South America pay little attention to prevention pollution policies, barely cope with control and pollution remediation and focus mainly on end-of-pipe solutions that entail investments that are usually unaffordable for McSEs (Thabrew *et al.*, 2009; Van Berkel, 2007; Montalvo, 2003; Ashton *et al.*, 2002). Environmental agencies may also support what is considered ideal for the social interest of environmental protection: the development of optimal solutions by the industry requiring radical innovation. Radical innovation, as opposed to gradual innovation can be problematic for these industries because it can work against the firm's own interests (profit and survival) (Moors *et al.*, 2005). Not surprisingly, McSEs end up in frequent conflicts with the environmental agencies.

The nature of the relationship of McSEs with environmental authorities can be considered a complex one (Mason and Mitroff, 1981, in Van de Kerkhof, 2004; Gummesson, 2007a) and, therefore, is characterized by

(1) scientific uncertainty as research is missing in strategic areas of prevention options like CP implementation;

(2) conflicting interests between organizations dealing with social matters and/or land issues, and those dealing strictly with technical and environmental concerns that aim ideally for radical innovation or end-of-pipe approaches;

(3) the urgency to formulate policy since it is recognized today that McSEs cause great collective damage;

(4) the strong linkage to other problems such as to the phenomena of informality and/or social inequity; and

(5) limited interdependency among actors such as the medium sized enterprises that compete for credit and for access to capacity building programmes with smaller sized enterprises but that may also offer McSEs, for example, opportunities to integrate horizontally in the production chain or to work on solid waste valuation programs in a collective manner.

As a consequence, the literature points to three issues:

a) The use of regulatory, market and persuasive policies often bypass McSEs because of their specific characteristics, and the only possibilities left to Government to deal with them are legal actions. Lawsuits against the polluters as well as the defaulting environmental authorities are more common (Blackman, 2010; Sánchez-Triana et al., 2007; Delli Priscoli, 2003). On the one hand, McSEs in the water sector are generally not included in the policy incentives targeted at the private sector and mainly end-of-pipe measures are often unaffordable leading to a cycle of exclusion, vulnerability and social unrest. The approaches can result in expensive end-of-pipe solutions, relocation without solving the polluting practices, and even industry shut downs with social implications (such as in Delhi and Agra in India, and Bogotá and Villapinzón in Colombia) (Soni, 2006; Van Hoof, 2005). On the other hand, it is difficult to work with the smallest industries that do not even have monitoring systems that can help them evaluate the benefits from implementing CP for example (Ashton et al., 2002; personal communication, 2009). It is being observed in developing countries as far apart as India (Soni 2006) and Colombia, that regulatory models and stringency of environmental policies are positively correlated with end-of-pipe techniques while costs savings, general management systems and specific management tools tend to favour cleaner production.

b) The implementation of cleaner production (CP) in McSEs in developing countries has, in general, been ineffective (Blackman, 2011; 2010; 2006; 2008; 2000; Montalvo and Kemp, 2008; Baas, 2007; Ashton et al., 2002; Blackman et al., 2012; 2009; 2007a; 2007b; 2006) even though CP has high potential benefits for industries. Cleaner production strategies reduce emissions and waste, and promote cost savings due to reduced resource use and reduced waste generation (Montalvo and Kemp, 2008). The mainstream approach to CP, based on quantitative approaches that correlate data but not context, can ignore crucial aspects related to the complexities of McSEs (Del-Río-González, 2009; Howgrave-Graham and Van Berkel, 2007). This approach has not focused on a profound understanding of their specific contexts and has not recognized specific needs (Van Berkel, 2011; 2010; 2007; Baas, 2007; Sánchez-Triana et al., 2007; Mitchell, 2006; Ashton et al., 2002; Baskerville, 1997). Besides, multiple actors are involved in cleaner production implementation. Bringing them to a common goal and commitments is especially difficult when there have been serious disputes between environmental authorities and polluters (Baas, 2007).

c) Although stakeholder participation may support the management of environmental problems, it does not by itself guarantee successful problem solving (Estacio and Marks, 2010; Hirsch et al., 2010; Jacobs, 2010; Reed et al., 2009; Terry and Khatri, 2009; Wiber et al., 2009; Ataov, 2007; Hayward et al., 2004; Van de Kerkhof, 2004; Karl, 2000; Berk et al., 1999; FAO, 1995; Grimble et al, 1995; Fiorino 1990), or in the specific case of vulnerable communities, that their interests will be respected (Cleaver, 1999).

1.2.4   Existing gaps in the theoretical fields of policy instruments, cleaner production and stakeholder participation

The above problematic issues suggest the existence of gaps in knowledge in the theoretical fields of policy instruments, cleaner production and stakeholder participation:

The theoretical field of policy instruments has not been successful in integrating the principles of sustainable development with rules of procedures that foster behavioural change and self-reliance in the context of McSEs (Blackman, 2010). It needs to deal with systemic and more comprehensive instruments (Wieczorek *et al.*, 2012; 2010) and focus at developing case study work for understanding change (Kemp and Pontoglio, 2008; White, 2008).

In the field of cleaner production, there is a gap in knowledge regarding the understanding and handling of the complex relationships between the industries and the environmental agencies; of the way different mechanisms support CP implementation and, hence, foster change; on how to build consensus; and on how the choice of technology is being implemented (Montalvo and Kemp, 2008; Montalvo, 2003). Authors like Van Berkel (2011), Altham (2007), Baas (2007), Van Berkel (2007), Mitchell (2006), Oosterveer *et al.* (2006), Frijns and van Vliet (1999) recommend focusing on context-specific strategies for CP implementation in McSEs aiming at building appropriate tools for behavioural changes and to pilot and evaluate the impacts. Within this scenario, the assessments on CP should no longer be limited to technical approaches. Instead, the social and psychological dimensions of organizational change are now being taken into account in the design and delivery of CP programs (Lozano and Huisingh, 2011; Van Berkel, 2011; 2007; Baas, 2007). By learning how to work in networks, with joint strategies and collective actions, these industries can acquire sustainability and improve their environmental status (Baas, 2007; Van Berkel, 2007; Frijns and van Vliet 1999).

Stakeholder participation has elaborated the virtues of participation (and drawbacks), and models of communication (Robinson and Berkes, 2011; Estacio and Marks, 2010; Geist, 2010; Hirsch *et al.*, 2010; Jacobs, 2010; Ataov, 2007; Van de kerkhof, 2004; Eversole, 2003) but does not have the tools for building consensus or for decision-making (Cleaver, 1999). For decades, technical elites have managed environmental problems without having to confirm community advice. Their decisions are still taken **for** and not **with** communities (Delli Priscoli, 2003).

This thesis discusses, hence, that since there are existing gaps, the above-mentioned theoretical fields cannot perform together the way they are needed in the context of McSEs in developing countries. A lack of integration and coordination between the theoretical fields of policy instruments, cleaner production, and stakeholder participation is then also seen as a major drawback in the context of McSEs in developing countries **(Chapter 2)**.

1.3   Research questions and objectives

The overall research question of this thesis is:

How can Micro and Small Enterprises be effectively engaged to achieve national water and environmental policy goals and approaches?

Against this background, this dissertation focuses on Colombia as a single, but layered case study. It studies the national and three sub-national situations of micro-tanneries impacting water bodies.

The three sub-research questions are:

1. What is the current institutional framework in Colombia for integrating and supporting micro and small industries in their adoption of cleaner production?
2. Which perceived mechanisms support cleaner production and how do these influence the adoption of cleaner production by micro and small industries in Colombia?
3. (How) can action research methodology be developed and tested to help McSEs implementing cleaner production?

From the latter research questions, the overall objective is:

To investigate how the specific problems of McSEs can be taken into account in developing water and environmental policy

The three sub-objectives are:

1. To investigate the current institutional framework and the role of formal authorities and institutions, with particular focus on participation and social inclusion in the context of McSEs adopting CP in Colombia.
2. To identify and analyze how the perceived mechanisms involved in the adoption of CP operate in the context of McSEs in Colombia.
3. To design and implement a process that empowers the McSEs' community and raises consciousness on the part of the authorities with respect to the McSEs' needs, and promotes the socialization of CP among McSEs in Colombia.

1.4   Steps in the thesis

There are seven methodological steps in this thesis:

First, the overall research question and the sub-questions were initially dealt with through a thorough literature survey.

Second, a methodology was developed for addressing the research questions. Two methods were developed: the comparative case study method from Yin (2009) combined with the extended case method (Buroway, 1998), and an action research approach – created during the research - called SASI (Systematic Approach for Social Inclusion). The combined case study method aimed at (a) acquiring a deep understanding of the multiple forces operating at McSEs implementing CP, and (b) getting acquainted with the chain of events (like behavioural change on CP implementation) during a long period of time. SASI is based on five principles, includes six spiral steps, derives from methods and techniques from policy instruments, stakeholder participation, and cleaner production, and is inspired by complex negotiation (including conflict resolution) (Holman, 2010; Holman *et al.*, 2007; Fisher *et al.* 1991; Lax and Sebenius, 1991) and managed learning (Burnes, 2004; Lewin, 1946; Schein, 1996) through an actor-oriented approach. SASI is characterized by the highest level of stakeholder participation in policymaking and scientific practice (Van de Kerkhof, 2004), and was inspired by approaches on sustainability. Developing action research in this community of micro-tanneries provided an ideal opportunity because it allowed: 1) to respond to an urgency that existed because of a crisis situation, 2) to deal with complexity and, maybe, 3) to innovate.

Third, the first 'what' sub-research question was addressed through content analysis of policy and legal documents, in-depth interviews with decision makers, and analysis of newspaper clippings.

Fourth, the second 'which' and 'how' sub-research questions were answered through the combined case study methodology at the local level where two case studies were chosen (Yin 2009; Buroway, 1998). In the two tannery cases: San Benito and Cerrito in which the involvement of the researcher (i.e. myself) was limited, the research analysed over a period of six years the contextual conditions relevant for CP implementation.

Fifth, the third sub-research question was addressed by testing the SASI approach with the micro-tannery community of Villapinzón, which represented an extreme case of pollution and social exclusion of McSEs in the tanning sector. Once the first SASI steps were met, the CP implementation could start through different quantitative and qualitative methods such as doing material & balance analysis, establishing a monitoring system that started with indicators based on conflict assessment, defining the technical options, and doing composting from the de-hairing and grease residues. This approach was developed over a period of six years.

Sixth, conclusions were drawn on McSEs in developing countries.

Seven, the data collected through the case studies was analysed, compared and integrated in order to draw contributions for both theory and practice, and an elaboration of the shortcomings and future perspectives was undertaken.

1.5   Focus, strengths, risks and limitations

1.5.1   The reflexive approach

This research adopts a reflexive approach. Reflexive science does not set boundaries between science, policy-making and society. It aims to build knowledge by deploying multiple dialogues among stakeholders, policy makers, practitioners and scientists (Gummesson, 2007a; Van de Kerkhof, 2004). For the purpose of solving complex environmental problems, all actors involved are placed on an equal footing for their contribution to knowledge. It enhances theory reconstruction, intervention, process and structure, and engagement (Burawoy, 1998:4). Each one of these issues is developed in this section.

Theory reconstruction

The reconstruction of theory is cyclic. It goes through intellectual debate by being confronted by participants that adapt it, refute it, or even extend it, before integrating it back into the scientific debate. Theories do not arise from the data but are constantly being developed and revised through an established dialogue in specific contexts. Dialogue is the unifying principle in reflexive science (*ibid*: 16). Context is seen then as a point of departure contributing towards the growth of knowledge, but not as an end point (Burawoy, 1998:13).

Focusing on micro-tanneries allows for not only looking at the relationship between the researcher and the participants but at the local and extra-local forces at work such as those springing from social, economic and environmental issues; from informal and formal sectors; from rural and urban settings, and from the way these interact among each other in the context of Colombia. By identifying these interactions, a constant dialogue is expanded not only between the different forces involved but also with theory itself.

Intervention

By being grounded in reflexive theory, which does not set boundaries between the social and natural sciences or between policy-making, science and society in order to solve today's pressing and complex environmental issues, this research incorporates intervention from all fields.

The driving thesis of this research is that articulating multiple dialogues and interventions among stakeholders[10], policy makers, practitioners and scientists may help to arrive at better explanations regarding environmental empirical phenomena. Such explanations may hence, produce more effective results in dealing with the complexity and the interdependence of factors impacting on the environmental problems society is facing.

Process and structure

In-depth understanding of specific contexts is considered necessary and prone to improvement by setting multiple dialogues. Those multiple dialogues imply multiple *processes* and methods with regards to the growth of knowledge. Action research needs a structured and systematic approach that improves stakeholder participation towards effective results on impending water problems related to McSEs. The inspiration for the systematic approach was taken from negotiation theory.

Engagement

Scientists, policy makers, practitioners and stakeholders may internalize their own roles more effectively in terms of solving and facing their daily challenges once a high degree of engagement and participation is reached. Behavioural change is expected as a result. This work envisions an actor-oriented approach and *engagement as the road to knowledge* (Burawoy, 1998:5).

1.5.2   Strengths and risks

In-depth understanding of complex problems has been pursued and, hence, qualitative research through case study research and action research over a long period of time have been developed. Whenever needed, the appropriate quantitative research for CP implementation was also developed.

Based on qualitative approaches, this research seeks then to present, as accurately as possible, the facts as they happen in real life through the eyes of a multiplicity of actors and by taking note for example of the number of people assisting meetings and/or through follow-up of the commitments set by the tanners themselves and derived from the conflict resolution exercises developed at the study-site.

Despite the egalitarian purpose of this action research work based on authenticity, reciprocity and trust, the issue of representation entails the risk of abuse of power (Jacobs, 2010; Burawoy, 1998; Denzin and Lincoln, 1994). The "power effects" are domination, silencing, objectification and normalization. In fact, an intervening social scientist may not always be impartial or even avoid dominating others (Burawoy, 1998; Denzin and Lincoln, 1994). I was aware of these effects and diminished my own intervention, as the tanners became their own masters.

---

[10] Representatives of societal and environmental non-governmental organizations, business or any other citizen that has a stake, interest on the problem and to a solution of the problem (Van de Kerhof, 2004)

In the course of this study it must be acknowledged that the privileged social and educational position of the researcher (i.e. myself) did influence the results in the context of a system of exclusion based on the social, educational or economic level of individuals. In the past, even environmental NGOs and the most prestigious private academic centres had excluded the micro-tanners from having access to technical solutions. The director of the faculty of environmental engineering from one of the most reputable private universities in Colombia has even considered it useless to help them solve their problems. He believed that the micro-tanners belonged to the same category of the dinosaurs and were, hence, condemned to extinction (personal communication, 2006-1)[11].

This work proposes that the degree to which the research empowers a vulnerable community or the degree to which the research responds to the situation and the people who are the subject of study should at the end, be taken into consideration for the academic evaluation process (Denzin and Lincoln, 1994).

1.5.3   Limitations

Some issues relevant to the subject of this thesis are taken into consideration in order to set its boundaries:

(a) Considering the complexity of the subject and the economic, social and environmental issues at stake, this research will therefore be restricted to analysing on the one hand, the literature on stakeholder participation, policy instruments, complex negotiation (including integrative negotiation, large groups conflict resolution), and managed learning (action research) theories; and on the other hand, to understanding the aspects focusing on the socialization of cleaner production by micro-tanneries in Colombia. Other topics will be addressed only when needed to understand the subject being researched.

(b) SMEs are traditionally referred to as micro, small and medium sized enterprises from either the industrial, commercial or service sectors. In this research, the term McSEs will just refer to the micro and small enterprises, as they are defined in Colombia. The case studies deal specifically with the industrial McSEs which impact upon the environment.

(c) Working on the challenges McSEs face when they impact the environment in Colombia shall not focus on the main pollution problem of the Bogota or Cerrito rivers, but rather on the risks posed to public health due to industrial discharges, a challenge that has not yet being taken into consideration by the institutional framework itself.

(d) This thesis focuses not only on the theoretical and analytical levels, but also on the practical levels bringing theory to implementable grounds through a systematic approach. Achieving CP implementation will be limited by the time factor of this research.

(e) The case studies in this research will focus on Colombia; and thus the extrapolation to other countries will be limited. McSEs and developing countries entail by themselves different and diverse categories. Nevertheless, since the SASI approach is based on universal principles, this extrapolation will go as far as considering other open developing economies (like for instance in Latin America) that share certain characteristics with Colombia.

---

[11] This comment was done at a Directive Meeting at the NGO Rio Urbano, when looking for supporting partners for the SWITCH project

1.6   Structure of thesis

Against this brief background, this Ph.D. thesis has 7 chapters. It presents the literature review and a theoretical framework (Chapter 2), the methodologies to be used in the research and proposes an integrated methodological framework called SASI (Chapter 3). It elaborates on the layered case study by first explaining the national context in Colombia (Chapter 4), and then develops the two comparative case studies (Chapter 5), and the action research within Colombia (Chapter 6). It presents the conclusions and contributions for theory and for practice elaborating on the shortcomings and perspectives (Chapter 7).

1.7   Inferences

The micro and small sized enterprises (McSEs) constitute a complex challenge in the developing countries since despite being an important group in terms of numbers and environmental impacts; they have been traditionally overlooked by policies and approaches.

Working on the McSEs' concerns has practical and theoretical implications. The problems are presented on the one hand with respect to the McSEs' specificities and to the McSEs' relationships with environmental agencies, and on the other hand, with respect to the existing gaps on the theoretical fields of policy instruments, cleaner production and stakeholder participation in the context of McSEs. Dealing with McSEs defined as marginalized groups in the water sector in developing countries implies a conflict resolution dimension inherent to the nature of the water related relationships and to the unconnected formal and informal worlds and entails many times social exclusion, and asymmetries from their own specificities.

Understanding the environmental institutional framework and the characteristics of McSEs, the role of the mechanisms influencing CP adoption which seems beneficial to them, and actually implementing (intervening on) an innovative approach based on action research may contribute towards building knowledge on theoretical, methodological and practical-policy issues related to micro-industries in Colombia, and on fostering change for successful CP adoption in McSEs' contexts.

2 A framework embracing diversity

2.1 Introduction

The literature on sustainability has adopted terms and concerns on multidisciplinary and holistic visions, prevention, cross-sectoral planning, inclusion of the marginalized communities, empowerment, representativeness, power effects, long-term change, context-oriented research, bottom-up approaches, diversity, self-reflection, transformative participation, agents of change, values and purposes of society. They all suggest building a better world but the question remains how we can make them work together.

Through a proposed theoretical framework embracing diversity for marginalized communities impacting water bodies, the complexity of today's developing world is faced. This chapter supports the overall goal of this research, which is to contribute towards the integration of micro and small sized enterprises (McSEs) in environmental policies and approaches. This chapter builds the theoretical framework that supports the integration of industrial McSEs. It intends to improve the way the theoretical fields of cleaner production, stakeholder participation and policy instruments operate in the context of McSEs in developing countries by adding insights from the theories of negotiation/conflict resolution, and action research.

The McSEs' concerns are framed by three main components that inspired the three sub- research questions of this dissertation. The problems focus on understanding why these are overlooked by the environmental policies, which mechanisms best support CP adoption and how they operate, and how can the approach towards McSEs be improved in terms of their adoption of CP.

This chapter presents cleaner production (2.2), stakeholder participation (2.3), negotiation/conflict resolution (2.4), action research (2.5), policy instruments (2.6), comparative analysis (2.7), and implications (2.8).

2.2 Cleaner production (CP)

2.2.1 The place of cleaner production in sustainable development/industrial transformation theories

For water issues, sustainability means that the needs of people should be satisfied and that as a limited resource water should be used and shared wisely (Gupta and Lebel 2010). The wise use must include all kind of uses (even the water uses of the polluters belonging to underprivileged groups). McSEs need to solve the adverse environmental impact they cause by choosing the best approaches that suit their needs as underprivileged enterprises. Their challenges were underestimated in the past (Baas, 2007; Van Berkel, 2007; Siaminwe *et al.*, 2005).

There are two types of abatement measures to deal with industrial pollution and that can complement each other: (1) End-of pipe measures, and (2) prevention focused strategies such as cleaner production (CP). While both may mitigate the harm induced by the industrial activities,

- end-of-pipe measures function by adding treatment processes for, primarily, cleaning the wastewater without changing the production process, while

- prevention focused strategies such as cleaner production focus on preventing the generation of waste or reducing the unnecessary use of resources.

The latter entails changing the production process itself, modifying the management of resources, housekeeping, substituting materials, and using new technologies (Gutteres *et al.*, 2010; Montalvo and Kemp, 2008; Nazer *et al.*, 2006; Frondel *et al.*, 2005; Van Berkel and Bouma, 1999).

Prevention focused strategies such as CP may lower the costs of today's water management and support more integral environmental, social and economic solutions. Preventing and reducing contamination at the source has more advantages than controlling contaminated discharges or eliminating the negative effects of produced waste streams (Baas, 2007; UNEP, 2006; Van Berkel and Bouma, 1999).

Today, many case studies show the relevance and importance of CP (Lozano and Huisingh 2011; Van Berkel, 2010; 2011; Gutterres *et al.*, 2010; Montalvo and Kemp, 2008; Blackman, 2006; Frondel *et al.*, 2005; Visvanathan and Kumar, 1999). There is a growing number of case studies of CP implementation in a wide range of industry sectors and businesses of varying sizes worldwide (Howgrave-Graham and Van Berkel, 2007).

Cleaner production (CP) works in conjunction with the principles of sustainability. Ever since the Rio Declaration on Environment and Development (1992), CP has represented an opportunity to focus on the transition (a) 'from waste management policies and approaches to waste prevention or industrial innovation policies of waste minimization' (Baas, 2007: 1205), and (b) from working *at* and *for* the industry towards an integrated and social-oriented approach of working *with* all the stakeholders that have interests in CP in order to build commitment towards prevention (Thabrew *et al.*, 2009; Van Berkel, 2007; Dieleman and Huisingh, 2006; Montalvo, 2003; Huisingh, 2002).

2.2.2   What is cleaner production?

Cleaner production (CP) is defined as an integrated preventive environmental strategy to processes, products, and services to increase overall efficiency and reduce risks to humans and to the environment (UNEP, 1999).

CP has been defined as a strategy in which all kind of processes and activities carried out generate the lowest environmental impact. As a strategy, CP can be applied to processes, and products in any industry and to the different service sectors in society (Siebel and Gijzen, 2002).

2.2.3   History of CP

The first attempts towards implementing prevention approaches in an industry came in 1975 by the American company 3M through their 3P program 'Pollution Prevention Pays'. Inspired by this initiative, the American Dupont Company elaborated a manual on pollution prevention, which was adopted in 1988 by the Environmental Protection Agency (EPA). One year later two demonstration projects were conducted in Sweden and in the Netherlands. The Dutch project called PRISMA inspired in 1991 the United Nations Environment Programme (UNEP) to support CP. After the Rio Declaration in 1992, UNIDO (United Nations Industrial Development Organization) and UNEP created through a joint effort the National Cleaner Production Centres (NCPCs), which were targeting especially the developing countries (Dieleman, 2007; UNEP, 1991).

Despite the enthusiasm generated in the nineties, further evaluations have shown important delays and up-scaling problems in CP implementation in the small industrial sector (Dieleman, 2007; Visvanathan and Kumar, 1999).

2.2.4 Elements of CP

Even though this Chapter discusses the theoretical framework, since CP has been considered more a strategy than a theory, the different elements and analysis discussed by the literature are tied to the practice of the theoretical aspects of CP.

Principles and practices

Since cleaner production started as a strategy to be developed within each industry, four principles were considered essential for the successful implementation of a CP program in a company (Van Berkel and Bouma 1999): (a) CP activities must be supported and initiated by a management commitment; (b) employees involved in the daily operations and implementation of CP must be strongly involved; (c) cost awareness and proper cost information showing savings due to CP implementation are a convincing factor; and (d) a well-established project management on CP opportunities is essential. It is hence based on the fact that raising cost awareness in an industry is the convincing factor towards CP implementation and thus towards self-reliance on sustainable principles.

As a strategy with respect to the efficient use of natural resources and pollution prevention, CP entails five prevention practices (UNEP, 1994; USEPA, 1992, cited in Van Berkel and Bouma, 1999).

1. Good operational practices,

2. Chemicals' substitution,

3. Technological changes,

4. On-site reusing, recovering and recycling, and

5. Product modifications

Siebel and Gijzen (2002), proposed to expand those strategies beyond the industries themselves and to look at the final discharges and impacts. They proposed to bring the CP concepts to the level of not only industries but also of households. They presented three steps in the adoption of CP that entail looking at the productive processes through a different perspective. The three steps are: (a) pollution prevention, (b) treating for re-use, and (c) self-purification.

Without CP, all types of waste effluents (domestic, industrial and runoff) tend to be mixed together making the possibility of material recovery very unlikely; waste is discharged without thinking of its real value while re-using it can imply additional sources of profits to any industry; and water bodies do not normally get the chance to go through their own self-purification processes (Siebel and Gijzen, 2002; Gijzen, 2001; 1998).

Recently, over the last decade, an essential issue has been the recognition that as a basic innovation, CP implementation involves besides the industrial sector, other institutions and stakeholders and there are significant barriers, which are related to the social, economic, political and cultural milieus (Lozano and Huisingh, 2011; Van Berkel 2011; 2007; Dieleman, 2007; Worrell and Van Berkel, 2001; Dasgupta, 2000; Frijns and Van Vliet, 1999). In order to build a strategy for the diffusion phase of CP, researchers agree that three aspects are to be considered: (a) active and intense information, (b) build-up of adequate policy instruments stimulating CP, and (c) cooperation among all involved stakeholders (Huisingh, 2002; Worrell and Van Berkel, 2001; Dasgupta, 2000; Frijns and Van Vliet,

1999). CP asks hence, for context specific research aiming at building appropriate tools for behavioural changes (Altham, 2007; Baas, 2007; Van Berkel, 2007; Mitchell, 2006; Oosterveer *et al.*, 2006). The latter is addressed through the identification of the mechanisms supporting CP.

Building–up appropriate policy instruments and cooperation among stakeholders have been targeted in recent years. CP implementation normally uses voluntary agreements between environmental agencies and industries, public disclosures and technical manuals targeting specific industries (Blackman *et al.*, 2012; 2009; Sánchez-Triana *et al.*, 2007).

Mechanisms supporting CP implementation

Gunningham and Sinclair (1997) consider that the mechanisms for CP implementation could be either exogenous or endogenous to a given firm and pertain to (a) the regulatory framework, (b) access to technology, (c) economic incentives, and (d) sound information.

Other authors consider that other factors should be considered first. For Montalvo (2003), better outcomes on CP policies could be obtained if the structural determinants of the behaviour are taken into account. For him the technological and organizational capabilities should be ranked first as backbones of the willingness to adopt CP. In 2006, Mitchell emphasized that there are systemic problems with regard to incorporating pollution prevention concepts. He proposed that besides the classical mechanisms from Gunningham and Sinclair (1997), the attitudinal ones be considered the root causes of commonly cited barriers (Van Berkel, 2011; 2010; 2007).

Consultation: the mainstream approach to CP

The mainstream consulting approach has been in charge of the CP programs in most of the developing countries as its techniques and principles have been widely adopted by specialists (Dieleman, 2007; Baas, 2007; Mitchell, 2006; Sánchez-Triana *et al.*, 2007; Baskerville, 1997).

A definition of consulting is inspired by Steele (1975) and taken from Baskerville (1997):

Organizational consulting is a process of temporarily using external expertise in order to obtain objective analysis, specialist knowledge or the benefits of cross-organizational experience without permanently acquiring additional organizational members.

This process is seen as a linear sequence of 4 stages: engagement, analysis, action and disengagement (Lippit & Lippit, 1978; Kubr, 1986). Baskerville (1997: 34-37) highlighted its principles and characteristics (Table 2-1):

Table 2-1 Characteristics & principles of consulting

a.	Outsider view of a problem. No involvement from consultant
b.	Analysis in the context of similar situations
c.	Proposed solutions are expected to be successful
d.	Knowledge transfer expected to be applied in future organizational situations
e.	Planning (as part of analysis) not a collaborative phase

f.   Consultant committed only to the owner of the firm

g.   Learning from the process is kept on an internal document

h.   Motivation is towards commercial benefits & proprietary knowledge

Source: Adapted from Baskerville (1997).

Proposed methodology

With the above as a backdrop, the proposed methodology from the early period of CP by the US/EPA manual from 1988 presented five phases (Figure 2-1): 1. Recognizing the need to prevent pollution, 2. Planning and organization, 3. Assessment, 4. Feasibility studies, and 5. Implementation. This methodology implied a constant reassessment in order to look for new areas, targets and repetition of processes in the implementation (Dieleman, 2007).

Figure 2-1 CP Methodology

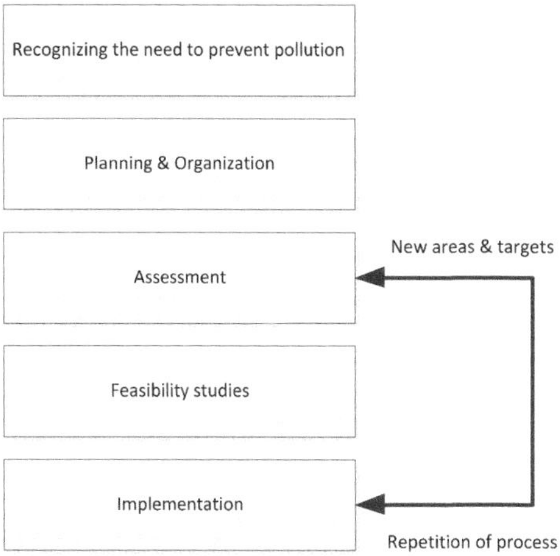

Source: US/EPA, 1998

2.2.5   Benefits, drawbacks and gaps in knowledge

The power of CP is presented in the definition of the concept itself (Table 2-2). It is a strategy that can be adapted to processes and products in any industry. As stated by Van Berkel since 1999, CP lowers the costs of today's water management and supports more integral and sustainable environmental, social and economic solutions. Hence, it can support the transition towards prevention policies and cross-sectoral planning. It is based on self-reliance. Preventing and reducing contamination at the source has more advantages than controlling contaminated discharges or eliminating the negative effects of produced waste streams. It introduced rules of procedure in industrial processes based on prevention and concepts such as waste valuation, re-use, and recycling. It is open to new concepts and innovations and is expected to offer a wide array of technical options (Baas, 2007; Dieleman, 2007; US/EPA, 1988).

Despite the mentioned benefits, literature reports that, in general, continuous evaluation, possibilities to repeat processes, implementation of multiple CP options are not becoming a daily chore as of yet. CP Consultants do not dig enough into the wide array of possibilities and do not work through consensus (Dieleman, 2007). The CP concepts have not been incorporated at the strategic stages of decision-making at the local, regional, or national levels. CP policies do not work in parallel with command-and-control (Frondel *et al.*, 2005; Visvanathan and Kumar, 1999) and instead, rely too much on quick savings to stimulate CP implementation (Van Hoof, 2005; Gunningham and Sinclair, 1997) (

Table 2-2).

Over the last decade, the efforts on CP implementation have proved to offer benefits to firms. The problem has been that especially in the developing countries, the efforts have privileged the larger firms (Montalvo and Kemp, 2008; Blackman, 2006; Frondel *et al.*, 2005). CP lacks larger scale recognition of McSEs in developing countries since these firms are not implementing the CP approach by themselves (Montalvo and Kemp, 2008; Blackman *et al.*, 2007; Dieleman, 2007; Soni, 2006; Frondel *et al.*, 2005). Since CP has basically focused on technical solutions, the social and economic components have not been targeted enough (Mitchell, 2006; Hamed and El Mahgary, 2004) and the difficult relationships between industries and agencies are not handled (Thabrew *et al.*, 2009; Van Berkel, 2007; Montalvo, 2003). Besides, the monitoring and evaluation processes are exclusively based on quantitative approaches that are usually unrealistic for McSEs (Del-Río-González, 2009; Howgrave-Graham and Van Berkel, 2007).

Table 2-2 Benefits and drawbacks of the interpretation and implementation of Cleaner Production

Benefits	*Source*
Can be applied to products & processes in any industry	Siebel and Gijzen, 2002
Introduced concepts such as waste valuation, recycling, re-using and steps to be followed (rules of procedure)	Gutterres *et al.*, 2010; UNEP, 1991; Dieleman, 2007; Van Berkel and Bauma, 1999
Based on the principles of sustainability, supports prevention approaches & policies	Baas, 2007
Based on self-reliance	UNEP, 1991
Potentially cost-effective	Van Berkel and Bauma, 1999
Open to new concepts & innovations	Baas, 2007; Dieleman, 2007
Can offer wide array of technical options	Dieleman, 2007
Successful results on large industries	Dieleman, 2007
Drawbacks	
Initially designed only with regard to technological solutions	UNEP, 1991; Frondel *et al.*, 2005
Only recently CP projects acknowledge social concerns, building consensus, and fostering behavioural change	Van Berkel, 2007; Dieleman and Huisingh, 2006
Regulatory instruments do not support CP. CP policies do not work in parallel with command-and-control	Frondel *et al.*, 2005
Relies too much on quick savings to stimulate implementation	Van Hoof, 2005
In reality CP lacks implementation of multiple options	Dieleman, 2007
Lacks repeating processes	Dieleman, 2007
Lacks larger scale recognition	Dieleman, 2007
CP programs are not successfully targeting McSEs: the monitoring and evaluation processes are exclusively based on quantitative approaches that are usually unrealistic for McSEs	Del-Río-González, 2009; Howgrave-Graham and Van Berkel, 2007
The conflicting relationships between environmental authorities and McSEs are not handled. Does not handle big groups undergoing technological change.	Thabrew *et al.*, 2009; Van Berkel, 2007; Montalvo, 2003

A literature analysis on the drawbacks of the interpretation and implementation of CP identified three major gaps in knowledge (a) the integration between regulatory, market and persuasive instruments, (b) the lack of vision of CP implementation as a change process and (c) the difficulties addressing McSEs.

(a) Since CP was always focusing on a company's perspective rather than on a regulatory perspective (Frondel *et al.*, 2005), the CP policies grew separate from regulation. They need to be integrated into the regulatory frameworks and incorporated at the strategic stages of decision making at the local, regional, or national levels. It has been observed in developing countries as distant as India (Soni, 2006) and Colombia (Van Hoof, 2005) that classical regulatory models and a stringency of environmental policies have been positively correlated with end-of-pipe techniques while cost savings, general management systems and specific management tools had favoured cleaner production. Siaminwe *et al.* (2005) in Zambia, Blackman *et al.*, (2012), Sánchez-Triana *et al.* (2007) and Van Hoof (2005) in Colombia, and Reijnders (2003) for example, consider that there are important weaknesses in the conventional regulatory frameworks that encourage industries to remain indifferent to production approaches based on prevention. Hamed and El Mahgary (2004) in Egypt, propose that the discharge limits should be fixed on loads and not just on concentrations in order to avoid the use of dilution and to stimulate prevention. For authors like Zhang *et al.* (2008), Montalvo and Kemp (2008) or Altham (2007), even though the government regulations are initially a determining factor for CP uptake, more attention should be given to market instruments, information instruments, and economic incentives regarding CP. For Baas (2007) and Frijns and Van Vliet (1999), policies should reflect the reality of organizations. For them, policies that do not take into consideration the McSEs' views and needs, work against building conditions for dialogue and trust which are essential for testing, disseminating and adopting new concepts like CP. Authors like Baas (2007), Van Berkel (2007), and Frijns and Van Vliet (1999) state that government initiatives that reinforce prevention, such as Malaysian programmes stimulating green champions, are showing positive results towards CP implementation. The same is observed in seven of the OECD countries where CP has become the preferred strategy chosen by the industry in general. Among them, Japan shows the highest percentage of CP adoption at 86.5% (Frondel *et al.*, 2005). Relying basically on financial win-win situations as one of the CP principles has brought frustrations to the diffusion of CP (Gunningham and Sinclair, 1997). In South America for example, CP has been handled through voluntary agreements and relied heavily on quick savings in the production processes to incentivize owners of the firms (Van Hoof, 2005). Such savings are not always easy to reach in the short term. Some countries' policies do not have fees for water use or for polluters, others do not have access to substituting materials at lower costs and/or low-cost clean technologies like grease traps for example. Even though CP entails environmental improvements, this does not automatically result in savings to firms because such savings vary from a precise particular circumstance of a given industrial sector, individual firms or environmental issues and costs cannot always be precisely calculated with respect to the smaller firms.

(b) In order to respond to the theoretical foundations of CP that emphasized continuous evaluation, repeated processes, and implementation of a wide array of CP options, Dieleman (2007) and Van Berkel (2007) propose to see CP implementation as a comprehensive change process that needs to focus on the specific contexts with multiple stakeholders. The latter responds also to the needs of working not just with the traditional mechanisms supporting CP but the organizational components as well. The latter gives a whole new social dimension to the field of CP. Through this vision, the CP implementation process entails aside from technical assessments; elaborated social and institutional assessments, feasibility studies, and elaborated interventions aiming at behavioural change like Community-Based Social Marketing that aims at working specifically to break barriers towards clean attitudes (Van Berkel, 2007); at consensus building (Hamed and El Mahgary, 2004; Mitchell, 2006); or at focusing on social systems (Lozano and Huisingh, 2011; Van Berkel 2011; Dieleman, 2007). This vision also entails disseminating CP through a consulting process that focuses more on the specific contexts and where the consultant becomes a change agent that gets involved in the implementation process itself (Dieleman, 2007; Dieleman and Huisingh, 2006). Historically, CP implementation has been disseminated through branch organisations, which rely on their preferred environmental management systems and less from academia (Baas, 2007). In Colombia, CP programmes have relied on private efforts from classical consulting bodies (Table 2-1) (Sánchez-Triana et al., 2007). Mitchell (2006) reports that in Vietnam CP knowledge, and the manner in which it is delivered is heavily dependent on foreign assistance. In general, the industry and the government find it easier to support environmental management systems that are mass-produced on the basis of end-of-pipe solutions rather than on CP implementation that demands systematic and time consuming reviews of each specific company and contexts (Baas, 2007). Besides, there is already inertia to keep on investing on existing high-cost end-of-pipe infrastructure. Authors report that the on-going higher costs needed to maintain this infrastructure prevents firms from venturing into innovative clean technologies (Baas, 2007).

(c) The McSEs' difficult relationships with authorities need to be handled and their interests recognized. In Vietnam and Colombia for example, (Mitchell, 2006; Van Hoof, 2005) cooperation and coordination of the various sectors and levels of government are lacking on CP efforts. With respect to monitoring and evaluating CP, innovative and flexible tools mixing qualitative and quantitative approaches may be more realistic (Del-Río-González, 2009; Howgrave-Graham and Van Berkel, 2007). Provided comprehensive and integral CP programs are implemented, better results with respect to the McSEs can be expected.

2.2.6　Lessons learnt for this thesis

This thesis considers that even though CP implied a major change for the industrial processes in the nineties, today's difficulties suggest that the field of CP needs to be complemented by other fields.

The CP approach cannot just be copied, as each particular situation is context specific; it calls for simultaneous changes in individual behaviour and is open to new concepts and innovations (Van Berkel, 2007). Within this scenario, the assessments of CP cannot be limited to technical approaches. Instead, the social and psychological dimensions of organizational change are now being taken into account in the design and delivery of CP programs (Lozano and Huisingh, 2011;Van Berkel, 2011; 2010; 2007; Baas, 2007). The analysis suggests hence, that the mainstream approach to CP has not been the right approach to McSEs. CP implementation seems to work in developed countries and in large industries but not in the case of McSEs (Lozano and Huisingh, 2011; Dieleman, 2007; Dieleman and Huisingh, 2006; Van Hoof, 2005; Gunningham and Sinclair, 1997). The classical consulting approach has been top-down, not focused on the consultant's involvement in order to thoroughly understand the specific contexts, to stimulate participation and learning processes for communities living within high complexity scenarios involving multiple social, economic and legal issues (Lozano and Huisingh, 2011; Del-Río-González, 2009; Baas, 2007; Howgrave-Graham and Van Berkel, 2007; Sánchez-Triana et al., 2007; Mitchell, 2006).

2.3　Stakeholder participation (SP)

2.3.1　The place of stakeholder participation in sustainable development

Meeting present needs without compromising the ability to meet future generations' needs entails unavoidably identifying the stakeholders involved, and stimulating participatory processes aimed at building agreements assuring that solutions on development include needs/interests from all kinds of stakeholders. As such, sustainable development (SD) is seen as a balancing act where different interests need to be taken into account (Morse, 2008; O'Hogain, 2008). It is argued that enhancing stakeholder and public participation in policy decisions improves the effectiveness of outcomes on sustainability (Robinson and Berkes, 2011; García and Brown, 2009; Reed *et al.*, 2009; O'Hogain, 2008; Reed, 2008; Selfa and Endter-Wada, 2008; Godard and Laurans, 2004; Van de Kerkhof, 2004; Lewis, 2001; Karl, 2000; Rowe and Frewer, 2000; World Bank, 1996a, 1996b; FAO, 1995; Grimble *et al.*, 1995; Fiorino, 1990). Participation in sustainable development has been considered essential especially when it deals with complex problems (Stringer *et al.*, 2007).

Participation stands nevertheless at the heart of the tensions between supporters of sustainable development and post-development. For authors like Escobar (1999) enhancing participation for the discourse on SD exacerbates the existing power inequalities and the 'western hegemony' (Morse, 2008). For these authors, participation is problematic because as there are power imbalances, representativeness of the different interests is at stake (Eversole, 2003), and the role of the expert in the process and aiming at consensus building are seen with suspicion.

2.3.2   What is stakeholder participation?

Stakeholder participation is based on participation theory that proposes that concerned people take part in collective decisions that affect them (Robinson and Berkes, 2011; García and Brown, 2009; Reed *et al.*, 2009; O'Hogain, 2008; Reed, 2008; Burawoy, 2007; Godard and Laurans, 2004; Rowe and Frewer, 2004; Van de Kerkhof, 2004). Stakeholder participation considers that stakeholders are the best judges of their own interest and that they can acquire the political skills needed to take part in governance (Godard and Laurans, 2004; Van de Kerkhof, 2004; Lewis, 2001; Fiorino, 1990) and that the way local members are included in a process defines the quality of the outcomes (Bakker *et al.*, 2012; O'Hogain, 2008; Kanji and Greenwood, 2001).

Based on the latter, defining participation may vary depending on the purpose: when participation is an end it supports empowering local people, inclusion and equity. When it is considered a means, it builds on specific methods aiming at implementing projects, policies or systems for wider goals (Kanji and Greenwood, 2001).

The World Bank sees participation as a process when stakeholders influence and share control over priority setting, policymaking, resource allocations and access to public goods and services (WB, 2006).

Specifically, for sustainable development, participation enhances the capabilities of the local people that have less chance to be heard through bottom-up approaches (O'Hogain, 2009; Burawoy, 2007); it implies a shift in power relations (Duraiappah *et al.*, 2005).

2.3.3   History of stakeholder participation

There is a vast literature on stakeholder participation going back for the last fifty years in the policy sciences, which reflects the increasing interest in participation at the national level in many western countries.

Late in the 60s, there was increased awareness of the benefits of participation with respect to social issues (Arnstein, 1969); in the 70s, local perspectives were incorporated into data collection and planning (Deléage, 1991); in the 80s, local knowledge was considered important and techniques on stakeholder (community) participation were developed as a reaction to traditional development processes (Morse, 2008; Duraiappah *et al.*, 2005; Kanji and Greenwood, 2001); in the 90s, public participation hit the international agenda in the Rio Declaration adopted at the UN Conference on Environment and Development (Deléage, 2002a; UNEP, 1992). Implementing Agenda 21 (A21) at the Rio Declaration entailed at the local level, to overcome the traditional consultation by stimulating a dialogue between communities and their local authorities (Coenen, 2009). Local authorities were considered the key agents in educating and building public capacity in Chapter 28 of A21 (UNEP, 1992). The principles contained in the Inter-American Strategy for the Promotion of Public Participation in Decision-Making for Sustainable Development (ISP) intended to lead to effective public policy-making (OAS, 2001). For researchers like Godard and Laurans (2004:1), even research on economic valuation of environmental issues should be anchored in participatory processes. They defended the thesis of the social process approach (SPA) where valuation itself is viewed as support for the search for legitimate agreements among actors. They stated that 'acknowledging complexity and conflicts in the decision making process calls for a new analytical perspective on valuation itself'. For other authors like Reed (2008) and Deléage (2002b; 2007), it was important to focus on inviting stakeholders to evaluate not just the outcomes of a participative exercise, but to evaluate the process itself. With respect to IWRM, Van der Zaag (2005), Van der Zaag and Savenije (2004) and Van der Zaag *et al.*, (2003) argue that the situation that is agreed is the most appropriate. In that sense, with effective participation, few surprises should emerge at the end of a water process and big changes of agreement on proposed actions are expected to take place (FAO, 1995). Despite such growing interest in participation, authors like Fiorino (1990) denounced lack of interest and low political awareness from communities because technicians have never been good at making agreements or arriving at consensus. He reports that, traditionally, defining and evaluating environmental risks has relied on technocratic rather than on democratic values. Fiorino (1990) and Pateman (1970) state that technicians have been good at setting standards but bad at building consensus. At the beginning of the century, there was growing disillusionment regarding participation. In the last decade there has been a reappraisal searching for best practices and learning from mistakes. It is even considered today that most of the best practices are developed in the developing countries and that the developed ones are now learning to involve stakeholders from those experiences (Reed, 2008; Burawoy, 2007).

2.3.4   Elements of stakeholder participation

Stakeholder participation has elaborated on the virtues of participation, on different methodologies stimulating discussions, and at identifying that there are different degrees/kinds of participation for different purposes.

8 key principles for best practice

Since stakeholder participation is focused today on searching for the best practices, Reed (2008) proposes the following features:

1. Focusing at empowerment, equity and trust learning. It is in fact important to determine whether the participants have the power to influence a specific situation (Bourdieu, 1998) or the technical capabilities to engage in any decision; to enhance building relationships; to support iterative learning not only between participants but also with researchers.

2. Introducing stakeholder participation as early as possible and all the time; setting clear limitations in order to avoid frustrations.

3. Identifying and analysing relevant stakeholders.

4. Setting objectives from the beginning, which are adapted to the needs and are negotiated.

5. Tailoring methods to the decision-making context.

6. Supporting highly skilled facilitation.

7. Integrating local and scientific knowledge.

8. Institutionalizing participation.

A wide array of methodologies is offered and is classified with respect to the objective sought (problem structuring, idea finding, enhancing awareness or mutual learning). Different methodologies include the dialectical debate, the decision seminar, the policy Delphi, group brainstorming or gaming simulation (Van de Kerkhof, 2004) or Social Network Analysis (Reed, 2008). Stakeholder analysis constitutes one of the most important tools. This tool focuses on the relevance of identifying stakeholders. It defines the different issues at stake of a social and natural system; identifies the individuals as well as the groups who are affected by a decision; and prioritises the actors (individuals or groups) to be involved in decision-making (Reed *et al.*, 2011; 2009). This analysis can work either top-down or bottom-up. It works with matrices regarding interest-influence, Venn diagrams or Knowledge Mapping Analyses (Reed, 2008).

Different kinds of participation

The quality of participation is considered by many authors as the starting point (Van de Kerkhof, 2004; Mayer, 1997; Argyris and Schön, 1996; Arnstein, 1969) to be taken into consideration for successful problem solving and decision-making. The quality of decisions is dependent on the nature of the process (Reed, 2008).

Participation can take place in different domains. It can be used for policy making or scientific practice. The difference between them is that in the policy domain, power is involved and short-term issues are usually focused on, whereas in scientific practice, the goal is more about 'learning' on the side-lines of the policy process, and working on long-term studies in a certain problem area. The problem that arises is that it is usually not clear why, how and for what purpose participation is being exercised and that different degrees of participation can be pursued (Engels, 2002). Three different degrees of participation are distinguished in both the policy-making and scientific domains (Van de Kerkhof, 2004, and inspired by Arnstein, 1969 and Mayer 1997): High, moderate and low.

Coordination, co-production and mutual learning are the highest degrees of participation in scientific practice in which the involved actors, in mutual interaction and deliberation, determine the outcomes of a process (

Figure 2-2). Those highest degrees of participation cannot be reached without two essential factors: (a) proper coaching, and (b) understanding that the highest degrees require mastering change processes. For Gummesson (2007a: 230), practitioners are suited to undergo both tasks. Coming from the management field, he advocates that practitioners have developed expertise in making decisions, initiating action and achieving results, and have the potential to contribute also to knowledge building. The moderate degree of participation implies that stakeholders are asked to give their input but scientists determine the outcomes. The lowest degree of participation is information in scientific practice.

With respect to policy-making, the highest degree of participation is when joint ventures foster change, and responsibilities are shared among all the actors involved. A moderate degree of participation implies that there is no follow-up to stakeholder participation and that there is no guarantee that the stakeholders' input will be reflected in the final outcomes. The lowest degree can entail manipulation by the policy makers or simply information given to the stakeholders. Such a degree implies only top-down approaches (Figure 2-1). The highest degree of participation is seldom reached or even sought for in policy making (Engels, 2002).

It is important to highlight that in both domains, classical consultation, which is the most standard practice in CP in developing countries belongs to a moderate degree of participation.

Figure 2-2 introduces some elements to the ladder of participation. It puts action research at the highest level for both domains simultaneously. It has also situated mutual learning, integrative negotiation and conflict resolution at the highest levels because they satisfy the requirements presented above where involved actors in mutual interaction and deliberation determine the outcomes of a process, and/or joint ventures are undertaken fostering change and responsibilities are shared among all.

Figure 2-2 Degrees of participation

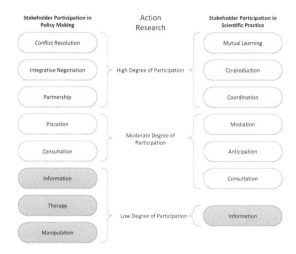

Source: Adapted from Arnstein (1969) and Van de Kerkhof (2004)

Taken from the analysis above, few approaches reach the highest level of participation and this is the reason why some scholars consider establishing a hierarchy of participative approaches or determining blueprints on the participation problematic (O'Hogain, 2008). They prefer first to identify that knowledge can be developed out of institutes of science and to set participative approaches that are best suited to develop knowledge that can be instrumental or reflexive and targeting academic audiences or extra-academic audiences (Table 2-3) (Mollinga, 2008; Burawoy, 2007).

Table 2-3 Division of disciplinary knowledge

	Academic audience	*Extra-academic audience*
Instrumental Knowledge	Professional	Policy
Reflexive Knowledge	Critical	Public

Source: Burawoy, 2007

Cornwall (2003) presents a matrix where the modes of participation depend on the reason to initiate a participatory process and the way participants are perceived (Table 2-4):

Table 2-4 Modes of participation

Mode of participation	Associated with...	Why invite/involve?	Participants viewed as...
Functional	Beneficiary participation	To enlist people in projects or processes, so as to secure compliance, minimize dissent, lend legitimacy	Objects
Instrumental	Community participation	To make projects or interventions to run more efficiently, by enlisting contributions, delegating responsibilities	Instruments
Consultative	Stakeholder participation	To get in tune with public views and values, to garner good ideas, to defuse opposition, to enhance responsiveness	Actors
Transformative	Citizen participation	To build political capabilities, critical consciousness and confidence, to enable to demand rights; to enhance accountability	Agents

Source: Cornwall, 2003

2.3.5   Benefits, drawbacks, and gaps in knowledge

Participation is seen through different lenses by the literature and can be considered to complement each other with respect to its benefits (Table 2-5). It can be a means and an end. It is a means when through collaboration a policy or a project can be implemented. It is also an end when it empowers people through skills that foster   self-reliance and self-management (O'Hogain, 2008; Selfa and Endter-Wada, 2008; Duraiappah et al., 2005; Kanji and Greenwood, 2001; Karl, 2000). Other authors like Reed (2008) consider it as playing a dual role: normative (implying democratic rights) or pragmatic (enabling technology adaptation, or the quality and durability of environmental decisions). Cornwall (2003) defines four modes of participation depending on the way participants are perceived and the reasons to initiate a given process. In consequence, participation offers a wide array of methods stimulating people's involvement (Reed, 2008; Van de kerkhof, 2004). It offers tools like the ladder of participation (Van de Kerkhof, 2004; Mayer, 1997; Argyris and Schön, 1996; Arnstein, 1969) or a matrix on four modes of participation (Cornwall, 2003). A useful element is stakeholder analysis that can successfully integrate bottom-up approaches and capture local knowledge (Reed et al., 2011; 2009). This work considers that for developing countries and for McSEs, all aspects and methods need to be considered.

Since stakeholder participation offers settings to discuss issues, it is recommended to handle complex problems such as water systems (Mollinga, 2008; Stringer et al., 2007; Van de Kerkhof, 2004) based on three claims (Van de Kerkhof, 2004): First, it may improve overall decision making by bringing new information that leads to better decisions. Decisions are enriched by legitimacy and accountability and as a consequence more public support is expected. Second, it may also improve scientific practice. The involvement of stakeholders in the evaluation and monitoring of scientific data can restore, to a certain extent, science's integrity and social recognition. The latter has been arguably lost because traditionally scientists were never in the habit of seeking public backing for their endeavours, and it has also been exposed to games of power and abuse as an integral part of society. Third, it may also finally support problem structuring in order to reach sound and more effective policy making. In fact, the definition of a problem on the basis of much conflicting information that is confronted, evaluated and integrated is considered of utmost priority. Errors may arise when analysts exclude relevant information about a problem, and this results in a limited perception of the problem.

Participation can stimulate cross-sectoral planning, which in Chapter 1 has been presented as one of the major challenges for sustainability (Thabrew *et al.*, 2009). It can potentially lower the costs of policy or technical implementation at the last stages (Grimble *et al.*, 1995).

Despite the above-cited benefits, literature reports drawbacks on stakeholder participation (Table 2-5). As important as it might be, however, increasing participation does not by itself guarantee problem solving, neither the implementation of policies, nor does it assure governance (Berk *et al.*, 1999). Stakeholder participation lacks rules of procedure for decision-making, for choosing stakeholders or for even choosing among the different methodologies (Reed, 2008; Gupta, 2004; Engels, 2002; Berk *et al.*, 1999). In fact, the literature reports delays on decision-making (Gupta, 2004); that there is no objective criteria in order to select stakeholders when their number can be very large and representativeness is at stake (Morse, 2008; Eversole, 2003; Berk *et al.*, 1999); that the average citizen is not capable of rational judgement on complex matters (Hisschemöller, 1993; 2005) and (Hisschemöller and Hoppe,1995); that stakeholders tend to mistrust scientific experts (Berk *et al.*, 1999); that interaction between stakeholders can be emotional and tend to aggravate conflict (Douglas and Wildavsky, 1983); that the choice of stakeholders is often problematic - inviting stakeholders implicitly implies excluding others, management of the stakeholder process can in effect imply greater bureaucracy; stakeholders within the system resist the inclusion of other stakeholders, and where the problems are complex, it is often very difficult to reach agreement (Morse, 2008; Gupta, 2004; Escobar, 1999)- ; and finally that methodologies do not suit the specific contexts (Reed, 2008). Most of the literature agrees that besides the latter, the major challenges that stakeholder participation faces are due to the fact that participation does not succeed in development: it does not deal with power imbalances, political engagement and co-learning (Wiber *et al.*, 2009; Morse, 2008; Reed, 2008; Burawoy, 2007; Duraiappah *et al.*, 2005; Eversole, 2003; Cleaver, 1999); and finally with change (Reed, 2008; Eversole, 2003). In fact, participation can reinforce the interests of the already powerful (O'Hogain, 2008; Morse, 2008; Berk *et al.*, 1999; Escobar, 1999; Cleaver, 1999); democracy can be threatened because participation can have the tendency of overlapping formal agreements (Cooke and Kothari, 2001) and stakeholders are most likely to defend their short-term interests versus collective ones (Coenen, 2009; Webbler and Renn, 1995). Since stakeholder participation can then be used for manipulative purposes, it can cause some resistance and be problematic (Escobar, 1999; Cleaver, 1999; Arnstein, 1969) and can raise expectations that cannot be met (Estacio and Marks, 2010; Hirsch *et al.*, 2010; Mehta *et al.*, 2001). There is a question whether stakeholder participation most likely works in egalitarian countries and not in totalitarian regimes. Even though there are worldwide positive experiences illustrating the benefits that stakeholder participation can offer regarding policy implementation, many agencies were thought to get suspicious about the motives and intentions of this participation (FAO, 1995). Finally, stakeholder participation imposes more up-front costs especially at the beginning of a process (Grimble *et al.*, 1995).

Table 2-5 Benefits and drawbacks of Stakeholder Participation

Benefits	Source
Can be a means (signing agreements) and an end (empowerment)	Karl, 2000
Supports democratic values	Reed, 2008
Offers settings to discuss complex issues	Van de Kerkhof, 2004
Potentially improves decision-making bringing new information and public support	Van der Zaag, 2005; FAO, 1995
Potentially improves scientific practice	Van de Kerkhof, 2004; Van der Zaag, 2005; Fiorino, 1990
Supports problem structuring for policy making	Van de Kerkhof, 2004; Godard and Laurans, 2004
Stimulates cross-sectoral planning	Thabrew et al., 2009
Offers a wide array of methods for people's involvement that can be bottom-up	Reed, 2008; Van de Kerkhof, 2004
Offers the ladder of participation	Van de Kerkhof, 2004; Mayer, 1997; Argyris and Schön, 1996; Arnstein, 1969
Offers 8 key principles	Reed, 2008
Offers stakeholder analysis	Reed, 2008; Reed et al., 2009
Potentially lowers implementation costs at the supervision stage	Grimble et al. 1995
Drawbacks	
Does not guarantee implementation of policies or governance	Berk et al., 1999
Cannot lead to problem solving	Berk et al., 1999
Lacks rules of procedure for decision making (delays occur), for choosing stakeholders and for choosing different participative methodologies	Berk et al., 1999; Reed, 2008; Gupta, 2004; Engels, 2002
Cannot avoid reinforcing the interests of the most powerful and does not know how to avoid conflicts	Gupta, 2004; Wiber et al., 2009; Reed, 2008
Raises expectations that cannot be met	Estacio and Marks, 2010; Hirsch et al., 2010
Cannot prioritize collective interests over short term individual ones	Webbler and Renn, 1995
Cannot face power imbalances, political engagement and co-learning	Wiber et al., 2009; Reed, 2008; Eversole, 2003
Cannot stimulate commitment and change	Reed, 2008; Eversole, 2003
Can be used for manipulation	Van de Kerkhof, 2004; Arnstein, 1969
Empowerment may have negative interactions with existing power structures	Cooke and Khotari, 2001
Cannot avoid overlapping formal agreements	Cooke and Khotari, 2001
Has more up-front costs	Grimble et al. 1995

The drawbacks of stakeholder participation presented above suggest some gaps in knowledge and have been identified in this thesis' analysis of the literature. Basically, the gaps are related to (a) the need to improve decision-making, to deal with conflicts, which seem natural in water issues; (b) the need to understand stakeholder participation as a process of change; (c) the need to identify the right participatory methods for different policy instruments; and (d) the importance of approaching capabilities for the marginalized supporting the on-going debate on the relevance of reaching consensus or not as a result of participative approaches.

In fact, (a) for authors like Godard and Laurans (2004:1) and Cleaver (1999) what lies at the heart of environmental complex problems is the need to acknowledge their social context and nature 'where decision making is related to complex negotiations and agreements among actors that hold diverse points of view, and science is affected by controversies and uncertainties'. There is a need to deploy effective dialogues between actors representing interests that may be in conflict (Cleaver, 1999). In this sense, the dialogues, which need to be reinforced, are the ones that originate at the local level from both institutions and ordinary citizens (Bakker *et al.*, 2012; Edelenbos *et al.*, 2009). The latter is especially missing with respect to water resources management as authors identify a gap between the international discourse and local realities (Mollinga, 2008). In the water domain it can be acknowledged that water by itself tends to build asymmetrical relationships, simply by the fact that it flows downhill and people downstream get affected by the upstream uses people give it (Van der Zaag, 2005). Stated this way, conflicts related to water seem inevitable; especially when even scientists are not always sure about the effectiveness of the available solutions in a complex and globalized world (Godard and Laurans, 2004; Vinke-de Kruijf *et al.*, (2010), Kolkman *et al.*, (2005) Cleaver, 1999).

Reed (2008: 2422) proposes (b) that in order to improve stakeholder participation, the definition itself should switch to 'a process where all parties build trust and learn from each other to negotiate potential solutions'. By looking at it as a process, holding long-term exercises aiming at learning and changing become most important (Vinke-de Kruijf *et al.*, 2010; Morse, 2008; O'Hogain, 2008; Reed, 2008; Cornwall, 2003; Eversole, 2003; Cleaver, 1999). The quality of decisions is dependent on the nature of the process. Goals are negotiated and outcomes uncertain (Reed, 2008). Participation is not only about the unveiled pursuit of power over the final decision (Ker Rault, 2008). Questions remain on how to enhance social learning, how to transform opposed relationships, how to arrive at more solid and long-term decisions, how to enhance public trust, and how to stimulate the diffusion and adoption of innovation. Over the last decade, researchers on stakeholder participation have been venturing in processes inspired by diverse approaches such as CDP (community development participation) based on RRA (rapid rural appraisal), PRA (participatory rural appraisal), PAR (participative action research), AI (appreciative inquiry), and PPA (participative poverty assessment), which ask for holistic, self-reflexive and inclusive thinking (O'Hogain, 2008; Burawoy, 2007). Robinson and Berkes (2011) recommend multilevel network participation approaches since the adaptive capacity to changes is a property of a social system.

Authors agree that there is (c) a research need to relate different kinds of participatory mechanisms to different kinds of policy problems and instruments. They even set the problem as one of setting prescriptive guidelines that would tell us how and when to use participatory approaches (Reed, 2008; Karl, 2000; Grimble *et al.*, 1995; FAO, 1995; Fiorino, 1990).

Authors like Duraiappah *et al.*, (2005) consider that (d) besides the challenges and gaps, the introduction to participatory approaches in the development arena over the last three decades have improved the capacity of poor communities to participate. They suggest that it is important to be able to assess the degree to which each approach supports capabilities of the marginalized and/or excluded. The literature on stakeholder participation finds the issue of reaching consensus as a result of participative approaches problematic because there are always risks of abuse of power (Cleaver, 1999). Morse (2008)'s advice is to embrace diversity and to understand that reaching consensus is not necessarily a target among many interests.

2.3.6   Lessons learnt for this thesis

Despite the fact that society recognizes the importance of participation in today's complex environmental problems, the field of stakeholder participation seems unable to face them. Much debate has been taking place on the nature and purpose of participation but approaches focusing on problem solving, decision-making, handling conflicts of interests, stimulating commitment and change, and avoiding manipulation have been relatively absent. Solving such challenge asks first of all for in-depth understanding of specific situations (Hayward *et al.*, 2004; Cleaver, 1999), and to target three main elements (Robinson and Berkes, 2011): (a) institutional environment, (b) inclusivity in decision-making, and (c) deliberation. Reed *et al.* (2011), Camargo *et al.* (2010), and Welp *et al.* (2006) propose deploying interdisciplinary teams for stakeholder dialogues, a variety of methods incorporating multiple and context-specific knowledge for complex systems and this will be implemented in this thesis.

The literature (Bernardini, 2008) agrees that the field is evolving and that such evolution is especially relevant with respect to underprivileged communities needing long-term comprehensive and complex approaches. Burawoy (2007) invites opening the social sciences to reflexive thinking, that encourages problematizing on values and purposes of society, and that concentrates on developing knowledge also at extra-academic audiences that point strongly at the underprivileged communities.

2.4   Alternative dispute resolution (ADR): Negotiation/Conflict resolution (N/CR)

2.4.1   The place of alternative dispute resolution in sustainable development

Meeting present needs without compromising the ability to meet future generations' needs entails necessarily identifying and prioritizing them in order to be able to arrive at legitimate problem solving and decision making for today's complex problems. Such a process entails complex negotiations, which can support cross-sectoral planning entailing multiple issues, levels and interests. ADR includes options for resolving disputes that can be alternatives to court proceedings and that can support sustainable solutions aimed at meeting all parties' needs.

2.4.2   What can be understood from ADR and from negotiation?

ADR is comprised of a wide variety of consensual approaches with which parties in conflict voluntarily seek to reach a mutually acceptable settlement. Those approaches can be highly diverse when they come from indigenous groups (Wolf, 2008; 2000). The most common are: arbitration, shared vision development, negotiation, mediation, facilitation and conciliation (Frenkel and Stark,

2008)[12]. The list can vary depending on the field, which can be law, political science or business. The focus can be power, rights or needs-based. They can go from the most adversarial to the least. This thesis aims at building sustainability and consensus with underprivileged groups, hence, the focus will be on a win-win, needs and interests-based options such as mediation and facilitation, or negotiation. Other dispute resolution options like adjudication focus more on win-lose situations where the legal rights determine final outcomes. Such decisions are often also acceptable because they are seen as just but will not be discussed here since these processes of conflict resolution may not be quite as effective for long-term commitments in the water field.

Negotiation refers to a process of communication in which the parties aim at influencing each other's decisions (TANDEM, 2010; 2005). Mediation/facilitation is initialed when a dispute or an issue asks for a third party. It is an assisted negotiation (Frenkel and Stark, 2008).

A negotiator's ability to exert influence depends upon a combined total of a variety of factors. These include: 1. Knowing and determining the people and the interests involved; 2. Having a good working relationship; and 3. Having a good alternative to a negotiated settlement (Ury *et al.*, 1993; Fisher *et al.*, 1991; Fisher and Ury, 1981). Basically, the first factor can be obtained through stakeholder participation and analysis; the second and the third are not necessarily assured unless there is a systematic approach aiming at win-win situations, decision-making, and establishing long-term commitments. Leading that approach entails, for environmental problems, constant facilitation inspired by the theories set forth below.

2.4.3   Elements of the theory of negotiation/conflict resolution

Distributive and Integrative bargaining

Most of the negotiation literature focuses on two attitudes, although these can be called by different names. The first one is positional (or distributive or competitive or traditional) bargaining, while the other is interest-based (or integrative or cooperative or new) bargaining. In their book on negotiation, *Getting to yes,* Roger Fisher and William Ury (1981) argue that there are three approaches: hard, soft, and what they call principled negotiation. Hard is essentially extremely competitive bargaining, soft extremely integrative bargaining (so much so that an actor gives up his own interests in the hope of meeting the other person's interests) and principled negotiation is supposedly somewhere in between. Negotiations are actually a combination of both approaches. They try first, to create value by enlarging the pie as much as they can (integrative, interest-based) and at some point in time most of the negotiations will then need to divide the pie (distributive negotiation). A negotiation is a combination of creating value and claiming value (Lax and Sebenius, 1991). The fundamentals of principled negotiations are: (a) separate people from the problem, (b) focus on interests, not on positions, (c) invent options for mutual gains, and (d) insist on using objective criteria (Fisher and Ury, 1981). Interests are essential for negotiation because they make people act and change. Interests cause people to take the positions they take. Positions are proxies of people's deepest needs, fears or concerns and are usually concrete demands and offers made in negotiation (Frenkel and Stark, 2008). The most powerful interests or positions may be anchored on fundamental human needs described by Burton (1990) on the Theory of Needs such as security, identity, self-esteem and recognition. He states that the most deep-rooted conflicts are usually caused by the denial of one or more of them.

---

[12] *palestineisraelresolutionscrt.blogspot.com/ on March 22 2012*

In order to understand the causes of human behaviour, Brett (2007) uses the figure of an iceberg where the visible part above the waterline is made up by behaviour and institutions, and below the waterline, two deeper psychological levels explain other attitudes like knowledge structures (values, beliefs and norms), and fundamental assumptions (see Figure 2-3). The figure explains why negotiations between different cultures can be more demanding (Ogliastri, 1999).

Figure 2-3 Culture is represented below the line in this iceberg representation of human behaviour

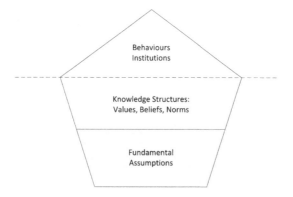

Source: Adapted from Brett, 2007

A systematic process

As Fisher *et al.* (1991) state, an essential key to negotiation is to concentrate on the process, not on the result. A systematic approach to negotiation starts by a preparatory or planning step, essential to all experts on negotiation, (Thompson, 2009; Frenkel and Stark, 2008; Brett, 2007; Saner, 2003; Raiffa, 2002) and entails the following steps (TANDEM, 2010; 2005):

> (1) Preparation: General data collection on the specific subject, information gathering about the BATNA (best alternative to a non-negotiation agreement), ZOPA (zone of possible agreement), goals, limits or walk away point, stakeholders' interests and preferences, and the initial perceived definition of the problem. Research on negotiation has shown that more than 50% of the outcomes derive from this step (Saner, 2003); (2) Building relationship: It basically implies trust building; (3) Sharing information; (4) Redefinition of the problem being dealt with: it can be different from the initial perceived definition of the parties; (5) Creation of options; (6) Agreement design; and (7) Implementation.

Negotiation occurs when the parties find that negotiating brings more benefits than not negotiating and that they believe they can work towards a mutual agreement zone (ZOPA). When parties have similar ideas about their BATNAs (alternatives in case there is no negotiation), it is said that they are ready to negotiate (Ripeness). The ZOPA depends most of the time upon values, interests and preferences (Fisher and Ury, 1981). If negotiation does not occur, the adjudicatory system is traditionally the only alternative left. If this option does not work, conflict arises, power and force are called upon unilaterally and the situation can be lead to violence and war (Ogliastri, 1999-2000).

Linear- systemic (complex)

Negotiation can be linear (simple) or systemic (complex), entailing multiple aspects, stakeholders and uncertainties (Thompson, 2009; Saner, 2003; Raiffa, 2002). Complex negotiations must be successful at integrating valuable data and designing solutions. In the water sector, there is a need to integrate multiple topics essential for decision-making (Van der Zaag, 2005). Water-related and, even more so, McSEs-related processes are complex, systemic ones because of their own specificities where conflicts can easily arise (Chapter 1). It is hence very common that interests, values, perceptions or even preferences are opposed to each other in this kind of negotiation. Fortunately, it is known how to bring the right parties to the negotiating table. It takes above all, the preparation of a conflict assessment by a neutral party like a mediator (Susskind *et al.*, 2003; Susskind and Field, 1996).

The theories of conflict resolution, which have evolved somehow independently from negotiation but are presented as a whole in this thesis, have advantages and disadvantages for each system of conflict resolution and each has elaborated rules of procedure regarding how solutions can be reached. Such rules are often missing in stakeholder participation, as shown in the previous section, and constitute preliminary and/or simultaneous steps for successful negotiations.

The integrative mediation process addresses conflicts. Its purpose is to invite an independent body to help the parties to work on a solution that satisfies their needs. Mediation follows the same steps as in negotiation except the implementation step (Step #7) as the mediator, always neutral, has no responsibility in that phase, and leaves the parties to work it out on their own. The mediator is impartial in the sense that he/she does not favour any party; he/she is also neutral because he/she has no personal preference regarding the outcomes (Frenkel and Stark, 2008). When conflicts involve communities, organizations or a multiplicity of actors, there is a need to engage people from all aspects of a system and increase their capacity to achieve what is most important, individually and collectively (Holman *et al.*, 2007; Holman, 2004; 2010; Holman and Devane, 1999). The latter is inspired by what could be considered for the purpose of a logical sequence of analysis as a 'complex integrative mediation'. These types of mediations are more likely to take place in the water domain and are based on participation theory that states that people support initiatives that they help create (Holman, 2004). However, dealing with a multiplicity of actors, issues and/or big groups in water conflicts can lead to more confusing and risky situations if they are not handled and managed with care by third parties called facilitators. Hence, social experts have been developing different processes and methodologies that deal effectively with large-scale conflicts (Holman and Devane, 1999). They are called Whole Scale Methodologies. Three such methods were chosen for their suitability in handling complex problems namely Appreciative Inquiry, Open Space, and Dialogue methods (Table 2-6). They all take advantage of the power of collective thinking and effective change.

Table 2-6 Comparative analysis of Appreciative Inquiry (AI), Open Space Technology (OST) and Dialogue

	AI	*OST*	*Dialogue*
Purpose/Outcomes	To enable full-voice appreciative participation that taps the organization's positive change	To enable high levels of group interaction and productivity, providing a basis for enhanced organizational function over time	To build capacity to think together, creating shared meaning To open communication channels, building trust and shared leadership
Process	Discovery	One law	Based on the power of

	Dream	Two principles	collective thinking
	Design		
	Destiny		
Types of Participants	Internal & External stakeholders Creators who hold images	Anybody who cares about the issue under consideration Diversity is a plus	Cross functions and cross management levels
Number of participants	20-2000 involved in interviews, large scale meetings	5-1000. No limits by using computer connected, multiple site, simultaneous events	5-100 people in circles
When to use	To create a positive revolution. Enhance strategic cooperation overcoming conflict	When time is pressing	To open communication channels and build trust based on deep inquiry
Creators	David Cooperrider & Suresh Srivastva 1987	Harrison Owen 1985	David Bohm 1985

Source: Holman (2004)

Using these methodologies would be useless if they do not aim at specific strategies for specific targets. The next section deals with the different strategies in conflict resolution and negotiation.

Strategies

There are four ways to deal with a conflict: (a) Evasion, (2) Give away, (3) Imposition or confrontation, and (4) Negotiation (Figure 2-4). Strategies are defined depending on the significance of the issue being discussed, the power involved, the short or long-term relationship involved, the principles and values involved and the level of interdependency of the concerned stakeholders (TANDEM, 2010; 2005). Blake & Mouton (1964) defined the latter as the 'double worry' model. Two axes define it. The Y- axis is how important it is for a person that his party gets a good outcome and the X-axis shows how important it is for the same person to get a good outcome for himself.

Figure 2-4 adapts the model for the purpose of this thesis to an environmental authority on axis X and to McSEs on axis Y.

It has been shown that reaching sustainable agreements in water issues entails integrative negotiations. In order to lead such negotiations, the interests from all parties must be met aiming ideally at win-win situations. The latter implies that in the figure below, the interaction should move towards (+) in both axes once behavioural change has been fostered. In the case of McSEs, the positive move on the model presented above is a complex endeavour, which may ask for specific strategies such as empowering. Empowering is based on the fact that 'people know best how they can be engaged in governance' (Fiorino 1990). The model shows that if the McSEs are powerless, the authority can impose. The latter is not ideal because in real life situations in developing countries the abuse of power is frequent and authorities can represent individual powerful interests. In the case of marginalized people, this can easily mean that their interests are not taken into account. In order to ensure respect for their interests, those communities need to be empowered through capacity building, for example, in best practices and financial programs adapted to their needs. By moving positively along the Y-axis, integrative negotiation can be possible and better outcomes are reached for the environment and for communities.

Figure 2-4 Integrative Negotiation through empowering.

Source: Adapted from Tandem (2005)

This model is based on interests but on a daily basis the strategies used, depending on particularities, can switch from one to another (evasion, imposition, etc). What really matters is that through the process there has to be trust building, positive appreciation of the other, true openness and rational emotiveness (Ury, 2007; Merlano, 2005).

2.4.4   History of the theory of negotiation/ conflict resolution

The first documented sophisticated system for negotiations ever found belongs to Ancient Greece. In the XVII century, under the French influence of Armand Jean du Plessis, Cardinal Richelieu, negotiations turned out to be seen through long-term perspectives. During that period, the western world's negotiations were conducted in French. In 1716, the diplomat Francois de Calliéres stated that a good negotiation should meet all the legitimate parties' interests. Despite this development, it was not before the onset of the two world wars that the fields of negotiation and conflict resolution really evolved (Saner, 2003). Peace research and non-violence movements were their early roots. Gandhi's ideas on developing a successful struggle to overcome injustice while being loyal to pacifist values were important. Needs, cooperation, or interests-based negotiations developed as a discipline following World War II and diverged from power-based negotiations that had been leading the political science's principles (Bradford, 2012). Social psychologists like Kurt Lewin considered conflict as a situation of tension where the needs are denied. Other important founders were Kenneth Boulding, Johan Galtung, and John Burton, who considered conflict not as a dysfunctional issue but as intrinsic in human relationships. These three scholars took the lead from the 60's at University College London, Yale University and Harvard University. In 1965 during the conflict between Malaysia, Singapore, and Indonesia, Herbert Kelman developed interactive conflict resolution, third party consultation, process-promoting workshops and facilitated dialogues. His thirty years' work can be summarized as (a) participants are influential, (b) facilitators can help determine the outcomes of the workshop, (c) meetings allow for creative thinking, (d) listening without judgement to people's needs is important, (e) focusing at clearing-up misperceptions and misunderstandings is a target, (f) conflict is jointly worked out, and (g) new knowledge is incorporated into policy-making. In the late 70s, the theory of principled negotiation was developed through the Harvard Negotiation Project (Fisher and Ury, 1981); new vocabulary was introduced as win-win, integrative or problem-solving approaches and BATNAS.

2.4.5   Benefits, drawbacks and gaps in knowledge

The power of negotiation/ conflict resolution is given by the fact that it entails a process of communication with specific rules of procedure in order to know when is the appropriate time for negotiation, how to conduct negotiation, and how to reach decision-making (Thompson, 2009; Frenkel and Stark, 2008; Brett, 2007; Saner, 2003; Raiffa, 2002). The field offers strategies depending on the kinds of stakeholders' relationship involved, the significance of the issue, and the nature of the relationship. Since the process of communication is private, it is prone to party autonomy, confidentiality and finality of awards. For cases where there is power imbalance and some parties may benefit from privileges under the applicable law and local politics, negotiation/conflict resolution may assure some neutrality for all parties and the protection of long-term objectives (WIPO, 2012; Merlano, 2005; TANDEM, 2005; Blake and Mouton, 1964). By promoting order and stability, legitimacy can be enhanced and the probability of future conflicts can be reduced (Giacomantonio *et al.*, 2010b). The field of negotiation/ conflict resolution creates and claims value based on interests (Thompson, 2009; Raiffa, 2002; Saner, 2003; Fisher *et al.*, 1991). It offers hence a wider range of solutions, flexibility, win-win opportunities and in-depth knowledge (WIPO, 2012). Some authors consider that it is cost-effective compared to court proceedings (WIPO, 2012). The theory focuses on processes to engage people towards outcomes; it is context-specific (Thompson, 2009; Frenkel and Stark, 2008; Brett, 2007; Raiffa, 2002; Saner, 2003, Fisher *et al.*, 1991). Lately, the field is focusing on the importance of emotions for training students (Shapiro, 2006), recognizing the importance of the emotional component (Barry, 2008), or focusing on the relevance of positive emotions to maximize results (Holman, 2010; Pietroni *et al.*, 2009).

Complex negotiation/conflict resolution deals with complex problems: big groups, multiple interests, multiple issues, and multiple stakeholders through methodologies such as AI, OST and Dialogue. It aims at engaging people for collective and individual best results (Holman, 2004) where specific benefits of group decision-making are (Raiffa, 2002): (1) ownership: people seem to support decisions that they have created; (2) arousal: people seem to work harder when others are around; and (3) improved resources: a group offers more manpower and expertise than individuals can.

The field offers the roles of negotiator, mediator or facilitator, which conduct the appropriate processes (Frenkel and Stark, 2008). The most important skills for such agents are: being a good communicator, skilled diagnostician, promoting creative problem solving, and being persuasive (Frenkel and Stark, 2008).

Despite the benefits and advantages presented above, the literature review highlights drawbacks. Since the processes involved in negotiation and conflict resolution are traditionally of a private nature, they are believed not to rely necessarily on rights, and not to be regulated by quality standards (WIPO, 2012).

Integrative negotiations entail more up-front costs because they require more preparation and information gathering regarding the parties' interests than simpler approaches and are hence, more difficult to conduct (Giacomantonio *et al.*, 2010a). Leading complex negotiations involving big groups such as in the environmental field imply greater efforts. Whole scale methodologies ask for facilitators with interdisciplinary and open orientations, which are not easy to find (Holman, 2004). Conflicts can be more costly and time consuming because their high complexity asks for a greater time investment. It is known that de-escalating a conflict requires more time than escalating it because there are usually wounds difficult to heal and there is the need to lead processes based on mistrust and suspicion towards processes based on trust (Frenkel and Stark, 2008).

The processes based on negotiation/conflict resolution have tools to deal with power imbalances but are rarely used on a long-term basis (WIPO, 2012; Moran and Ritov, 2007). The field of negotiation or conflict resolution do not support once implemented, long-term change processes (Moran and Ritov, 2007).

The mediator or facilitator, always impartial and objective, is not supposed to get involved in the concerned processes (Frenkel and Stark, 2008).

Table 2-7 Benefits and drawbacks of Negotiation and Conflict Resolution

Benefits	Source
Offers specific rules of procedures for decision-making based on specific criteria and structured approaches	Thompson, 2009; Frenkel and Stark, 2008; Brett, 2007; Saner, 2003; Raiffa, 2002
Promotes order & stability and feelings of self-efficacy; reduces probability of future conflicts; stimulates economic prosperity	Giacomantonio *et al.*, 2010b
Offers party autonomy, neutrality, confidentiality and finality of awards as being private	WIPO, 2012
Potentially cost-effective compared to court proceedings	WIPO, 2012
Offers a wider range of solutions, flexibility, win-win opportunities, in-depth knowledge	WIPO, 2012
Can handle complex problems through win-win approaches; can create and claim value based on interests	Thompson, 2009; Raiffa, 2002; Saner, 2003; Fisher *et al.*, 1991
It is context-specific and process oriented	Thompson, 2009; Frenkel and Stark, 2008; Brett, 2007; Saner, 2003; Raiffa, 2002; Fisher *et al.*, 1991
Can handle conflicts entailing big groups through whole scale methodologies that engage people for collective and individual best results	Holman, 2010; 2004; Holman *et al.*, 2007
Offers the roles of mediator and /or facilitator	Frenkel and Stark, 2008
Offers strategies that tell when and how; that may focus long-term	Merlano, 2005; TANDEM, 2005; Blake and Mouton, 1964
Focuses on emotions	Pietroni *et al.*, 2009; Barry, 2008; Shapiro, 2006
Drawbacks	
Does not necessarily clarify/ rely on rights	WIPO, 2012
Not regulated by quality standards	WIPO, 2012
Potentially entailing slow processes and more up-front costs	WIPO, 2012; Frenkel and Stark, 2008
Integrative negotiations are difficult to be conducted	Giacomantonio *et al.*, 2010a
Whole scale methodologies imply greater efforts: broader facilitators' minds, interdisciplinary orientations	Holman, 2010; 2004
Power imbalances can be dealt on a specific situation but are not commonly used on a long-term basis	WIPO, 2012; Moran and Ritov, 2007
Does not support on a long-term and change processes	Moran and Ritov, 2007
Mediators or facilitators do not get involved, do not handle	Frenkel and Stark, 2008

change processes

The drawbacks of negotiation/ conflict resolution presented above suggest some gaps in knowledge. They relate to (a) the need to break with existing cultural barriers in order to lead successfully in integrative negotiations, (b) the need to consider negotiation/conflict resolution processes and theories deriving from the field of change, and (c) the need to tie them to legal rights and frameworks.

The literature reports (a) that leading integrative negotiations constitutes still a difficult endeavour (Thompson, 2009; Frenkel and Stark, 2008). They consider that there is a reluctance to change the leading adversarial culture among most lawyers; that integrative attitudes are seen as weak; and that people are overly cautious about sharing information. Moran and Ritov (2007) consider that assessing the opponents' interests is not enough; assessing their gains should also be considered essential. There is the need to identify what makes people prefer fairness to personal gain, or mutual benefits to personal victory (Giacomantonio *et al.*, 2010b) or deal with social dilemmas (Raiffa, 2002). More research is suggested on the psychological mechanisms that drive people to focus on interests or on issues, on the emotions that build integrative behaviour (Giacomantonio *et al.*, 2010b; Pietroni *et al.*, 2009; Barry, 2008; Moran and Ritov, 2007). The more abstract the analysis of a negotiation, the more prone stakeholders are to focus on pro-social values and fairness and to adopt long-term perspectives (Giacomantonio *et al.*, 2010b). Multiparty interactions are reported to do poorly because of their high complexity (Raiffa, 2002). Complex negotiations need to take advantages of the adaptive capacity of specific social systems (Robinson and Berkes, 2011), of the power of collaborative learning, of the social capital implied (Wiber *et al.*, 2009) or even more precisely of the relational capital (Welbourne and Pardo, 2009) within groups, which make learning and co-production possible. The latter seems highly useful for McSEs that are commonly grouped.

For Moran and Ritov (2007) (b) understanding the determinants of integrative attitudes is not enough; through negotiations there is the need to transfer learning, to make negotiations a real change process. Negotiators, mediators or facilitators, who are expected to be neutral in dealing with power imbalances, are not trained to foster and support change processes in the long run.

Since (c) negotiation, mediation, and/or conflict resolution have been mainly private initiatives, they are not necessarily bound to rights or legal frameworks (WIPO, 2012). Institutionalizing them as formal policy instruments could contribute towards providing legitimacy and legality to the processes as these are regulated by quality standards (WIPO, 2012).

2.4.6   Lessons learnt for this thesis

In contrast, to the field of stakeholder participation, negotiation/ conflict resolution offers rules of procedures, roles, strategies and methodologies that may potentially lead to decision-making and to conflict management with respect to McSEs in developing countries. Bringing negotiations based on interests opens possibilities towards better outcomes; people support initiatives that they help create. By focusing on big groups going through change, it may be possible to work with people on common grounds while respecting the individuals by using AI and OST. Through AI, visioning is based on positive insights of problems. By using OST, discussions are framed and the number and kind of people attending are not a limiting factor and representativeness is not at stake. Consensus building is then seen through a different lens than stakeholder participation because it is important to work on common grounds in order to meet common goals but to also point at respecting individuals' diversity through specific methodologies.

This thesis supports authors like Shmueli *et al.* (2009) that state that the negotiation field can enhance civic capacity among community leaders that draw from their heuristic knowledge. The concerns on McSEs nevertheless entail demanding processes that can ask for long-term social change processes including power imbalances, which the field of negotiation/conflict resolution does not handle and negotiators/facilitators are not trained for. Empowering entails that McSEs walk through a long-lasting change process and maybe also the authorities, provided they are integrated into the field of policy instruments.

2.5 Action Research (AR)

2.5.1 The place of action research in sustainable development/industrial transformation theories

Complex and contested circumstances that urgently demand change, such as dealing with a marginalized community with highly polluting practices, require from the researcher long-term commitment and willingness to become involved in the process of change in order to aim at sustainability. AR adheres to such processes of change where the sustainability concerns have gone from mere availability of natural resources and pollution control to broader visions that put human development at the heart of solutions (Quental *et al.*, 2011) and ask for institutions that may support social justice, environmental protection and poverty eradication (Dobers and Halme, 2009).

Because AR constitutes the main axis of the theoretical and methodological framework, the elaboration on its definition, history, and elements are presented in chapter 3. This section will just elaborate the analysis on its benefits, drawbacks and gaps in knowledge for the purpose of this chapter, in order to avoid repetition.

2.5.2 Benefits, drawbacks and gaps in knowledge

AR allows learning about a system as it goes through change and can, hence, face today's impending complex challenges asking for action and change (Burnes, 2004; Ravetz, 1999; Lewin, 1946). It offers a cyclic and structured process aiming at leading and at understanding the change process and the natural resistance to change (Schein, 1996; Lewin, 1946). It deploys multiple contextual dialogues enabling the understanding of how the forces implied in those dialogues interact (Levin and Greenwood, 2011; Arieli *et al.*, 2009; Martínez *et al.*, 2006). It increases the skills and confidence of people that work in real life (Kevany, 2010; Kevany and Huisingh, 2013; Jønsson *et al.*, 2009; Terry and Khatri, 2009; Lincoln and González, 2008; Dick, 1999). It focuses on in-depth knowledge. Learning and changing a system are bound together and are characterized by the highest degree of participation, which offers legitimacy or, for authors contesting hierachies, by the mode of participation that is transformative (Levin and Greenwood, 2011; Arieli *et al.*, 2009; Van de Kerkhof, 2004; Cornwall, 2003). The learning process is based on trial-and-error (Schein, 1996).

AR claims validity by being grounded on theory and by constantly reconstructing it (Van de Kerkhof, 2004), and rigour by using multiple approaches to gain knowledge (Denzin and Lincoln, 2011). It focuses on building socially robust solutions (Levin and Greenwood, 2011; Van Herk *et al.*, 2011; Bodorkós and Pataki, 2009; Gummesson, 2007a; Argyris, 2004; Dick, 1999; Burawoy, 1998; Greenwood and Morten 1998; Argyris and Schön, 1996; Funtowicz and Ravetz, 1993; Whyte, 1991; Whyte, 1998).

The role of the change agent is one of being an expert on how and when to foster change. It can switch from helper, to mediator, to facilitator (Schein, 1996; Lewin, 1946). It assesses the risk of intervening through a first low-key inquiry step (Burnes, 2004; Schein, 1996; Lewin, 1946). It puts scientists, policy-making, and practitioners on equal footing with respect to knowledge (Van de Kerkhof, 2004).

Despite the benefits presented above, AR does not seem to build enough upon participative relationships, handling conflicts and pressures, and accounting for multiple interests (Aziz *et al.*, 2011; Jacobs, 2010; Arieli *et al.*, 2009). Despite being a superior way to link teaching, research and the real world, AR is not dominating research in universities (Levin and Greenwood, 2011; Bodorkós and Pataki, 2009; Mejía *et al.*, 2007; Hansen and Lehmann, 2006; Martínez *et al.*, 2006; Steiner and Posch, 2006). AR takes longer, is more demanding, and publishing is not that common (Levin and Greenwood, 2011; Bodorkós and Pataki, 2009).

The issue of representation entails the risk of abuse of power (Jacobs, 2010; Fals Borda and Mora-Osejo, 2003; Burawoy, 1998). Supporters of participatory AR aiming at egalitarian purposes sit uneasily with this because they state that paradoxically, many AR projects would not happen without the initiative of someone that inevitably is a member of the educated group that has historically oppressed the others (Salazar, 1991). The latter reports how participatory researchers in the Colombian society end up being involved in a long chain of transmission of authoritarian actions that may reproduce the usual domination patterns. The role of the researcher (change agent) is not always clear and easy to play (Jacobs, 2010; Arieli *et al.,* 2009).

Table 2-8 Benefits and drawbacks of Action Research

Benefits	*Source*
Allows learning about a system as it goes through change. Faces impending complex problems asking for action	Burnes, 2004; Ravetz, 1999; Lewin, 1946
Offers a cyclic, structured process and understanding of change. Learning based on trial-and-error.	Schein, 1996; Lewin, 1946
Allows for understanding the various forces through multiple dialogues	Denzin and Lincoln, 2011; Arieli *et al.*, 2009; Martínez *et al.*, 2006
Increases the skills and confidence	Jønsson *et al.*, 2009; Terry and Khatri, 2009; Lincoln and González, 2008; Dick, 1999
Focuses on in-depth knowledge. Learning and changing a system are bound together. High degree of participation offers legitimacy	Denzin and Lincoln, 2011; Arieli *et al.*, 2009; Van de Kerkhof, 2004
Offers the versatile role of the change agent	Schein, 1996; Lewin, 1946
Assesses the risk of intervening	Burnes, 2004; Schein, 1996; Lewin, 1946
Puts scientists, policy-making, practitioners on equal footing with respect to knowledge	Van de Kerkhof, 2004
Claims validity by being grounded on theory and rigour by multiple data collection processes	Denzin and Lincoln, 2011; Van de Kerkhof, 2004
Focuses on socially robust solutions	Van Herk *et al.*, 2011; Levin and Greenwood, 2011; Bodorkos, 2009; Gummesson, 2007a; Argyris, 2004; Dick, 1999; Burawoy, 1998; Greenwood and Morten 1998; Argyris and Schön, 1996; Funtowicz and Ravetz, 1993; Whyte, 1991; Whyte, 1998

Drawbacks	
Entails the risk of abuse of power	Jacobs, 2010; Burawoy, 1998
Has difficulty engaging multiple interests and conflicts	Aziz *et al.*, 2011; Jacobs, 2010; Arieli *et al.*, 2009
Entails more up-front costs and takes longer	Levin and Greenwood, 2011; Bodorkos, 2009
Publishing is not common	Levin and Greenwood, 2011; Bodorkos, 2009
The role of the change agent is difficult	Jacobs, 2010; Arieli *et al.*, 2009
Not prioritized as research	Levin and Greenwood, 2011; Bodorkos, 2009; Mejía *et al.*, 2007; Hansen and Lehmann, 2006; Martínez *et al.*, 2006; Steiner and Posch, 2006

The drawbacks of AR presented above suggest some gaps in knowledge. They are basically related to (a) the need to deal with conflicts, different interests, and to lead to decision-making on structured processes, (b) the need to prioritize real-life problems and multidisciplinary approaches in academia, and (c) the difficulties of playing the role of change agent.

Despite the fact that Levin and Greenwood (2011) consider AR as the superior way to link teaching, research and real world engagement, (a) the field would benefit from other fields such as the field of negotiation and conflict resolution, presented previously, that focuses on handling conflicts, decision-making and/or multiple interests. In this regard Faure *et al.* (2010) point at the importance of moving from AR to action research partnership (ARP), as it is needed to reach collective and individual objectives and to lead negotiations. Even in the engineering field, academicians recognize that in order to foster coordination and cooperation, there is a need for collaborative work among disciplines and with stakeholders (Martínez *et al.*, 2006). Kisito *et al.*, (2009) point at focusing on structuring (moving towards consensus) the problems prior to action to avoid frustration; Kangas *et al.* (2010) at in-depth knowledge of stakeholders' profiles and at supporting also more structured processes than what is currently implemented from AR approaches; and Estacio and Marks (2010) and Hirsch *et al.* (2010) at evaluating more precisely the risks and reactions to actions from underprivileged communities.

AR (b) has not reached the position in academia that it deserves because multidisciplinary and collective research is commonly discouraged and also because real-life social concerns are not always a priority at higher education centres (Levin and Greenwood, 2011; Bodorkós and Pataki, 2009; Mejía *et al.*, 2007; Hansen and Lehmann, 2006; Martínez *et al.*, 2006; Steiner and Posch, 2006).

The role of the change agent (c) has stimulated substantial debate because it is quite complex and entails power effects (Jacobs, 2010; Arieli *et al.*, 2009; Morse, 2008; Fals Borda and Mora-Osejo, 2003). Being able to assert when to be a helper, a mediator, a facilitator, or an observer asks for special training (Faure *et al.*, 2010; Burnes, 2004; Schein, 1996). Developing criteria to assess the risk of intervening, of being an expert in helping and of being self-reflexive has provoked profound debates at the very heart of the meaning itself of the social sciences (Morse, 2008; O'Hogain, 2008; Burawoy, 2007; Eversole, 2003).

2.5.3   Lessons learnt for this thesis

This thesis has found that dealing with marginalized groups may be better targeted, understood and complemented by AR. Authors like Burawoy, (2007), Cornwall (2003), and Reason (1994) state that supporting marginalized groups requires transformative participation in the creation of instrumental or reflexive knowledge, long-term change, empowering, and in-depth understanding. Moreover, in the case of McSEs implementing CP, conflict resolution and integrative negotiation in developing countries may complement AR by giving it structure and strategies dealing with multiple interests in order to be successful.

The theory of change (Lewin, 1946) gives insights into the nature of change for McSEs, on the need to have a change agent that is above all a helper, that has a holistic view, that leads approaches suiting the specific culture, that stimulates trial-and-error as the learning path, and that asserts the risk of intervening through a first low-key inquiry step.

2.6   Policy Instruments (PI)

2.6.1   The place of policy instruments in sustainable development/industrial transformation theories

Meeting present needs without compromising the ability to meet future generations' needs entails necessarily dealing with the impending problems arising from human impact through appropriate tools. These tools are expected to address the challenges posed by sustainability, which are its: unstructured character, path dependency and technological lock-in, institutional and cultural embeddedness, and knowledge dependence (Wieczorek et al., 2010). Within this scenario, the traditional instruments do not seem to be successful, especially on a large-scale use (Blackman, 2011; 2010; 2009; 2005; Raven et al., 2010; Wieczorek et al., 2010; Kemp and Pontoglio, 2011; Kemp and Martens, 2007; Sterner, 2003).

2.6.2   Definition

This thesis refers to the field of environmental policy instruments (PI). Policy is commonly defined as the method or mechanism used by government, political parties, businesses or individuals to achieve a desired effect through legal or economic means (ULeeds, 2012). Instruments pertain to the tools used to achieve the desired effect (OECD Directorate, 2012). The rationale for government involvement is market failure[13] in the form of externalities[14] due to the free-rider problem[15] or the tragedy of the commons[16] (Sterner, 2003).

2.6.3   History of environmental PI

The 1970's are considered the decade of the beginning of the modern environmental policy. On January first 1970, President Nixon signed the National Environmental Policy Act (NEPA) from which, the United States of America was considered a world leader in conservation. In July 1973, Europe adopted its first Environmental Action Programme at the first meeting of the Council of Environmental Ministries. Since then, considerable legislation has developed in all areas of environmental protection (Andrews, 2006).

---

[13] Conditions under which the free market does not produce optimal welfare

[14] Non-market side effects of production

[15] It is when the private marginal cost of correcting an impact is greater than the private marginal benefit the individual can obtain, and subsequently at the social level, and the social marginal cost is less than the social marginal benefit.

[16] By the commons, it is understood goods such as public goods that no one owns and that are commonly overused.

2.6.4   Elements of environmental PI

The theory of policy instruments has a set of principles that are described by Beder (2006).

1. Sustainability principle: points at establishing commitment on a long-term basis regarding the environmental, social, and economic dimensions;

2. Polluter pays principle: targets the polluter's responsibility;

3. Precautionary principle: states that where the cause-effect relationship is uncertain, the possibility of irreversible impacts if the effect is caused is reason enough for action;

4. Equity principle: argues that individuals deserve equal opportunities in a safe and healthy environment and should be treated equitably;

5. Human rights principle: is based on the United Nations Declaration from 1948 and promotes basic rights; and

6. Participation principle: Those who are affected by a decision have the right to get involved in the decision-making.

The literature on policy instruments focuses traditionally on regulatory, economic or persuasive instruments and the characteristics of each instrument is summarized below (see Table 2-9). Recently, due to the high complexity of today's environmental challenges, a set of systemic instruments has been proposed (Wieczorek *et al.*, 2010). Systemic instruments face structural and non-structural issues and take an integral view of the problems. It is believed that ideally, several instruments can be combined in a way that they match the specific needs of the problem being faced without undermining individual original frameworks. Command-and-control approaches may, for example, be complemented by persuasive instruments and capacity building processes, which require a high degree of participation. The goal is, hence, to allow flexibility in terms of compliance with the law while reducing uncertainty of reaching sustainability (Beder, 2006; Sterner, 2003).

Table 2-9 Some relative features of the traditional policy instruments

Characteristic	*Regulatory*	*Economic*	*Persuasive*
Route for influencing decisions	Alteration of the set of available options	Alteration of the cost-benefit ratios of available options (direct, indirect by fiscal means, or market creation)	Alteration of actors' sensitivity or preferences
Potential incentive (for actors)	Avoid penalty or action (judicial/administrative procedures)	Economic incentive Flexibility of choice	Avoid social isolation Project 'green' image Flexibility of choice
Potential advantage (from regulator's point of view)	Effectiveness May use existing regulatory framework (decreased administrative costs)	Efficiency Dynamic May have fiscal benefits May be easier to administer	Acceptability Feasibility (less costs)
Potential disadvantage (from regulator's point of	Are static Enforcement difficult or	Macro-economic implications	Uncertainty Free riding

view)	costly (esp. for many actors)	Distributive & equity considerations Uncertainty Reduced control over industrial operations	
Examples of instruments (national level)	Behavioural, technology standards (e.g. for emission or energy efficiency) Bans Mandatory energy audits	Taxes (energy, carbon, emission) Permits or quotas (national, sectored) Subsides (reduction/ temporary allocation) Financial (grant, loan)	Awareness programs Eco-labelling Voluntary agreements (VA)

Source: Soni (2006).

Sterner (2003) considers that the field of policy instruments faces the most impending challenges in the developing world. The latter explains that relevant PI research is conducted in those regions of the world (Blackman, 2010; Jordan *et al.*, 2005).

First, the main problem faced by PI is that they are usually not used adequately especially on a large-scale. Second, since poverty is inter-related with environmental degradation, tailoring the appropriate instruments to the specific contexts and essential needs has become especially important. Third, selecting a proper mix of policies in developing countries is not always reachable as, for example, certain market-based instruments cannot be implemented. Fourth, voluntary agreements are widely used but there is no consensus as to whether they have been effective (Sterner, 2003). Fifth, in developing countries there is the risk that the long sought environmental governance competes with weak government structures as in the absence of state involvement, private measures try to provide economic efficiency (Jordan *et al.*, 2005). Five contextual elements may affect the effectiveness of the policy instruments in developing countries: 1. Knowledge, 2. Trust in governance, 3. Design of participatory processes, 4. Women's status, and 5. The status of minorities.

Overall, in this scenario what has been observed is that the use of regulatory, market and persuasive policy instruments (PI) often bypasses the McSEs. The outcomes of using these PIs have not been successful whereas only a small number of McSEs are effectively reached (Baas, 2007; Montalvo and Kemp, 2008; Frondel *et al.*, 2005). Traditionally, in the developing countries the effort has been applied to the large-scale industrial sector, which has the resources and often the capacity to develop and use new technologies. Within the McSEs, the enforcement of policies has frequently resulted in costly end-of-pipe solutions, relocation without solving the polluting practices, and even shutdowns with social implications (such as in Delhi, Agra in India and Bogotá and Villapinzón in Colombia) (Soni, 2006; PROPEL, 1995).

2.6.5   Benefits, drawbacks and gaps in knowledge

PI aims at allowing flexibility in terms of compliance with the law while reducing the uncertainty of reaching sustainability (Beder, 2006; Sterner, 2003). PI is framed within six principles and offers rules of procedure (Beder, 2006). For complex environmental problems, systemic policy instruments have been recently framed (Wieczorek *et al.*, 2012). Systemic instruments aim at understanding a problem from the perspective of structural and non-structural dimensions. Addressing systemic problems entails the participation of actors, the learning process, appropriate hard and soft instruments, physical and knowledge infrastructure, and interfaces among heterogeneous actors.

The field of policy instruments has already been largely put in practice, especially in the developed world; weak enforcement of policy instruments does not help to trigger the change towards clean industries (Beder, 2006; Soni, 2006). Policy instruments today offer a wide array of different tools, which need to be combined in such a way that they are congruent with the specific contexts (OECD directorate, 2007). There is more than one way to stimulate the diffusion of cleaner technologies including national policies compatible with regional and local ones, and community initiatives (Blackman, 2010; 2009; Baas, 2007; Steiner and Posch, 2006; Soni, 2006). As such, there is consensus in the literature that the design of effective policies must be built on the basis of understanding the complexity and the diversity of the social capital of McSEs in developing countries (Blackman and Kildegaard, 2010; Soni, 2006). Even though the effect of persuasive instruments like voluntary agreements (VA) might be overestimated, VAs are supporting the institutions' transition towards law enforcement and improving regulatory regimes (Blackman *et al.*, 2012; Blackman 2011; 2010).

Despite the above-cited benefits, PI does not offer consensus regarding the efficiency and effectiveness of specific policy tools. Some support the idea that economic instruments provide better incentives for technology adoption and innovation than direct control (Jung *et al.*, 1996; Milliman and Prince, 1989). However, Adelle *et al.* (2010), and Nentjes (1988) show that pollution taxes or an emissions trade system provide less incentive for innovation. For Blackman (2010), Adelle *et al.* (2010), Opschoor and Turner (1994), the issue is to choose, design and implement suitable instruments for the particular circumstances as the techno-economic and political contexts are considered relevant. For Bohm and Russel (1985), regulatory instruments are technology forcing. They tend to be enforceable to the extent that technological possibilities exist or are likely to be developed (Barde, 2000). Kemp (1995) considers that innovation waivers and tradable quotas may be better than direct regulation and taxes promoting innovation. Most recently, Wieczorek *et al.* (2010) consider that since systemic instruments stimulate innovation systems, there is a win-win situation and sustainability is better supported.

In the developing countries and specifically in the contexts of McSEs, where complexity is high, the theory of policy instruments has not shown positive results (Blackman, 2010; Frondel *et al.*, 2005; Jordan *et al.*, 2005; Montalvo and Kemp, 2008). The accent has been on regulatory instruments whereas command-and-control is usually weak (Blackman, 2000). Recently, the use of persuasive instruments has been analysed and preliminary results are not suggestive of successful long-term results (Blackman, 2010).

Table 2-10 Benefits and drawbacks of Policy Instruments

Benefits	*Source*
Large experience reported	Beder, 2006; Soni, 2006
Aims at allowing flexibility & reducing uncertainty	OECD directorate, 2007
Is framed by six principles and rules of procedure	Beder, 2006
Offers wide array of tools to be adapted to different contexts	Soni, 2006
Offers new integrative instruments called systemic instruments	Wieczorek *et al.*, 2012
Drawbacks	
Does not offer consensus regarding the efficiency of the tools	Adelle *et al.* 2010; Barde, 2000
No positive results with respect to McSEs	Blackman *et al.*, 2012; Blackman, 2010

| Persuasive instruments are overestimated in the developing countries | Blackman, 2010 |
| Main lesson is that weak enforcement does not support change | Beder, 2006; Soni, 2006 |

The drawbacks of policy instruments (PI) presented above suggest some gaps in knowledge. They relate to the need (a) to develop further the fields of climate change and innovation in the developed countries supporting in-depth study and systemic instruments; (b) to reach effectiveness without putting aside the formal regulatory frameworks in developing countries also through more integral approaches by systemic instruments.

In the developed countries, (a) research on PI is being oriented towards facing today's complex problems such as climate change or towards innovation. The Institute of European Environmental Policy suggests looking at the global impacts, and at going beyond the EU's borders (Adelle et al., 2010). The literature reports that raising environmental awareness has brought positive outcomes even for the EU McSEs, which seemed reluctant to accommodate environmental concern in the recent past and considers that they need to be constantly stimulated (Revell et al., 2010). Kemp and Pontoglio (2008) and White (2008) state that case studies have proven to be relevant sources of empirical evidence leading to in-depth knowledge and that as such they should be supported. The evidence brought by case study research could eventually contribute towards identifying the effectiveness of the existing policy and systemic tools.

In the developing countries, (b) due to the ever-existing complexity, the need to fill the gaps in knowledge on PI is always present (Blackman, 2010; Sterner, 2003). Blackman (2011; 2010) reports that in the last decade, persuasive instruments like public disclosures, eco-certification, or voluntary agreements have been widely used and are sidestepping the institutional and political constraints. The latter is not without risks: agencies lacking funding, expertise and personnel; a majority of informal firms lacking monitoring and support; and lack of political will to allocate the few resources to address environmental concerns (Blackman, 2010). The problem is that overestimating these persuasive instruments can be easy and non-regulatory tools may not be able to motivate the firms because they are known to be weak (green markets are small, stock markets are thin); informal regulation needs at the end strong formal regulation in order to be effective; small-scale firms are prominent and more difficult to be reached by regulators; and despite the fact that public disclosures seem to work, information does not flow freely. Overall, most authors agree that besides voluntary regulation being widely used, in parallel the formal regulatory frameworks are to be strengthened (Blackman et al., 2012; Blackman, 2010; 2009; 2005; Jordan et al., 2005; Sterner, 2003). In Colombia, for example, the discharge fee program has developed not only as a disincentive for polluters but also for regulatory authorities to improve monitoring, permit procedures and enforcement (Blackman, 2010; 2005). The voluntary environmental agreements have contributed towards improving the regulatory regimes, to improve communication and the flow of information, and to support capacity building (Blackman et al., 2012; 2009; 2006; Blackman, 2008). In short, persuasive instruments may open a road towards transitioning to sustainability and more data is needed.

Since solving organizational challenges have become key issues in the developing countries, organizational matters have turned out to be fertile grounds for future PI investigation. Blackman and Kildegaard (2010) consider that clean technological change does not really depend on the firm size or on the regulatory pressure but rather on the firm's human capital. Other authors suggest the firm's relational capital as the key issue and that as such, policy tools are advised to focus on stimulating firms that are successful negotiators to develop collaborative agreements (Barcellos et al., 2009; Welbourne and Pardo, 2009). Steiner and Posh (2006) suggest focusing on building mutual learning tools, supporting learning environments (Merrey *et al.*, 2007; Van der Zaag and Bolding, 2005), and learning through experimentation (Raven *et al.*, 2010) whereas trade associations are important sources for technological input and capacity building (Blackman and Kildegaard, 2010). Barcellos *et al.* (2009) invite to build networks for decision-making and to focus on leadership, change and learning processes. Best *et al.* (2005) points at building linkages and enhancing trust between small-scale producers and buyers in growing markets. Van Berkel (2011) states that in order to be more effective, the existing national cleaner production centres (NCPCs) ought to establish solid links with the specific market needs. All the latter invite to develop knowledge on systemic instruments (Wieczorek *et al.*, 2012). The developed world is realizing that structural analysis, which has been traditionally used to evaluate innovation systems, has proven to be insufficient (Schmoch *et al.*, 2006). Instead, also focusing on functional approaches may lead to an enhancement of the processes for the best performance of innovation systems. Processes are seen as functions of innovations like market formation or knowledge development and linking structural and functional approaches may at the end lead to sustainable technological change. Policy instruments are thought to stimulate system change (Wieczorek *et al.*, 2012; 2010). For these authors, discouraging negative influences, stimulating the positive ones and improving capabilities on a broad spectrum seems to be the path towards sustainability.

2.6.6   Identification of issues and mechanisms that influence CP

CP policies operating in parallel to command-and-control ones are suggested in the context of McSEs in developing countries. The attitudinal, organizational and systemic mechanisms are key for successful CP implementation. The list below suggests issues and mechanisms influencing CP derived from the analysis of PI in 2.6.5, 2.6.6, and 2.6.7 and from the analysis on CP in 2.2.7:

1. Attitudinal & Organizational (systemic) instruments

- − Focus on social and economic issues,

- − Emphasis on highly participative and trial-and-error processes,

- − Consensus building,

- − Regional authority's leadership and commitment to CP, and

- − Consultant seen as a change agent.

2. Regulatory, economic, and persuasive instruments

- − CP policy enforced in parallel with command-and-control's measuring loads and not just concentration,

- − Discharge limits according to local situations, and

- − Market incentives.

3. Access to financing

4. Choice of technology

5. Access to information

2.6.7   Lessons learnt for this thesis

The literature review has shown that the policy instruments' field is moving towards broader concepts, analyses, processes and applications, and towards taking into account contexts and specific characteristics in order to reach effectiveness. This thesis supports Raven (2010)'s proposition to elaborate a thorough diagnosis of hidden problems in order to be able to design suitable instruments, and Wieczorek *et al.* (2012)'s thesis to link the structural and the functional components in order to aim at sustainable technological change.

So far there is worldwide consensus on not exclusively targeting economic growth but also focusing on environmental protection, social development, and change (Stiglitz, 2012; Kemp and Martens, 2007; Cleaver, 1999). The latter entails then, building strategies for complexity, which may influence positively the study field of McSEs in developing countries, since it is considered a complex problem and has been disregarded by Economics or Business Management (Audretsch, 2009). Within this scenario, in the developing countries where McSEs are so prevalent, integral instruments need to be reinforced. The latter is even true among the highest performing countries such as Chile that offer good developed banking systems and/or subsidize innovation-based programs between private firms and universities as Aagosin *et al.* (2009) have reported.

2.7   Comparative Analysis

2.7.1   Introduction

The theoretical fields of policy instruments, cleaner production and stakeholder participation do not address the specific characteristics of McSEs in developing countries. AR and negotiation/conflict resolution (N/CR) that handle transformative participation may contribute towards building a better theoretical and methodological approach towards McSEs implementing CP because AR supports change and long-term endeavours, and Negotiation/Conflict resolution (N/CR) offers structure, rules and approaches towards decision-making and problem solving based on multiple interests.

2.7.2   How the theories fit together

The aim is to build a robust theoretical framework for McSEs implementing CP in developing countries whereas the gaps in some theories are addressed through the strengths of the others.

Tables 2-11 & 2-12 integrate the benefits, drawbacks, gaps in knowledge, and elements of the different theories.

Table 2-11 Combining Cleaner Production, Stakeholder Participation, Policy Instruments, Negotiation /Conflict Resolution and Action Research: benefits.

Benefits	CP/ SP	CP/ SP/ N/C R	CP/SP/ N/CR/ AR	CP/SP/ N/ CR/AR/ PI	*Reference*
CP					
Can be applied to products & processes in any industry	●	●	●	●	Siebel and Gijzen, 2002
Introduced concepts such as	●	●	●	●	Gutterres et al., 2010; UNEP, 1991; Dieleman, 2007;

waste valuation, recycling, re-using and technical steps to be followed (rules of procedure)				Van Berkel and Bauma, 1999
Based on the principles of sustainability, supports prevention approaches & policies	● ● ●		●	Baas, 2007
Based on self-reliance	● ● ●		●	UNEP, 1991
Open to new concepts & innovations	● ● ●		●	Baas, 2007; Dieleman, 2007
Can offer wide array of technical options	● ● ●		●	Dieleman, 2007
SP				
Offers settings to discuss complex issues that can be bottom-up improving scientific practice and policy making.	● ● ●		●	Reed et al., 2009; Mollinga, 2008; Reed, 2008; Stringer et al., 2007; Van der Zaag, 2005; Van de Kerkhof, 2004; Godard and Laurans, 2004; Fiorino, 1990
Stimulates cross-sectoral planning	● ● ●		●	Thabrew et al., 2009
Can be a means (signing agreements) and an end (empowerment)	● ● ●		●	O'Hogain, 2008; Selfa and Endter-Wada, 2008; Duraiappah et al., 2005; Kanji and Greenwood, 2001; Karl, 2000
Offers the ladder of participation	● ● ●		●	Van de Kerkhof, 2004; Cornwall, 2003; Mayer, 1997; Argyris and Schön, 1996; Arnstein, 1969
Offers 8 key principles	● ● ●		●	Reed, 2008
Offers stakeholder analysis	● ● ●		●	Reed, 2008; Reed et al., 2011 &2009
Potentially lowers implementation costs at the supervision stage	●			Grimble et al., 1995
N/CR				
Offers specific rules of procedures for decision-making based on specific criteria and structured approaches.		● ●	●	Thompson, 2009; Frenkel and Stark, 2008; Brett, 2007; Saner, 2003; Raiffa, 2002
Promotes order & stability and feelings of self-efficacy; reduces probability of future conflicts		● ●	●	Giacomantonio et al., 2010b
Offers a wider range of solutions, flexibility, win-win opportunities, in depth knowledge		● ●	●	WIPO, 2012
Can handle complex problems through win-win approaches; can create and claim value based on interests		● ●	●	Thompson, 2009; Saner, 2003; Raiffa, 2002; Fisher et al., 1991
It is context-specific and process oriented		● ●	●	Thompson, 2009; Frenkel and Stark, 2008; Brett, 2007; Saner, 2003; Raiffa, 2002; Fisher et al., 1991
Can handle conflicts entailing big groups through whole scale methodologies that engage people for collective and individual best results		● ●	●	Holman, 2010; 2004; Holman et al., 2007
Offers the roles of mediator and /or facilitator		● ●	●	Frenkel and Stark, 2008
Offers strategies that tell when and how; that may focus long term		● ●	●	Merlano, 2005; TANDEM, 2005; Blake and Mouton, 1964

	CP/SP	CP/SP/N/CR	CP/SP/N/CR/AR	CP/SP/N/CR/AR/PI	Reference
Potentially lowers implementation costs at the supervision stage		●	●	●	Grimble *et al.*, 1995
Focuses on emotions		●	●	●	Pietroni *et al.*, 2009; Barry, 2008; Shapiro, 2006

Benefits	CP/ SP	CP/ SP/ N/C R	CP/SP/ N/CR/ AR	CP/SP/ N/ CR/AR/ PI	*Reference*
Allows learning about a system as it goes through change. Faces impending complex problems asking for action			●	●	Burnes, 2004; Ravetz, 1999; Lewin, 1946
Offers a cyclic, structured process and understanding of change. Trial-and-error			●	●	Schein, 1996; Lewin, 1946
Allows understanding the various forces through multiple dialogues			●	●	Denzin and Lincoln, 2011; Arieli et al., 2009; Martínez et al., 2006
Increases the skills and confidence			●	●	Kevany, 2010; Kevany and Huisingh, 2013; Jønsson et al., 2009; Terry and Khatri, 2009; Lincoln and González, 2008; Dick, 1999
Focuses on in-depth knowledge. Learning and changing a system are bound together. High degree of participation offers legitimacy			●	●	Denzin and Lincoln, 2011; Arieli et al., 2009; Van de Kerkhof, 2004
Assesses the risk of intervening and the role of Change Agent			●	●	Burnes, 2004; Schein, 1996; Lewin, 1946
Puts scientists, policy-making, practitioners on equal footing with respect to knowledge			●	●	Van de Kerkhof, 2004
Claims validity and rigour			●	●	Denzin and Lincoln, 2011; Van de Kerkhof, 2004; Burawoy, 1998
Focuses at socially robust solutions			●	●	Van Herk et al., 2011; Levin and Greewood, 2011; Bodorkos, 2009; Gummesson, 2007a; Argyris, 2004; Dick, 1999; Burawoy, 1998; Greenwood and Morten 1998; Argyris and Schön, 1996; Funtowicz and Ravetz, 1993; Whyte, 1991; Whyte, 1998

PI					
Large experience. Main lesson is that weak enforcement does not support change				●	Beder, 2006; Soni, 2006
Aims at allowing flexibility & reducing uncertainty				●	Beder, 2006; Sterner, 2003
Persuasive might support transition towards formality				●	Blackman et al., 2012
Is framed by six principles and rules of procedure				●	Beder, 2006
Offers wide array of tools to be adapted to different contexts				●	Soni, 2006
Offers new integrative instruments called systemic instruments				●	Wieczorek et al., 2012

CP = Cleaner Production, SP = Stakeholder Participation, PI = Policy Instruments, N = Negotiation, CR = Conflict Resolution, AR = Action Research

Table 2-12 Combining Cleaner Production, Stakeholder Participation, Policy Instruments, Negotiation /Conflict Resolution and Action Research: drawbacks.

CP = Cleaner Production, SP = Stakeholder Participation, PI = Policy Instruments,

Drawbacks	CP/SP	CP/SP/ N/CR	CP/SP/ N/CR/AR	CP/SP/N/ CR/AR/PI	
CP					
Only recently CP projects acknowledge fostering behavioural change and social concerns	●	●			Lozano and Huisingh, 2011; Van Berkel 2011; Van Berkel, 2007; Dieleman and Huisingh, 2006
Regulatory instruments do not support CP. CP policies do not work in parallel with command-and-control	●	●	●		Frondel *et al.*, 2005
Lacks repeating processes, multiple options, larger scale recognition	●	●			Dieleman, 2007
Relies too much on quick savings to stimulate implementation	●	●			Van Hoof, 2005
The conflicting relationships between environmental authorities and McSEs are not handled. Does not handle big groups.	●				Thabrew *et al.*, 2009; Van Berkel, 2007; Montalvo, 2003
SP					
Has more up-front costs	●	●	●		Grimble *et al.* 1995
Cannot lead to problem solving	●				Berk *et al.*, 1999
Lacks rules of procedure for decision making (delays occur), for choosing stakeholders, for choosing different participative methodologies	●				Morse, 2008; Reed 2008; Gupta, 2004b; Eversole, 2003; Engels, 2002; Berk *et al.*, 1999
Cannot face power imbalances, conflicts, political engagement and co-learning	●				Wiber *et al.*, 2009; Morse, 2008; Reed, 2008; Eversole, 2003; Escobar, 1999
Raises expectations that cannot be met	●				Estacio and Marks, 2010; Hirsch *et al.*, 2010; Mehta *et al.*, 2001
Cannot prioritize collective interests over short-term individual ones	●				Webbler and Renn, 1995
Cannot stimulate commitment and change on long-term basis	●	●			Reed, 2008; Eversole, 2003
Can be used for manipulation	●				Van de Kerkhof, 2004; Arnstein, 1969
Empowerment may have negative interactions with existing power structures	●				Morse, 2008; Reed, 2008; Burawoy, 2007; Duraiappah *et al.*, 2005; Cooke and Khotari, 2001
Cannot avoid overlapping formal agreements	●	●	●		Cooke and Kothari, 2001
N/CR					
Does not necessarily clarify/ rely on rights		●	●		WIPO, 2012
Not regulated by quality standards		●	●		WIPO, 2012
Integrative negotiations are difficult to be conducted		●			Giacomantonio *et al.*, 2010a
Whole scale methodologies imply greater efforts: broader facilitators' minds, interdisciplinary orientations		●	●	●	Holman, 2010; 2004
Power imbalances can be dealt on a specific situation but are not commonly used on long-term basis		●			WIPO, 2012; Moran and Ritov, 2007

2.7.3   Gaps in the theories

Table 2-13 lists the gaps presented for each theory for McSEs. An integrative theory that incorporates the strengths of each may provide better solutions.

Table 2-13 Comparing the gaps in theories on Cleaner Production (CP), Stakeholder Participation (SP), Negotiation/Conflict Resolution (N/CR), Action research (AR) and Policy Instruments (PI)

Theory	Gaps	Source
	Was designed initially exclusively towards technological solutions. Needs rules of procedures to build consensus, to deal with the social concerns and interests, to handle conflicts, to foster behavioural change for big groups, not to rely just on quick savings.	Dieleman, 2007; Van Berkel, 2007; Gunningham and Sinclair, 1997
CP	CP policy has a voluntary nature that does not work in parallel with regulatory policies. CP needs to be integrated into the regulatory and formal frameworks.	Blackman, 2010; Frondel *et al.*, 2005
	CP programs need to address the McSEs needs and interests; do not have strategies to overcome barriers, do not offer opportunities of focusing on trial and error as learning processes, of mixing qualitative and quantitative approaches.	Del-Río-González, 2009; Howgrave-Graham and Van Berkel, 2007
	Needs to stimulate dialogues originating at the local level	Bakker *et al.*, 2012; Edelenbos *et al.*, 2009)
	Needs to set participation as a process of change	Morse, 2008; O'Hogain, 2008; Reed, 2008; Cornwall, 2003; Eversole, 2003
SP	Needs holistic, self-reflexive and inclusive thinking, to embrace diversity and to understand that reaching consensus is not necessarily a target	O'Hogain, 2008; Burawoy, 2007; Morse, 2008
	Needs prescriptive guidelines that say how and when dealing with participative approaches in the technological arena	Reed, 2008
	Needs rules of procedures leading to decision-making or to solving conflicts, or strategies for claiming and creating value.	Robinson and Berkes, 2011
	The facilitator needs to learn to handle power imbalances, interests, decision-making, co-learning, conflicts, or change processes.	
N/CR	Needs to be institutionalized as formal policy instruments	WIPO, 2012
	Needs research on motives for searching mutual benefits	Giacomantonio *et al.*,

		2010b
	Needs specific rules for dealing with change and especially with empowering groups before leading them to successful negotiations. The mediator or facilitator 's role needs to engage in change processes	Moran and Ritov, 2007
	Needs to focus on networks, collaborative learning	Wiber *et al.*, 2009
AR	Needs to learn to handle conflicts, multiple interests, and/or negotiations	Faure *et al.*, 2010
	The change agent needs to know how to lead negotiations or how to deal with conflicts	Faure *et al.*, 2010; Jacobs, 2010; Arieli *et al.*, 2009
	Needs to motivate academia to work on a multidisciplinary and collective way to stimulate real-life research	Denzin and Lincoln, 2011; Bodorkos, 2009
PI	Needs to face complex problems/conflicts and to stimulate innovation	Adelle *et al.*, 2010; Kemp and Pontoglio, 2011
	Needs to focus research on organizational matters, to stimulate negotiations, decision-making and collective agreements	Barcellos *et al.*, 2009; Welbourne and Pardo, 2009
	Needs to stimulate learning through experimentation	Raven *et al.,.* 2010
	Needs to foster change at long term basis and prove a systemic framework	Wieczorek *et al.*, 2012
	Needs to deal with specific local contexts	Wieczorek *et al.*, 2010

Table 2-14 presents the elements of the different theories that can be tested in this research.

Table 2-14 Elements from Cleaner Production (CP), Stakeholder Participation (SP), Negotiation/Conflict Resolution (N/CR), Action research (AR) and Policy Instruments (PI) selected

Theory	Elements selected for testing
CP	Principles on sustainability
	Mechanisms
	Self-reliance
	Options on prevention
SP	Principles
	Stakeholder analysis

	Ladder/purpose of participation
	Role of facilitator
	Principles based on interests
	Strategies for decision-making
N/CR	Structured process
	Role of negotiator/mediator
	Inclusive processes AI, OST
	Principles based on change/empowering
	Cyclic processes and concepts of change
AR	Role of Change Agent
	Learning based on trial-and-error
	First low-key inquiry step
	Kinds of instruments: regulatory, economic, persuasive and recently systemic
	Rules of procedures
PI	Principles
	Institutional acceptance

2.7.4 Recapitulating

Chapter 3 integrates this analysis further in the SASI model. Merging and integrating the theories targeted in this thesis lead to reducing the drawbacks of the individual approaches towards underprivileged groups in developing countries. The comparative analysis highlights key concepts of the different theories that when interacting together can make the McSEs' kite fly towards sustainability. In Figure 2-5, the kite, the main interactions are presented on a one to one basis with respect to each field under a, b, c, d, e, f and g and are explained below. Interactions between PI and SP (h) and between PI and N/CR (i) are implicit and presented despite the fact that the design of the kite does not show a direct contact between them.

The figure of the kite symbolizes the fact that complex endeavours like McSEs may be solved by being targeted through combining theories, adopting integral concepts, raising social awareness, building networks and supporting open attitudes.

Figure 2-5 How CP, Stakeholder Participation, Negotiation/ Conflict Resolution, Policy Instruments and AR fit together.

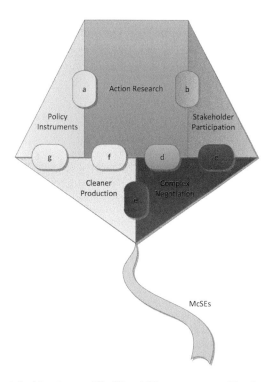

On the one hand, the original barriers on CP, SP and PI are overcome: The theoretical approach to CP starts by recognizing that conflicts can be handled, and by not having resistance to problematize in big groups once it starts to merge with N/CR (e). An intense dialogue between CP with AR results and fostering behavioural change becomes the central axis, as well as the social concerns; since AR is based on trial-and-error, repeating processes & generating multiple options are adopted, and confirms an original principle of CP; relying too much on quick savings to stimulate CP implementation is instead discouraged because there is commitment and engagement building at early stages; thanks to AR, theories are reconstructed (f); finally, once working also with PI, systemic instruments may well work in parallel with command-and-control and strengthen institutions in developing countries (g). In the same line of thought, SP may face problem solving; may incorporate rules of procedure for decision-making, for choosing stakeholders, and tailored-made participative methodologies; may face power imbalances and conflicts; may put limits to expectations; may prioritize collective interests over short term individual ones and solve the debate over the relevance of consensus; may avoid manipulative purposes and may support empowerment by merging with N/CR (c). Through a dialogue with AR, commitment and change are stimulated on a longterm basis (b). By including PI, SP would not overlap with formal agreements (h). Finally, the field of PI deepens into the world of change, into developing integral instruments for the underprivileged with AR (a); into empowerment, decision-making with N/CR (i). On the other hand, the theories of N/CR and AR can also grow stronger. N/CR also gains by a dialogue with CP with respect to working on sustainability (e), with AR by moving comfortably on change processes at long-term basis (d). Finally, AR learns to structure and to handle conflicts with N/CR (d), to target sustainability with CP (f) and to lead formal approaches with PI (a). Despite all those improvements, the drawbacks left are overestimating persuasive instruments, the up-front costs, the power effects, the fact that publishing is not common, and that academia does not prioritize AR and interdisciplinary approaches. The role of the change agent implies the most demanding role since it needs to ascertain when to be a helper, a facilitator, a researcher or a mediator, to evaluate the risk of intervening in a specific context, and to lead multidisciplinary approaches.

2.8   Implications

A robust theoretical approach towards McSEs in developing countries was developed by incorporating the benefits of different theories, by identifying that the gaps can be overcome with the support from other theories, and finally by adding-up elements from all of them. The next Chapter elaborates on the methods and implementation derived from the latter analysis which will attempt to empower McSEs.

3   Approaches for the marginalized: Methods

3.1   Introduction

Former Colombian candidate to the presidency of the World Bank in 2012, Jose Antonio Ocampo has said when asked why environmental sustainability is not yet a main core of the teachings in the faculties of Economics: *"One of the main reasons is the lack of solidarity....I would like to see microeconomic models based on solidarity"* (EnDeuda, 2012). This Chapter builds on this value, on the need to focus on the marginalized.

This thesis aims at contributing to helping McSEs implement environmental policy. The challenges facing McSEs in adopting CP in Colombia demands a methodology that captures this complexity. Engaging the concern of McSEs impacting water bodies implies looking at both the local and extra-local forces at work. Furthermore, these forces have social, economic and environmental drivers that originate in multiple sectors be they informal or formal, rural or urban, more to less developed settings. Dealing with this high complexity asks for in-depth understanding, context-orientation and, hence, for qualitative detailed casework over time.

The layered case study approach was chosen focusing on three cases embedded in the larger case of Colombia. Two of the cases are more typical examples of a classical combined case study while the third was elaborated as an AR.

This chapter first presents the national case study implemented (3.2); it then focuses at the combined case study method developed to compare two tannery cases (3.3). This is followed by an explanation of the AR method and the systematic approach for social inclusion (SASI) created for the third tannery case (3.4). An elaboration is done on the CP implementation for SASI (3.5). After identifying the lessons learnt from the application of the methods (3.6), the conclusions are presented (3.7).

3.2   The National Case Study Method

3.2.1   In-depth interviews and micro-cases

This case study is based on a content analysis and, in-depth interviews at national level, which were conducted before starting the local case studies in 2003. Valuable data was also taken from observations and comments taken from strategic actors at seminars, debates or local meetings and informal talks and is presented as *personal communication*. Since this research has had an inter-subjectivity focus allowing a more intense and continuous data collection process in the long term, some of the in-depth interviews themselves evolved into intense interaction (micro cases) with these stakeholders over the years 2005-2012.

Table 3-1 shows the stakeholders interviewed at the national level and the questions asked are presented in Annex 4.2.

Table 3-1 In-depth interviews at national level

Date	Institution	Position	Interviews (#)
Public National			
2002	MADS	Former 1st minister	1
2003	MADS	Former vice-minister	2
2003	MADS	Economic analysis dept	3
2003	IDEAM	Researcher Hydrology	4[1]
2003			4[2]
2004	Chamber Representatives	Cundinamarca region	5
2006	MADS	Head of tanneries	6
2010	Presidency	Former Secretary Legal Office	7
Public Regional[17]			
2003	SDA (former DAMA)	Manager Bogotá River	8[1]
2003			8[2]
2003	CAR	Director	9
2007	CAR	Responsible Tanneries	10
2007	CAR	Director (time of conflict)	11
Academia			
2003	UNAL	Director IDEA Institute	12
International Org			
2003	CECODES	Executive Director	13
2003	WB	Head of Environment	14
NGO			
2003	Conservation International	Executive Director	15
2003	Quinaxi	Director (former vice-minister)	16

Thirteen in-depth interviews were conducted in 2003, one in 2004, one in 2006, two in 2007 and one in 2010. Micro-cases (mc) resulted with the stakeholders as shown in Table 3-2 and in Annex 4.1.

---

[17] Even though these interviews are regional, the focus was on national matters and concerns.

Table 3-2 Micro-cases at national level

Date	Institution	Position	#
2004-2012	Senate	Head of Senate (Former Representative in Table 3-1)	*mc*1
2004-2012	Ministry	Head of tanneries	*mc*2
2004-2010	Regional Court	Magistrate on the Bogotá River Court Order	*mc*3

3.3   The Combined Case Study Method

3.3.1   Introduction

This section deals with the case study method addressed in the two comparative case study areas (Cerrito and San Benito) in which the involvement of the researcher was limited. The research analysed over a period of six years the contextual and complex conditions relevant for CP implementation.

This section presents an overview of the classical case study approach (3.3.2), the combined case study method (3.3.3), the design and process of the combined approach (3.3.4), and inferences (3.3.5).

3.3.2   An overview of the classical case study approach

Case study work allows for looking at social phenomena from multiple platforms and developing in-depth understanding of today's complex problems. Complexity asks for explanations that cannot be described only through statistical analysis or determined by the frequency of appearance of certain issues or matters. Scholars focusing on the challenges faced by complexity in the fields of CP and PI suggest more case study work for in-depth understanding (Kemp and Pontoglio, 2008).

Definition, elements and criteria for a good quality case

The classical case study is an empirical inquiry that usually addresses the 'how' or 'why' research questions, that focuses on a contemporary phenomenon within some real-life context, and where the researcher has limited or no control over events (Yin, 2009). Clearly case defined borders are argued to be essential.

For Yin (2009), the elements of a good case study are (a) to cope with more variables of interests than can be expected if the method was an experiment, which isolates variables from daily life, (b) to rely on multiple sources of evidence with data meant to converge, and (c) to benefit from a dialogue with theory to guide data collection and analysis. The criteria for judging the quality of research designs are usually based on constructing validity and reliability mainly in the phase of data collection, internal validity of the data analysis, and external validity of the research design (Yin, 2009). The types of research case studies depend on the type of question being asked: they can be explanatory, exploratory, illustrative, and evaluative.

Even though Yin's contribution to developing the case study method is unequalled and has been tremendously valuable in social science research, his approach is quite positivist. He sees the researcher as a neutral bystander and the case as having clear boundaries. Cases according to Yin are examined and compared as if they exist separate from the region they are embedded in, as if the activities beyond the case do not 'leak' into the case or the other way around. Rather than clear boundaries others argue that these boundaries are really porous and forever changing (Denzin and Lincoln, 2011; 1994; Gummesson, 2007a; Flyvbjerg, 2004; Burawoy, 1998, Reason, 1994).

Since this research commits with reality and with building actionable knowledge, it supports a more reflexive approach to the classical case study method, such as the extended case method. This approach recognizes that cases are contextually embedded in larger scales, and the inter-subjectivity of the researcher and the subject of study (Burawoy, 1998). The latter implies that the (qualitative) 'researcher is not an objective, authoritative, and politically neutral observer and his research is historically positioned and locally situated' (Denzin and Lincoln, 1994 citing Bruner (1993:1)).

Methods for implementing a case study can be qualitative such as interviews, participant observation, focus groups, stakeholder analysis, but can also use quantitative approaches depending on the specific needs of the research. In-depth research is problem-driven and not methodology-driven and is not expected to support an artificially built dilemma between qualitative and quantitative approaches as they can complement each other with respect to a specific problem (Gummesson 2007a, and Flyvbjerg 2004). Gummesson (2007a) proposes qualitative, quantitative, social and natural work in symbiosis and not through conflicting relationships. Freire (1982: 30) argues that the concrete reality is the connection between subjectivity and objectivity never objectivity isolated from subjectivity.

Case study research constitutes a difficult challenge because it produces the context dependent knowledge. It emphasizes building the skills for doing good research and does not restrict to just rule-based research, which is mainly conducted by beginners (Flyvbjerg, 2004). Case studies also face problems related to legitimation (researcher needing to make certain that voices are heard), representation (there must be ways that best describe and interpret the interests of others), dealing with a multiplicity of forces, and unmanageability because of the existence of thick descriptions (Denzin and Lincoln, 2011; 1994). It is also important to differentiate between theories and tools from early stages of the research (Sayer 2010; 1992).

3.3.3   The combined case study method

In sum, the case study method developed in this thesis for the 2 cases privileges an actor-oriented approach where underprivileged communities are put on equal footing with respect to policymakers or officials from the environmental authorities. The method applied uses Yin's (2009) structure and criteria, is inspired by the extended case study methodology from Burawoy (1998), and fed by insights from qualitative social scientists like Denzin and Lincoln (2011/1994), Gummesson (2007a) and Flyvbjerg (2004). The combined method gives better understanding of the complexity involved by giving the opportunity to interact more closely with the actors over time, and by unveiling the multiple forces at work at local, regional and national levels.

The extended focus

Observations at the national level unveiled the interactions between the regional and the national levels at the National Tannery Committee at the Ministry (Gummesson, 2007b). The cases on micro-tanneries were extended in relationship with the national case (Burawoy, 1998) where archival research, in-depth interviews, and a micro-case on the head of the Tanneries at the Ministry were developed. Section 3.3.5 exemplifies the intense interactions that resulted from studying the local cases.

The role of the researcher

The inter-subjectivity focus was meant to allow the data collection process to be more intense and continuous. The process was expected to develop multiple opportunities for data collection and to get acquainted with the change process. Micro-cases were developed for strategic stakeholders and preferred over isolated interviews.

3.3.4   Design and process of the combined approach

The case studies of Cerrito and San Benito address sub-research question #2:

Which perceived mechanisms support Cleaner Production and how do these influence the adoption of Cleaner Production by Micro and Small Sized Enterprises in Colombia?

The question aimed at discovering the complexity of McSEs facing CP implementation through the eyes of the actors themselves and, hence, the case study method needed (a) acquiring in-depth understanding of the multiple forces operating on McSEs implementing CP, and (b) getting acquainted with the chain of events (like behavioural change on CP implementation) during a long period of time.

This research was conducted at the national, regional and local levels. Basically, steps 1 (Preparation) and 3 (Follow-up) were developed at the national levels. Steps 2 (Getting to know each other) and 4 (Final scope) were conducted at regional and local levels (Table 3-3).

At the national level, (a) archival research was compared with people's perceptions and with the researcher's observations; interviews at national level were conducted; (b) direct observations at the national tannery committees took place twice a year at the Ministry of the Environment; these were analysed to illustrate the relationship between the tanners and their regional environmental authority from 2005-2008.

Among the tannery communities in Colombia, the two tannery cases were chosen because:

1) Following years of no effective results in terms of mitigating their environmental impacts, they started participative and apparently successful CP implementation projects (as it was presented at the National Tannery Committee in 2004);

2) Micro and small sized tanneries represented more than 70% of their industrial communities in terms of the total numbers, and

3) The pollution caused by the tanneries was substantial in both cases.

At the regional and local levels, both communities were visited twice: first in 2004, when each community was starting the CP implementation project[18], a stakeholder analysis wasdeveloped, and in-depth interviews used *exploratory* questions (which and what) – identifying mechanisms supporting CP; second in 2009, after 5 years of CP implementation, the questions used were *explanatory* (how) –

----

[18] Since the creation of the CP policy in 1997, some regional authorities were trying to stimulate CP agreements.

determining how those mechanisms operated - and accounting for the chain of events during that period of time (Yin, 2009) (Table 3-3).

Observations, focus groups (dialogue groups), micro-cases, and in-depth interviews were conducted on each visit. In 2009, the focus group (N=6 and mixed F/M)[19] targeted its questions with the actors on the lessons learnt from their own CP projects.

Table 3-3 Multiple opportunities for data collection and intervention

Steps	Methods	Year
1. Preparation	Archival & legal data, discussions	2004
2. Getting to know each other (Researcher & Stakeholders)	Stakeholder analysis, exploratory in-depth interviews & focus groups, micro-cases	2004
3. Follow-up	Observations, discussions, archival & legal data, micro-cases	2005-2008
4. Final scope	Explanatory in-depth interviews & focus groups, micro-cases	2009-2010

The answers to the questions were drawn on radar graphs (like). A score of 0 was given when an issue was not mentioned, of 1 when it was just considered by some and of 2 when it was considered relevant by all interviewees.

As written above, the question asked in 2004 was exploratory: Which mechanisms support CP implementation? (illustrates the answers in 2004). With that question the objective was to identify, before the CP implementation was starting, the ideas people had with respect to CP. In 2009, as CP was being implemented, it was important to identify which instruments were useful and how did the successful instruments operate. The question was then explanatory and was digging into the processes. Radar graphs were able to give insights on the similarities and differences on the mechanisms between groups such as the tannery communities and information on the existence of consensus for CP implementation. Scores of 1, which meant lack of consensus were important to suggest deeper analysis in subgroups such as tanners or officers or even deeper among the different groups of tanners themselves.

---

[19] N stands for number of participants, F for female and M for male.

Figure 3-1 Radar graph to assess mechanisms supporting CP

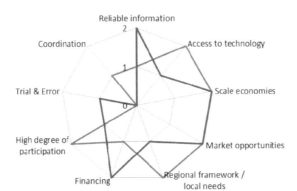

Table 3-4 Stakeholders interviewed in Cerrito and San Benito between 2004 and 2009

Type of Stakeholder	Cerrito	San Benito
Head of the national tannery committee (Nt)	1	1
Head of the CP project (HdCP)	1	1
Technician in charge of the CP project (Tech)	1	1
Officer at the regional environmental authority (Rg)	1	1
Local environmental authority (Lc)	1	1
Tanners	4	9
Total number of interviewees	**9**	**14**

For each case, in-depth interviews with the Head of the national tannery committee (Nt) (same as Head of Tanneries in Annex 4.1) and the Head of CP projects (HdCP) developed into micro-cases and offered a more continuous and on-going opportunity for data collection (Annexes 5.1 and 5.2).

The chosen number of interviews with tanners was not meant to be statistically representative. The choice of the tanners was intended to represent the different groups of interests once a stakeholder analysis was initially developed in the different local cases; as such it became a stratified sample. As a result, 4 tanners were chosen in Cerrito where a more homogeneous group was found, and 9 tanners were chosen in San Benito representing different associations, size of industry, and stages of CP implementation (Chapter 5).

3.3.5   Inferences

Since data collection and analysis were continuous and for 5 years, it was possible (a) to acquire in-depth understanding of the multiple forces operating at McSEs implementing CP, and (b) to get acquainted with the chain of events (like behavioural change on CP implementation) during that period of time. In fact, the results would have been very different if the research had stopped after two years of fieldwork because the tanners in Cerrito had been slow in the CP implementation process and the improvements in the water quality standards were not yet possible to be calculated (see Chapter 5).

Through this continuous process, a certain degree of trust and commitment was developed between all the actors (including the researcher). The fact that the local cases are in real life embedded in other administration levels was expressed by the multiple interactions that resulted from studying them through this process: (a) The case of Cerrito itself inspired the tanners in Villapinzón to implement CP, once they discovered that they shared common problems and characteristics with the tanners in Cerrito. (b) The head of the Tanneries and of the National Tannery Committee at the Ministry, who had a global knowledge of the problem, became the key person at the Ministry to support the process in Villapinzón. (c) The tanners from Villlapinzón inspired the tanners from San Benito belonging to the group of 80 tanners with respect to the adaptation of CP options to the local needs (see corresponding Chapters). (d) The head of the CP project in Cerrito supported initially the process in Villapinzón by offering credibility to authorities in conflict with the tanners.

The combined methodology developed to study the two comparative cases was meant to fit the context being targeted and was fine-tuned during the research.

3.4   Action research methods and SASI

3.4.1   Introduction

The third tannery case dealt with the extreme pollution and social exclusion case of the tanneries of Villapinzón. Here, the involvement of the researcher was more continuous and intense. The research designed, implemented and analysed over a period of six years a new methodology that I named as SASI (Systematic Approach for Social Inclusion). SASI is based on AR and inspired by approaches from complex negotiation/conflict resolution, policy instruments, stakeholder participation, and cleaner production. This section presents the AR approach (3.4.2), the SASI method (3.4.3), the design and process of SASI (3.4.4), and inferences (3.4.5).

3.4.2   Action Research (AR) Approach

Definition

AR develops knowledge while acting. In the environmental field, it recognizes that complex and impending environmental problems can be solved effectively with action accompanying the learning processes (Vinke-de Kruijf *et al.,* 2010; Burnes, 2004; Ravetz, 1999). For McTaggart (1996) AR is not a method or a procedure but a series of commitments to observe and define problems through implementing principles for social enquiry. "It is a form of collective enquiry undertaken by participants in social situations in order to improve the rationality and justice of their own social or educational practices, as well as their understanding of those practices and the situations in which the practices are carried out" (Kemmis and McTaggart 1988).

AR facilitates action through joint learning and research with those who will carry out the actions, through participation in research by stakeholders as well as specialists. Research is experienced within the real world in which it is applied (Levin and Greenwood, 2011; Hansen and Lehmann 2006; Burnes, 2004; Ravetz, 1999; Schein, 1996; Lewin, 1946). As a constructivist approach, AR does not claim truth because a fact accurately reflects nature but because those who are involved in the issue certified it as true (policy-makers, stakeholders and scientists) (Van de Kerkhof, 2004). Here stakeholders can be as much experts as scientists or policy-makers are (Mitroff *et al.*, 1983) and the domains of science, policy and society are not isolated entities (Dick *et al.*, 2009; Gummesson, 2007a; Fisher, 2000; Wynne, 1994). AR asks whether data or descriptive information is actionable or implementable or socially robust (Levin and Greenwood, 2011; Van Herk *et al.*, 2011; Bodorkós and Pataki, 2009; Gummesson, 2007a; Argyris, 2004; Dick, 1999; Burawoy, 1998; Greenwood and Morten 1998; Argyris and Schön, 1996; Funtowicz and Ravetz, 1993; Whyte, 1991; Whyte, 1998). It is, hence, used for increasing the skills and confidence of individuals needing to affect change and for solving difficult or complex problems (Kevany, 2010; Kevany and Huisingh, 2013; Jønsson *et al.*, 2009; Terry and Khatri, 2009; Lincoln and González, 2008; Dick, 1999).

Instead of isolating few variables from real life as mainstream research approaches do, AR recognizes complexity and the need for in-depth understanding of contexts (Gummesson, 2007a). Context is seen as a point of departure and theory is constantly being reconstructed (Burawoy, 1998:13). It works from an actor-oriented approach, on the real life constructs. Understanding the dynamics entails learning how the different forces interact with each other.

Some history

AR authors initially proposed this approach for underprivileged groups to solve their own problems in the post-war years (Gummesson, 2007a). Lewin (1946) designed AR as a research approach that states that there is no better way to learn about a system than in its own context, and changing it (Lewin, 1946; Schein, 1996). The latter is not without risk because for Lewin (1946) 'everything you do with a system is an intervention' (Schein, 1996) and the researcher needs to develop criteria that balance the risk from that intervention by designing a first low-key inquiry step thought to assess the risk. The researcher must engage in a managed learning process and his/her role was defined as a change agent that needed to be an expert on how and when to foster change and not precisely on the technical problems inquired about, and on how and when to be able to alternate the roles of helper, facilitator or mediator.

By the mid 20th Century a liberationist movement that responded to the political agendas and social contexts in the 'South' and more specifically in Latin America (Freire, 1993; 1970; Chambers, 1994; 1981; Fals Borda and Rahman, 1991) developed first rapid rural appraisal (RRA) thought to reduce the bias against research on the poorest communities and participatory action research (PAR) where the main target was the enlightment and awakening of common people. For Freire, (1970) learning was seen as an act of culture and freedom. Dialogue becomes the path to build a more profound understanding of a situation between the tension of the formally educated people and popular knowledge (Fals Borda and Rahman, 1991); Chambers (1981) integrates non-formal and experiential education with community-participatory appraisal techniques. Co-operative inquiry is a strategy for groups that want to explore their practice together. Action inquiry focuses on building the individual skills for valid inquiry (Reason, 1994). Another branch of AR has been post-normal science (mostly used in Climate Change research), which is seen by some authors as a family of research methodologies used for increasing the skills and confidence of individuals needing to change and for solving difficult or complex problems (Levin and Greenwood, 2011; Dick *et al.*, 2009; Dick, 1999; Baskerville, 1997). More recently, Faure *et al.* (2010) promoted moving from AR to action research partnership (ARP), as it is needed to reach collective and individual objectives and to lead negotiations. Managed learning is then a kind of AR that has been targeted in the business schools. As learning and changing turn out to be bound together (Bodorkós and Pataki, 2009), this thesis considers the theory of managed learning from Schein (1996) and stemming from Lewin (1946) as the reference theory on AR to build its own approach.

AR has continued to evolve into multiple combinations, meanings and processes (Brydon-Miller *et al.*, 2011) but all focus at the importance of experiential learning (Reason, 1994). The cases on AR can come from multiple fields: research, education, medicine or psychology, and more recently climate change.

Elements

Data collection process and principles

AR is oriented towards achieving two outcomes simultaneously: Action (change) and research (understanding). It has four characteristics: action-oriented, emergent (learning is reached in cycles of action-research-action), participation (validity is given by the quality of the participation) and qualitative research (as language is the vehicle by which participants communicate and that communication constitutes a large part of the data for understanding the situation) (Denzin and Lincoln, 2011; Dick, 1999). AR aims to enhance the actors' capacity to transform thoughts into action; to generate actionable knowledge; to make sure that the decision-makers' participation is assured in action planning (Ataov, 2007).

AR is done in cyclic processes. Each cyclic step is critically reflective and includes planning, acting, observing, and reflecting (Kemmis and McTaggart, 1988). The reflection at the end of each cycle fits into the planning of the next cycle. Data collection and data analysis are developed in parallel through cyclic processes. The researcher plans before acting and reflects on the findings and the methods after observing (

Figure 3-2). The approach demands from the researcher long-term commitment, self-reflexivity and willingness to dialogue and to work towards change.

Figure 3-2 Cyclic processes in Action Research.

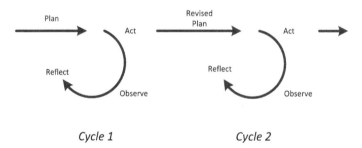

Cycle 1   *Cycle 2*

Source: Kemmis & McTaggart 1988

The seminal principles of AR are (1) that everything you do with a system is an intervention, (2) that you cannot understand a system until you try to change it, (3) that having the correct knowledge does not lead to change per sé, (4) that there is nothing as practical as a good theory (constantly reconstructing it), and that (5) the action researcher (agent of change) must be an expert on how and when to foster change more than on the specific problems of the subjects (Burnes, 2004; Schein, 1996; Lewin, 1946). Stemming from these principles, it becomes clear that learning about a system and changing it is the same. AR is then called managed learning and it is based on Lewin's change theory.

Change theory

Since change theory is based on the fact that knowledge does not lead to change per sé; attention must be paid to the matrix of cultural and psychological forces through which the subject is constituted (Winter, 1987). Lewin's change model encompasses three concepts – (1) unfreezing, (2) changing and (3) refreezing (figure 3-3):

(1) Unfreezing is based on the fact that human behaviour is based on quasi-stationary equilibrium. Complex psychological conditions must operate in order to break the status quo. It involves three processes: (a) Disconfirmation. The known data does not seem to help the people as it used to, (b) Induction of guilt or survival anxiety: At this stage, the disconfirming data is still considered valid and relevant but because of a crisis, people are having anxiety and the need to be offered new options. At this stage people ask for help. (c) Creation of psychological safety: Without sufficient psychological safety, the disconfirming data will not be denied, no survival anxiety will be felt and no change will take place.

(2) Changing is the next step. Once there is motivation to learn by unfreezing, the direction of the learning is not necessarily assured. Change occurs then thanks to a cognitive redefinition. Two mechanisms are available for the adoption of the new information, namely (a) learning through identification with available role models; and (b) learning through a trial-and-error process constructing the personal solution. The second mechanism is the most stable solution since the learner has invented it and it fits the learner's culture. This mechanism contrasts with the mainstream approach of consulting (Baskerville, 1997) and seems relevant for the particular context of McSEs that are undergoing technological change.

(3) Finally, there is personal and relational refreezing. In order to enter the state of equilibrium again, new knowledge must be compatible with the rest of the behaviour and culture of the learner. New knowledge is 'refrozen' and long-term change can be possible. If the new knowledge is not compatible with the culture, new cycles of crisis come along.

Figure 3-3 Kurt Lewin's Change Theory

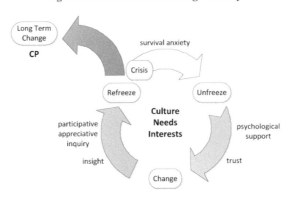

Source: Adapted from Schein, 1993.

Change agent
Change requires an agent. In order to be helpful, and understand where a system needs input, from the onset a researcher needs to design a very low-key inquiry oriented diagnostic intervention, designed to have a minimal impact on the processes being inquired about in order to have the best initial picture of it. The latter is based on Lewin's statement that says that any approach to a system is always an intervention. The flow of change or 'managed learning' process is one of continuous diagnosis as a change agent is continuously intervening (Schein, 1988). This approach to knowledge is called process 'consultation' and acknowledges that the researcher is not an expert on anything but how to be helpful and starts with total ignorance of what is actually going on in the targeted system (Schein, 1988; 1968). For the reflexive social sciences, training the change agents means placing emphasis more on the clinical criteria of how interventions will affect systems rather than on the techniques of how to gather scientifically valid information (Schein, 1992; 2010). During the last decade, supporters of reflexive approaches facing the social challenges have worked on the researcher's role as change agent rather than expert, on considering experts as just as important as other stakeholders in extra-academic settings, and as considering participants on transformative processes, agents living a change process (Morse, 2008; O'Hogain 2008; Burawoy, 2007).

In the water sector, since IWRM requires the integration of the institutional capacity, there is the need to train and educate the new water manager that would, in terms of the change theory, become a change agent (Van der Zaag, 2005; Van der Zaag *et al.,* 2003).

Lewin's change theory (1946) is considered relevant for this dissertation because (a) it is based on three concepts anchored in empirical reality and where people are seen as having a natural resistance to change, (b) it gives the framework for managed learning stating the importance of a first low-key inquiry step when a community is going through a crisis, and a need for disconfirmation, survival anxiety and to build psychological safety. At this time vulnerability is high and minimum intervention is of outmost importance as people ask for help and choose what they consider their best-suited solutions. Since the direction of change is not assured, communities can be prone to manipulation if the process is handled without respect and without setting high values and purposes of society (Morse, 2008; O'Hogain, 2008; Burawoy, 2007; Escobar, 1999). Change theory is also considered relevant (c) for building knowledge for specific communities because it is based on trial-and-error approaches that have proved to build more long-term learning fitting specific contexts. The arguments explained above set an important framework for research endeavours with marginalized communities.

The elements of a good case on AR are then to follow the principles, characteristics and processes. The approach necessarily needs researcher and stakeholder involvement at all levels: visioning, planning, decision-making, policy-making, implementation, and evaluation (Godard and Laurans, 2004). The AR claims validity by being grounded on theory and by constantly reconstructing the theory (Van de Kerkhof, 2004). It also claims rigour by following multiple methodologies, multiple sources of information, and multiple data collection and analysis techniques (Denzin and Lincoln, 2011; Arieli *et al.*, 2009; Martínez *et al.*, 2006; Van de Kerkhof, 2004; Dick, 1999). Since relationships are seen as an essential part of AR, their nature and quality are critical to both the outcomes and the process (Arieli *et al.*, 2009).

AR has been criticized as being too qualitative and lacking scientific rigour and objectivity (Levin and Greewood, 2011; Smith, 2001; Wynne, 1994). AR is in fact heavily context-oriented and the basic principles of reliability and generalizability used in the quantitative sciences do not hold. In fact, quantitative research is successful at answering questions of how frequent, how much and how many, but not at facing complex phenomena that need to be understood in-depth and that usually entail change as a major force (Gummesson, 2007a; Flyvbjerg, 2004). What brings legitimacy to AR is the high degree of stakeholder participation or the use of transformative participation, which improves decision-making in policy-making and enhances scientific practice (Denzin and Lincoln, 2011; Arieli *et al.*, 2009; Van de Kerkhof, 2004). Some scholars go as far as proposing that AR should be the main kind of research because it links academic knowledge to impending complex challenges from real-life (Levin and Greenwood, 2011).

AR has its weaknesses. First, the role of the researcher is not always clear. According to Van de Kerkhof, (2004) the researcher's main target is not to solve others' problems but to make others learn how to solve theirs' and to internalize their own understanding and solutions. However this could be perceived as placing the researcher above the dynamics of learning and change that takes place within AR. Authors like Jacobs (2010) and Arieli *et al.* (2009) find themselves in a paradoxical position when as AR-ers, they support democratic values but end-up imposing participatory methods and, hence, abusing power (Burawoy, 1998). AR does not adequately deal with the participative relationship itself and the complexity of handling conflicts and the multiplicity of interests (Aziz *et al.*, 2011; Arieli *et al.*, 2009).

On a different note, since AR is rather time demanding, researchers employ less time to publish in peer reviewed journals and results from AR are less visible (Bodorkós and Pataki, 2009; Levin and Greenwood 2011). Even though AR constitutes a *"superior way to link teaching, research and real-world engagement"* (Levin and Greenwood 2011), it has not been successful in dominating research because most AR activities are discouraged by the structural criteria of universities regarding good quality research. Pertinent societal challenges are usually not targeted (Levin and Greenwood, 2011; Bodorkós and Pataki, 2009; Burawoy, 2007; Mejía *et al.*, 2007; Hansen and Lehmann, 2006; Martínez *et al.*, 2006; Steiner and Posch, 2006).

In summary, this thesis designs a methodology on AR that develops the role of the researcher in a more specific way in order to deal with marginalized communities (committing with real-life concerns); that uses conflict resolution methodologies, which does not impose or control the flow of change; and that handles strategies from negotiation in order to improve decision-making within a multiplicity of interests. This thesis aims to contribute to the need to fill the gap between solving relevant practical problems and to elaborate a fine-tuned theoretical and methodological framework (Levin and Greenwood, 2011). Building such an approach calls for facing what Holman (2007) presents as the cycle of Mastery: to recognize that leading methodologies targeting change require putting skills, knowledge and intuition at the same level.

3.4.3   The SASI method

The SASI method was developed for the extreme case of pollution and social exclusion at the micro-tanneries in Villapinzón. Choosing an extreme case was interesting in order to design SASI because extreme cases deploy more essential elements from a system (Flyvbjerg, 2004). This case asked for urgent action in a conflict ridden situation and to foster behavioural change (Chapter 6).

AR seemed like an opportunity to empower the tanners' community and to foster change. Dealing with big groups in conflict with the environmental authority, aiming at building agreements and at improving the outcomes for both parties was, nevertheless, not assured by traditional AR methods as shown in Chapter 2. The Systematic Approach for Social Inclusion (SASI) was then inspired by AR and complex negotiation, and grew from insights from CP, PI and SP.

SASI is meant to allow contribution to knowledge and reconstruction of theory, as well as a successful change in behaviour of McSEs in developing countries. SASI focuses on the necessity to face the social exclusion of McSEs and to help them adopt appropriate CP to cope with environmental policies in developing countries. The approach is based on empowering through internal strengthening of the target communities, and on building strategic alliances in order to address the issues at stake identified with the actors in their context. It aims to boost the negotiation power of the underprivileged groups, on stimulating them to learn to solve their own problems and on raising awareness on the part of the authorities about the specific characteristics and problems of the McSEs.

Such a process entails placing the researcher inside the community of micro-tanneries in order to be able to dig into the deeper causes of their problems and their consequences rather than to just describe symptoms and the frequency of appearance of the problems (Oosterveer *et al.*, 2006). As a result, the methods used are mainly qualitative such as stakeholder analysis and Strengths, Weaknesses, Opportunities and Threats (SWOT) analysis, micro-cases (continuous data collection process to strategic actors), in-depth interviews, participant observation and focus groups. Since they are meant to deal with exclusion and change, methodologies handling conflicts for big groups – called Emergent Change Processes (Holman, 2010) - are privileged such as Appreciative Inquiry (including visioning) or Open Space Technology (OST). The latter "engage the diverse people of a system in focused yet open interactions that lead to unexpected and lasting shifts in perspective and behaviour" (Holman, 2010: 201). As authors like Burawoy (2007) and Holman (2010) state, such processes need to focus on values, principles and creativity (intuition).

Elements of SASI

The elements of SASI are (a) based on sustainability, which enhances self-reliance, prevention, regulation and participation. SASI builds on (b) five principles stemming from:

Negotiation (including conflict resolution)

   (1) People support initiatives that they help create (Holman, 2010) and participation increases commitment (Robinson and Berkes, 2011; Geist, 2010; Ataov, 2007; Dick, 1999);

(2) When focusing on large groups, conflict resolution should build common grounds within those groups while respecting individual autonomy (Holman, 2010; Bodorkós and Pataki, 2009; Holman *et al.*, 2007; Holman, 2004; Holman and Devane, 1999); and

(3) Negotiation based on interests and not on positions will open up possibilities towards creative outcomes that generate better results for all stakeholders involved (Thompson, 2009; Raiffa *et al.*, 2002; Fisher *et al.*, 1991).

Managed Learning (AR)

(4) There is no better way to know a system than trying to change it (Reason, 2006; Burnes, 2004; Lewin, 1946); and

(5) The learning process has better results when it works through trial-and-error (Bodorkós and Pataki, 2009; Schein, 1996; Lewin, 1946).

SASI is (c) a systematic process with six spiral steps focusing at multilevel and multidisciplinary interventions (Figure 3-4). The spiral nature implies cyclic processes that can give rise to further steps. Each spiral step is critically reflective involving observing (Gummesson 2007b), planning, acting, observing and reflecting (

Figure 3-5). The reflection at the end of each step fits into the observations of the next step. Data collection and data analysis are developed in parallel through spiral processes. The different issues being focused have their own pace and can follow their own spiral.

Figure 3-4 Six basic spiral steps of SASI

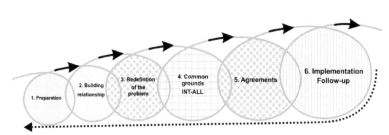

Figure 3-5 Spirals of SASI

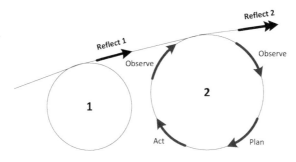

Source: Author

Negotiation theory inspired the six basic steps of this process: Preparation; Building a relationship; Redefinition of the problem; Building common grounds first among the target community and then between all stakeholders; Agreements; and Implementation and follow-up.

The discontinuous arrow in Figure 3-4 indicates that the process can go back any time, as these social processes need to be constantly rebuilt (*e.g.* new stakeholders in a scenario can push towards working again step 2 focusing at building relationship).

Since this research focuses on working with large groups experiencing social exclusion, step 4 targets building common grounds first among the underprivileged community and then between all the stakeholders. SASI works on common interests respecting individuals. It works with small groups on the first three steps and with large groups on the final three steps of the process. SASI promotes simple negotiations as well as complex or even big groups entailing conflict resolution approaches (Figure 3-6).

Figure 3-6 How negotiations may evolve

Table 3-5 explains the steps of SASI in detail, their aims, the theories and elements needed to reach these aims, and the expected results. Most of the steps are framed by N/CR and AR and are based on prevention from CP, on participatory ideals and on targeting accurate rules of procedure, which can formalize the McSEs' activities. The expected results column in Table 3-5 tells when a situation is ready to move to the next step e.g. Lewin (1946) stated that it was ideal that real intervention (step 2) starts once the community itself asks for help (end of step 1).

Table 3-5 The six steps of SASI

#	Step	Aiming at	Insp by*	By means of	Exp Results
1	Preparation	Obtaining the initial picture of the conflict	N/CR	In-depth interviews-observations. Identifying interests, possible allies, BATNAs (Best Alternatives to Non negotiated Agreements)	Initial picture allowing accurate diagnosis & possibilities for solving conflict
			N/CR	Initially perceived definition(s) of the problem	
			N/CR	Defining the nature of relationships	
				Setting indicators	
			N/CR	Defining issues at stake	Subjects ask for help
				Allowing minimum intervention	
			AR		
2	Building relationship	Supporting psychological safety, trust building	N/CR	Sharing reliable information through dialogue meetings	Help is provided on impending issues (not just technical)
			CP, SP	Principles of CP & SP	
			PI	Rules of procedure PI	Subjects ask for the need to clarify misunderstandings
			AR	Prioritizing impending issues, helping	
				Trail & error	
3	Redefining the problem	Building consensus on the initially perceived definition(s) of the problem	N/CR	An as large as possible representation of each of the stakeholders in the conflict.	One definition of problem agreed by all actors
				Big group methods	
		Visioning and building an initial strategy with allies	N/CR		
				Principles	An initial strategy is suggested with marginalized leaders
			PI, SP, CP	Mechanisms, options	
4	Building common grounds; first internally, then	Empowering the marginalized communities, creating a win-win	AR	Respecting culture & context	Leaders present strategy to all group

	among all stakeholders	situation for all stakeholders, building realistic and accurate solutions to the conflict	N/CR	Big groups methods as open space technologies initially with the concerned community, then with all stakeholders	Strategy is shared with all stakeholders
		Feedback to the strategy & alliances			
5	Building agreements	Establishing commitments	PI, SP	Principles	Agreements are set among stakeholders
			CP	Options	
			N/CR	Big groups methods & small committees	
6	Implementation & follow-up	Implementing the technical options and monitoring the implementation	AR	Developing solutions by trial-and-error	Implementation starts on technical issues
			CP	Technical advisory	
			PI, SP	Principles & rules	
				Constant feedback into a dynamic process (evaluation from subjects)	
			N/CR		

(d) Furthermore, a system of indicators meant to assess the initial situation as a low key inquiry step is needed (

Table 3-6); and (e) the role of the researcher as agent of change that entails commitment on a long-term basis, self-reflexivity, and independency.

Table 3-6 Indicators for assessing the initial situation

Indicators	Relevant issues	Values
H_{S1}	Access to participation	2 optimal
		1 fair
		0 non-existent
H_{S2}	Quality of relationship among stakeholders	2 harmonious
		1 tense
		0 aggressive
H_{S3}	Existence of consensus on perceived problem definition	1 consensus
		0 lack of it

3.4.4   Design and process of SASI

Developing SASI at these micro-tanneries provided an ideal opportunity because it allowed to: 1) respond to an urgency that existed because of a crisis situation, 2) deal with complexity and 3) innovate.

This research dealt with sub research question #3 and lasted six years: (How) can AR (methodology) be developed and tested to help McSEs implement Cleaner Production?

Diverse methods were applied at the national, regional and local levels. Some initial in-depth interviews at national, regional and local contexts developed into micro-cases informing continuously the researcher. Table 3-7 shows the opportunities for observations and interventions through the process. The techniques for big groups were chosen from a whole array of techniques from conflict resolution called Whole Scale Methodologies or Emergent Change Processes (Holman, 2010). They were chosen for their suitability and purpose in the context of McSEs.

Table 3-7 Opportunities for observations and interventions at Villapinzón

Level	Observations & Interventions
National	2 Court appeals
	National tannery committees at Ministry - twice a year- 2004-2007
	Two Presidential Councils - President Uribe- 2004-2005
	Four meetings with the Office of the Presidency (2004-2005)
	Meetings with the Public Prosecutor (2004-2009)
	Meetings with the National Comptroller (2009)
	Elaboration of a micro-case with the president of the Senate (2004-2010)
	Elaboration of a micro-case with the Magistrate (2004-2009)
	Elaboration of a micro-case with Head of Tanneries (2004-2010)
	7 Denunciations and publications in main newspapers (2004-2010)
Regional	Meetings with the regional authority CAR (2004-2010)
	Follow-up Court Order ruling on the Bogotá River recovery (2003-2005)
	Joint work with Governorship of the province surrounding Bogotá (2004-2010)
	Visits to the tannery community of Cerrito that was implementing successfully CP
	Elaboration of a micro-case with the responsible of CP Centre for Cerrito and region (2004-2010)
	Elaboration of a micro-case with head of the technical department at regional authority CAR (2004-2007)
Local	Meetings in the municipality of Villapinzón/ mayor and local actors like the big tanner and the small tanners (2004-2006)
	12 in-depth interviews with stakeholders
	6 OSTs and AIs to handle time pressing issues among all stakeholders (2004-2010)
	Elaboration of a micro-case with the new leader in Villapinzón (2004-2013)
	Follow-up committees (2004-2010)

Table 3-8 Time line for Action Research in Villapinzón

#	Steps	Dates
1	Preparation	March 2004 - April 2004
2	Building Relationship	April 2004 - May 2004
3	Redefining the problem	May 2004 - June 2004
4	Common grounds internally & all	June 2004 - June 2005
5	Agreements	June 2005 - December 2006
6	Implementation & Follow-up	December 2006 - January 2010

Table 3-9 Stakeholders interviewed in Villapinzón in 2004

Type of Stakeholder	#
CAR officer in Bogotá	1
Regional CAR officer	1
Public Prosecutor	1
Big tanner	1
Mayor	1
Tanners	3
Leader time of conflict	1
Officer Governorship	1
Ministry Head of tanneries	1
Judge (Magistrate) Court order	1
Total number of interviewees	**12**

3.4.5   Inferences

The AR method and hence, SASI respond to complexity and to the urgency to act in the context of underprivileged communities. They both offer the opportunity to better understand McSEs by supporting change and confirming the Change theory from Lewin (1946). Since SASI offers strategies to improve the outcomes of needed agreements, to follow a systematic process of negotiations, to implement tools for solving conflicts, and for handling big groups, it can be a comprehensive approach in the context of McSEs in developing countries needing to leapfrog into sustainability.

3.5   CP implementation for SASI

Once the initial SASI phases of building relationship, trust and defining the problem definition and common grounds were met, the regional environmental authority accepted the environmental plans on CP elaborated by the Chamber of Commerce of Bogota (CCB) and the technical implementation of CP could start supported by the EU project SWITCH[20] and the institute IDEA from the Universidad Nacional (step 5) (Table 3-8).

---

[20] SWITCH stands for Sustainable Water In tomorrow's Cities's Health (2006-2011). An EU project on Sustainable Water Management with 33 institutions around the world. University Nacional worked in Villapinzón supporting this thesis

Quantitative and qualitative approaches were set with respect to CP implementation:

1. Researchers and tanners developed selection criteria for choosing 12 tanneries that were going to be used as pilot industries before scaling-up CP to the whole community.

2. Material and energy balances were elaborated for the 12 tanneries involved in the project

3. The best CP options for the specific characteristics of the tanneries and of the Bogotá River in the upper basin were defined and implemented after a literature review, trial-and-error practices, and individual agreements with the tanners

4. 30 practical workshops on CP implementation were conducted

5. A CP monitoring system was developed initially on the Hs indicators presented in Table 3-6 and afterwards on technical, economic and social indicators

6. Solid waste valuation options were studied; composting of the de-hairing and grease residues was tested first, at UNAL laboratory, second, at a tannery and third, at a tanners' landfill bought by the tanners and meant to implement collective solutions for 30 tanneries.

3.6   Lessons learnt from the application of the methods

3.6.1   Combined case study method

The combined case study method used in this investigation was different from most classical approaches. It involves a greater engagement in time and resources on the part of the researcher involved in the case study community. Although this appears to be more expensive and inefficient as a research method, it is more effective because (a) it gives the researcher a much deeper understanding of the context of the problem and of the chain of events that leads to a change in behavior, and (b) makes the results of the research more reliable and relevant. In fact, the results from this research would have being very different if this work had lasted only for one year. In 2005, when the CP programs were on the verge of being cancelled, the main elements of a process supporting CP would not have been unveiled and conclusions would have been misleading. The smallest tanneries in Cerrito had been slow to meet the standards, needed more flexible deadlines but finally succeeded two years after. On a different note, it is important to highlight that permanent contacts with the subjets of study constitute a source of reliability. The flow of communication between the researcher and the heads of CP bodies was continuous in both case studies. At any time, when the specificities of each situation asked for the researcher's understanding and physical contacts were not at hand, communication was possible through telephone conversations or email contacts.

3.6.2   SASI

SASI is an improvement on existing AR methods because it creatively combines insights from different theories in order to (a) follow a systematic process with specific targets per step inspired by integrative negotiation for decision-making between McSEs and authorities, (b) offer methodologies to overcome social exclusion and conflict for big groups such as Open Space Technologies (OST) and Appreciative Inquiry (AI) as visioning, and (c) boost the negotiation power of the marginalized groups through strategies aiming at internal strengthening and at building alliances in order to improve the outcomes of needed agreements.

The role of the researcher in the SASI approach confirms the role of the change agent from Schein (1996) and from Lewin (1946) in the sense that it needs to assess the risk from intervening, to switch from helper to facilitator or researcher as needed, and to focus on being an expert on helping, more than on specific matters. Through the SASI approach, this role evolves into the Agent of Inclusion (AoI) that needs to focus holistically on the complex world of McSEs entailing social, economic and environmental issues at stake needing urgent attention; emphasizes the willingness to commit on a long-term basis; assumes self-reflexivity; and essentially keeps its independence from the wide array of the possible conflicts of interests involved. The latter was, on the one hand, especially meaningful in three circumstances in Villapinzón. First, when there was a need as part of the strategy for public denunciations when the spatial planning issues were hampering the CP implementation. The politics of knowledge did not allow Universidad Nacional to confront the regional authority, which was delaying the spatial planning decisions. In that specific situation, the researcher made the denunciations individually as an international PhD student. Second, at earlier stages of the process, the tanners considered the researcher's independence as an important asset entailing trust. When the Chamber of Commerce tried to run the process and set aside the researcher, the tanners argued that they would continue the process, provided the researcher was kept as the facilitator of the process. Third, even before the process started officially in Villapinzón, the independence was also considered important by the magistrate who gave the court order on the Bogota River. She argued that the researcher's independence helped her to rely upon the research (personal communication, 2004). On the other hand, that independence is not without risks. It was initially seen with suspicion by the regional authority, which was reluctant to trust the researcher. Bringing academia to the picture, which the authority believed in, turned out to be essential for the process (personal communication, 2009).

SASI was successful in respecting stakeholders' own decisions while supporting the common good (supporting consensus on strategic common issues and solving the debate on consensus on stakeholder participation); understanding the situation of the McSEs and privileging contextual understanding over generalizations (Baskerville, 1997); undertaking top-down or bottom-up analysis - it can descend to the level of the local actors and identify details of the environmental technologies, internal and external elements of an industry, and the social and economic relationships that are relevant to explain the technological change that cannot be captured by aggregate quantitative analysis (Del-Río-González, 2009); promoting the culture of participation based on positive and not on claim oriented approaches; promoting the dynamic creation of knowledge to deal with changing situations; ensuring medium-term trust and commitment; promoting behavioural change (*e.g.* the tanners were willing to clean the river once they felt their interests were taken into consideration, and win-win situations were possible through CP implementation[21]); stimulating third party involvement and commitment that goes beyond mediation; using Appreciative Inquiry which led other manipulative actors in the field to lose their influence and chances for corruption; innovating on technical processes once people get out of the vicious cycles; and developing a comprehensive holistic, self-enforcing, prevention-based solution. The solution based on CP was able to work on semiquantitative methods because of the characteristics of McSEs and also because the initial measures of the discharges were difficult to be taken because the regional authority only re-opened the tanneries and allowed them to work two years after SWITCH

---

[21] Researchers and CP technicians had difficulties monitoring and evaluating CP implementation in the context of McSEs because it is hardly difficult to keep track of the implementation through exclusive quantitative approaches. Monitoring through semi-quantitative innovative approaches can balance these weaknesses. SASI was successful at supporting CP implementation because it worked towards building common grounds between stakeholders understanding each other's interests and towards focusing on integrated solutions.

had started. CP implementation worked participatively and based on trial-and-error. Researchers, officials, and tanners learned from each other through experiential learning. Actors grew in knowledge in their own specific contexts and needs.

SASI is, however, a more expensive and time-consuming process because it requires building common grounds and long lasting change; entails longer periods for reporting, writing, and publishing scientific papers (Bodorkós and Pataki, 2009; personal communication, 2011-1)[22]; requires skills for organizing and making multidisciplinary teams work efficiently; requires commitment from the participants; needs constant feedback and reflection; and implies a risk for intervening that must be assessed at the beginning of the process and the derived power effects.

The experience from applying SASI showed that committed ARers who can switch from researcher, helper, to facilitator or negotiator as needed (Bodorkós and Pataki, 2009; Burnes, 2004; Lewin 1947; Schein 1996) must maintain independence; assume self-reflexivity; commit on a long-term basis; be open to discuss all issues and be willing to take a holistic approach; and play a leadership role in empowering people to find appropriate solutions. Clearly as the target group becomes empowered, the role of the ARer is minimized and the process becomes sustainable.

3.6.3   How to choose methods for complex environmental inquiry

This thesis shows that choosing appropriate approaches towards complexity asks for creative and integral initiatives from multiple disciplines, which may appear eclectic to   mainstream academics.

Yin (2009) has elaborated a comparison of the traditional social methodologies like survey and archival analysis, and a case study based on the conditions demanded by the specific research situation. Dealing with complex problems may be considered an extension of such analysis. Complex problems do not require control of behavioural events and focus on contemporary situations like case studies and are opposed to experiments and historical analysis. The research question is the first condition for choosing an appropriate method to address complex challenges. This thesis suggests three additional conditions: (1) to identify whether there is a need to interact more closely with actors, (2) to determine whether there is a need to intervene, to produce actionable knowledge, and (3) to specify whether there is a need to focus on relational social interactions like negotiations. The table below presents the relevant situations for different research methods that can be used to address complex problems such as McSEs.

Table 3-10 Relevant situations for different research methods

Method	Form of research question	Need for intense interaction	Need to act	Need to focus on relational interaction
Survey	Who, what, where, how much, how many	No	no	no
Archival analysis	Who, what, where, how much, how many	n/a	n/a	n/a

---

[22] This fact was constantly brought up in the first steps of the SWITCH research because publishing was getting delayed with respect to original plans. SWITCH was an Action Research.

Case study	How, why	No	no	no
Combined case study	How, why	Yes	no	no
AR	How, why	Yes	yes	no
SASI, ARP	How, why	Yes	yes	yes

The kind of social inquiry, the role of the researcher and of the participants can change depending on the purpose of participation. Table 3-11 was inspired by Cornwall (2003). It does not establish a hierarchy or a ladder on methods for social inquiry. It determines instead the appropriate methods as complexity and specific demands increase, and systemic inquiry becomes a priority. The researcher's role implies also that as the research methods get more complex, the researcher's role becomes also more demanding and specific characteristics are needed.

Table 3-11 Social inquiry/modes of participation/roles of researcher

Modes	*Aiming at*	*Social inquiry*	*Participants*	*Researcher*
Functional	Who, How many	Surveys, questionnaires	Objects	Positivist
Instrumental	How to conduct, What	Interviews Workshops Case study	Instruments	Inquirer
Consultative	How to get response In the long term	Participative methodologies Workshops Case study Combined Case Study	Actors	Facilitator
Transformative	How to foster change How to support it in the long term How to empower	AR Change methods: AI, OST SASI	Agents	Agent of Change Agent of Inclusion (AoI)

SASI and the combined case study methods are based on producing context dependent knowledge, which brings about experts. They follow Yin's recommendations (2009): to cope with (a) more variables of interests than can be expected if the method was an experiment, which isolates variables from daily life, (b) to rely on multiple sources of evidence with data meant to converge, and (c) to benefit from a dialogue with theory to guide data collection and analysis. Both methods offer better understanding of the complexity involved by giving the opportunity to interact more closely researchers and actors over time.

In the event that a specific situation only asks for understanding of a problem related to McSEs, a methodology based on long-term reflexive case study research is suggested. This methodology called combined case study constitutes a more reflexive approach to the classical case study.

3.7   Conclusions

Both methods allowed to follow a process of change (combined case study) or to engage on it (SASI). Both methods used, the combined case study methodology and SASI, may go futher from the methodology that inspired them. They nevertheless must face the challenge set by the high up-front costs due to the long lasting time employed compared with mainstream approaches. Such up-front costs can be compensated by the fact that these approaches can keep track of the chain of events leading to change (combined case study method) or can even generate behavioural change (SASI) and can then imply better and more reliable results than the classical methods for complex environmental problems. In the specific case of SASI, the costs of decades of conflict, non-action and misinformation in Villapinzón, overrun its high up-front costs. Once empowered and CP aware, the smallest enterprises pushed for prevention adoption not only by the environmental authority but also by the biggest enterprises that were not involved in the change process whereas both had exclusive end-of-pipe orientations.

SASI has confirmed the individual theories and has improved the approach for McSEs by testing them acting together. CP was 'socialized' for McSEs through SASI.

SASI was successful in bringing a highly relevant social case on social exclusion to higher education. This outcome was possible because the SWITCH project was an AR by itself and adapted its objectives to the needs of McSEs. As Universidad Nacional in Colombia had an institute dealing with the environmental challenges through a multidisciplinary approach, different faculties supported the research on McSEs. The application of SASI fell nevertheless short of motivating the financial system meant to deal with informal industries. No strategic people were found to support the process in the banking system. If strategies were to be designed again with the banking system, direct support from the ministry of finances would be the target from the beginning of the process. (Vice-ministers and president of the official bank dealing with farmers were contacted at later stages). SASI has basically supported to open the social and environmental sciences to reflexive thinking (Burawoy, 2007) or to revitalize universities in order to commit with a core responsibility in bringing about the social challenges (Levin and Greenwood, 2011: 28).

4   Colombia: Needing its own identity as a water sensitive and inclusive country

4.1   General Background

4.1.1   Introduction

"Poverty is either cause or consequence of the environmental degradation...We need to build a just society which lives within the limits set by nature" (Manuel Rodriguez, 2008:69 Former Minister of Environment of Colombia).

This Chapter deals with the first sub-research objective, which is to investigate the current institutional framework and the role of formal authorities and institutions in the context of industrial McSEs adopting CP in Colombia. Fig. 4-1 presents the map of Colombia and the case study regions.

Figure 4-1 Map of Colombia, Bogotá, Cerrito, San Benito and Villapinzón

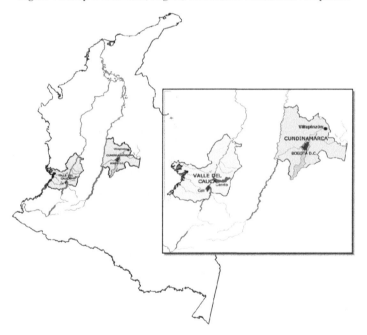

This case study is based on a literature survey, content analysis, and interviews. The content analysis included analysing available official documents, newspapers, reports, presentations, and debates related to the National Water Policy, the National SME (maybe McSE) Policy, and the National Cleaner Production Policy. It also implied carrying out in-depth interviews and whenever possible micro-cases with strategic actors. This Chapter presents a general background (4.2), the national water-environmental policies (4.3), the national SME policy (4.4), the Cleaner Production policy (4.5), inferences (4.6), and conclusions (4.7).

4.2   A general background

4.2.1   Political context

Colombia is a unitary constitutional republic with 32 departments. The executive branch dominates the government structure. The country became independent from Spain on July 20th 1810. Its present constitution dates from 1991 and has been amended several times. The Spanish and the French civil codes have influenced the country's civil law system.

At the executive branch, the President is chief of state and head of government. The president and vice-president are elected by popular vote for a four-year period and are eligible for a second period. The last elections (May 2010) brought Juan Manuel Santos to power.

The legislative branch is constituted by a bicameral congress where the Senate holds 102 seats and the Chamber of Representatives has 166 seats. Both senators and representatives are elected for a four-year term.

The judicial branch has four co-equal supreme judicial organs: The Supreme Judicial Court that handles civil and penal matters, the *Consejo de Estado* (State Council) that deals with administrative issues, the Constitutional Court that focuses on issues that can potentially violate the Constitution and the *Consejo Superior de la Judicatura* (Superior Council) that deals with the judges' disciplinary sanctions (CIA, 2012).

Within the judicial branch, the Office of the *Fiscal* (Attorney General) has administrative autonomy and was designed to prosecute offenders, investigate crimes, review judicial processes and accuse penal law infractions against judges and courts of justice.

Colombia has also two control institutions. The *Procurador* (Inspector General) who is the chief public prosecuter, and the *Contralor* (Comptroller General) who acts as the highest form of fiscal control.

Some political history

Colombia has faced in the past hundred years, long lasting conflicts between the two political parties, the Liberal and the Conservative parties. Different ideals inspire both. The Conservative party has supported development by stimulating and privileging private investments and the free market without prioritising the social role of the government and promotes a Catholic state. The Liberal party instead believed in a stronger commitment from the government with respect to the under-privileged and on fixing stronger limits to private initiatives. It was responsible for declaring a secular state.

In the past four decades, two additional forces, outlawed armed groups whose leaders were inspired by Marxian thoughts and the drug cartels, have fuelled the above-cited conflicts. To give some numbers regarding the generalized violence at the beginning of the 21st Century, in 2002, 70-75% of the municipalities had insecurity caused by the guerrillas; between 1998 and 2002, 17 mayors had been kidnapped, 554 threatened and 54 killed; and 300 mayors were forced to work far away from their municipalities (Velásquez and González, 2003). Over the last nine years, there has been a continuous effort to improve the security of the country. The Colombian army has improved military education and technological aspects (US dept, 2010). Over the same period, a left democratic party has flourished to the point that Bogotá, the capital, has been steadily ruled by it for the last ten years.

Two characteristics distinguish the Colombian political system: authoritarianism and clientelism.[23] As a consequence there has been high distrust towards participation from all actors (Velásquez and González, 2003). The Head of the Senate highlighted this issue (mc1 in Table 3-2).

4.2.2   Socio-economic data

Colombia is the 3rd most populous country in Latin America after Brazil and Mexico (46, 927,125 inhabitants) (WB, 2011a). The urban population constitutes 74% of the total population (CVNE, 2012; USdept, 2010). Its land area is 1,144,394 m^2.

Main Colombian economy

The Colombian natural resources are: coal, petroleum, natural gas, iron, nickel, gold, silver, copper and emeralds. Mining represents 7.8% of the GDP. The main Colombian exports are petroleum, coffee, coal, nickel, emeralds, apparel, bananas, and cut flowers. They were valued at approx. USD 39 billion (2010). The major markets are USA (38%), EU (15%), China (3.5%) and Ecuador (3.4%) (2010). The imports are machinery, grains, chemicals, transportation, equipment, mineral products, consumer products, and paper products. The major markets are USA (25%), China (15%), Mexico (11%), Brazil (5%) and Germany (4.1%) (2010) (Indexmundi, 2012; MCIT, 2012).

Agriculture represents 7.1% of GDP mainly with coffee, bananas, cut flowers, cotton, sugarcane, livestock, rice, corn, tobacco, potatoes and soybeans. Services represent 46% of GDP through government services, financial services, commerce, transportation and communication. Industry represents 14.4% of GDP mainly through textiles, garments, footwear, chemicals, and metal products (DANE, 2012a; USdept, 2010).

The Colombian economy faces contrasting situations. On the one hand, figures reveal that the Colombian industrialization process has been literally halted since the 70s. Compared to recently industrialized nations (Taiwan, South Korea and Singapore among others), the Colombian economy did not evolve from the labour-intensive industry to superior process levels entailing a knowledge intensive industry (Pinto, 2006). Colombia never reached 30% of its total GNP[24] through the manufacturing industry, as did other recently industrialized economies. Colombia went as far as achieving 23.5% in 1975, and this percentage has been declining ever since. Since 1975, the growth rate of the manufacturing sector has dropped from 6% to 1.1% per annum (Pinto, 2006). Other figures from the National Department of Planning (DNP by its name in Spanish) report 2.6% growth per annum between 1980 and 2010 and are still far from being good (Hommes, 2011). Colombia is relying basically on extracting natural resources and has not been successful in stimulating labour intensive industries (Kalmanovitz, 2012; Personal communication, 2012-07/20[25]; Ocampo, 2010). 65% of 2011's exports were oil and coal; foreign investment was represented in 2011 by 60% on mining and fuel, 10.3% on commercial activities and hotels, 4.6% on financial services, and only 4.4% on manufacturing and industry (Kalmanovitz, 2012). The industrial performance has gone from August 2011 to August 2012 from a growth of 9.9% in 2011 to a decrease of 1.9% in 2012 (DANE, 2012a). On the other hand, for the last two decades, Colombia has been opening its doors as a more open

---

[23] serving particular interests.

[24] Gross National Product: the total value of goods and services produced over a year by the people of a nation including the value of income earned in foreign countries.

[25] Comment made by Ocampo at a meeting on development Club El Nogal

economy. It has signed free trade agreements with countries such as Canada, the EU, Israel, Turkey, South Korea and with the USA.

The economy's performance

Colombia's GDP is USD 193 831 million (USD from 2005) and has been growing steadily since year 2000 (Banrepublica, 2012). The annual growth rate is at 4-6% (Banrepublica, 2012). In 2011 the inflation rate was 3.7% and unemployment 10.8% (one of the highest in the region) (Indexmundi, 2012). Due to the democratic security strategy of former President Uribe, which was based on regaining regional state presence and strengthening institutions and the military, the GDP grew 4.4% on average for the first 4 year's presidential term and 8.0% on average for the last two years of his second presidential term in office because private investment was on the rise (DNP, 2007a). The indicator of Total Investment/GDP has also surpassed 20%, the highest rate in Latin America since the year 2005. This total investment was mainly 90% represented by the private and foreign component (DNP, 2007a). Nevertheless, during that period, stagnation of the small industrial and the rural sectors and in wealth redistribution was underlined (Echeverry *et al.*, 2011; Stiglitz, 2003; Fedesarrollo, 2006). Some critics considered that relying on foreign investment for economic growth is risky if it is not considered a step while the level of domestic savings provides the necessary resources to reach a stable growth (Carrizosa, 2008; Fundación Agenda Colombia, 2007). Foreign investment is considered capricious and not always disposed to finance minorities like SMEs/McSEs that embody greater financial risks (Rocha, 2007).

Inequality

Economists are observing inequality from a different perspective over the last few years. The tendency today is not to target growth primarily but to see how the resources are distributed (DNP, 2011; The Economist, 2011; PNUD, 2010). Wilkinson *et al.* (2010) report that even in the developed countries, there is a positive correlation between a low level of income differences and a good performance on health and social problem indexes.

Over the next few years, Colombia faces the risk of becoming the most unequal country in the region (PNUD, 2010). The country has fallen in the Human Development Index over the past decade (from position 54 to 73) (Fundación Agenda Colombia, 2007). While in the last decade the middle class has doubled in countries such as Chile, Argentina and Costa Rica, in Colombia it has increased by 30% (WB, 2012). In spite of stable growth over the last nine years, 37.2% of the population lives below the poverty line and the Gini co-efficient is 0.56 with a tendency to increase (PNUD, 2010).

Inequality is especially important in the rural sector. The rural Gini coefficient of 0.87 shows the huge inequalities that coexist (Hoyos, 2012; Echeverry *et al.*, 2011). Colombia can be even divided into six areas, which have very different development levels. The central region with the main cities, is the only one considered on the path to modernity (López, 2011). In the other regions, the lack of social mobility lies at the heart of the difficult conflict (Posada-Carbó, 2012; Gaviria, 2011; Reyes, 2009). While the richest 10% of the population absorbs half of the GDP, the poorest 10% receives merely 0.6% of the GDP (Semana, 2011); land property has traditionally been in the hands of the few, especially in the rural sector, where 41% of the land belongs to 3.8% of the owners (ElTiempo, 2012; Machado, 2012), where 64% of the population is below the poverty line, and over the last twenty years a great number of peasants (more than 3 million 400 thousand representing 10% of the Colombian population) have been violently displaced by extremists (García *et al.*, 2010). More than five million hectares belong today to either guerillas or mafia cartels (Molano, 2010).

Large income disparities support inequality. Informal workers earn half the salary of formal workers[26]. Informality entails also a reinforcement of gender inequality: an informal woman worker earns on average 13.2% less than a man in similar conditions while in the formal sector the difference is 9.2%. In the formal sector, the scenario of income disparities is also worrisome: only 27% of the salaries surpass two minimum salary wages (El Tiempo, 2011a).

Echeverry *et al.* (2011) has identified three factors causing Colombian inequality:

(a) Difficult access to higher education. Considering that the Colombian society is divided into five groups in terms of its socio-economic levels, 18% of the youngsters from the lowest level groups 1 and 2 have access to graduate education while 100% of the highest group 5 do. Since the biggest job opportunities are for the most qualified, the gap between all social groups continues to increase. Between 1993 and 2009 modern job opportunities for the most educated have increased by 38% while for the less educated the job opportunities have decreased by 45% (Echeverry *et al.*, 2011).

(b) Informality is high at 50.4% in 2009, 57% in 2010 and 51.3% from June to August 2012 (DANE, 2012b). Despite stable growth, Colombia has resilient informality (WB, 2010). In 2009, the highest levels of informality were found among the lowest qualified workers, entrepreneurs, young workers between 15-24 years of age, and microenterprises. For all those groups, productivity is lower and workers are more vulnerable in terms of social security than in the formal system (Echeverry *et al.*, 2011). The Ministry of Commerce, Industry and Tourism (MCIT) reported that in 2006, informal labour accounted for as much as 4 million 778 thousand jobs, while the formal sector was responsible for just 3 million 351 thousand jobs (Caro and Pinto, 2007).

(c) Difficult access to loans and to the financial system. Access to financial services such as loans, deposits, or payment instruments brings benefits to households. The classical mechanisms that bring income and consumption stability to households are nevertheless insufficient and ineffective for the most vulnerable people; credit opportunities are almost non-existent for micro-enterprises (personal communication, 2009-06)[27]. While only 20% of the households in Colombia uses formal banking, the rest looks for informal instruments like friends, family, or individual moneylenders that are usually highly expensive but reachable (Echeverry *et al.*, 2011). The ineffectiveness of those informal instruments does not only affect the households' consumption patterns but their access to nutrition, health and education as well. The income instability creates persistent poverty (Echeverry *et al.*, 2011).

Even though Colombia has increased the social spending by more than 60% between 2001 and 2008, poverty has diminished but inequality has remained (Echeverry *et al.*, 2011; Semana, 2011). Economic policies being implemented have been inappropriately focused, the public finance management has been ineffective, and there have been institutional weakness, lack of political will, and/or unnecessary bureaucracy on a long-term basis in Colombia for wealth redistribution (DNP, 2011; Gaviria, 2011). The situation is not promising with respect to indicators evaluating the performance of institutions favouring progress in terms of per capita income (Fundación Agenda Colombia, 2007). The institutional framework is dysfunctional in coordinating and addressing all the actors involved in the country's economic development and the latter poses a very serious challenge for Colombia's

---

[26] Informality in Colombia is still over 50% of the potential working population (El Tiempo, 2011a).
[27] Comment made by the Colombian representative at IMF

leadership. Solving conflicts through the institutional conduits is held back by lack of a clear definition regarding property rights and by ineffective institutions (Posada-Carbó, 2012; Gaviria, 2011; Reyes, 2009; Merlano and Negret, 2006). The tax policy system constitutes an example of the lack of focus. The Colombian Government has limited capacity to use it for equality purposes. Individuals with the highest incomes get tax incentives that drop their tax tariff from 33% as a nominal national tax rate to an effective 14%. With a population of around 45 million people, only 1.3 million declare taxes. In contrast, in Chile, with a population of 16 million, 4.5 million contribute to the treasury (El Tiempo, 2011b).

In response to the above scenario, the Government has been creating tools to ease a transit to a more inclusive society. It has implemented a decentralized strategy since the 1991 Constitution; it is working on a tax reform aiming at increasing the taxes for the richest population; the creation of the Law on Land Restitution (2011) aims at restoring the damage caused to the under-privileged and to return the lands, estimated at more than two million of hectares, to their original owners[28] (García *et al.*, 2010); the law 1508 on APP (Public Private Alliances) (2012) established the possibility for the public and the private sectors to work on a cooperative basis provided the initiative is inspired by social responsibility and public interest.

Despite efforts, which have put Colombia as one of the top three business climates of South America, the WEF (World Economic Forum) considers that corruption constitutes the main issue affecting Colombia's competitiveness (AnteaGroup, 2012). This statement is supported by in-depth interviews at national level to former vice-minister (1) and to the representative (5) in Table 3-1. Because of the high inequality, the process of decentralization of the past two decades has not only resulted in social and political innovation (Duque, 2012) in regions but has also meant decentralized corruption (López, 2011). As powerful local lords aim primarily at meeting their private interests, local political and control authorities have been prone to manipulative processes (Semana, 2012; El Tiempo, 2012; López, 2011) and the law 1508 from 2012 on Public-Private Alliances, which was thought to stimulate investment based on public interests could be losing track.

4.2.3   Micro, Small and Medium Sized Enterprises

Within the scenario presented above, the prevalence of micro, small and medium sized enterprises (SMEs) is particularly marked in Colombia. Even though there are difficulties in accurately counting them, an effort was made in 2006 and 2007 by the National Planning Department. The study identified over 684,000 such enterprises, meaning over 19.6 per one thousand people (Saini, 2006). SMEs represent 99.9% of the enterprises, offer 80.8% of national employment, and are responsible for 37.7% of the national production (El Tiempo 2005b; DNP, 2007a). McSEs represent 99.4% of the total number of enterprises, and 67.9% of the work force. Of all the enterprises in Colombia, 96.4% are microenterprises and they employ 50.3% of the total Colombian workforce (DNP, 2007a) (

Figure 4-2 & Figure 4-3).

---

[28]www.congresovisible.org/agora/post/la-restitucion-de-tierras-en-colombia/155
    www.accionsocial.gov.co/estadisticasdesplazados/DinamicaGeneral.aspx

Figure 4-2 Percentage of Enterprises by category in Colombia.

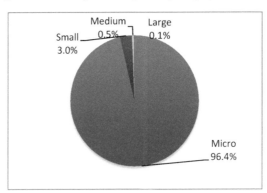

Source: DNP 2007a

Figure 4-3 Percentage of workforce in micro, small, medium or large enterprises.

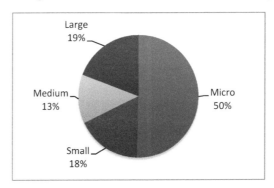

Source: DNP 2007a

In 2003, a survey aiming at filling the information gap on the Colombian SMEs was performed on 687 SMEs in the Colombian biggest cities and included 63 officers working in the public sector and in NGOs (Rodriguez, 2003). The results unveiled their main barriers to development, which were pertaining to macro and micro origins and divided in three categories:

- Main barriers: (1) Economic situation of Colombia, (2) Access to financing, (3) Tax system, and (4) Internal market.

- Secondary barriers: (1) State support, (2) Public security, (3) External markets, and (4) Labor legal framework.

- Internal barriers: (1) Limited human capital, (2) Effective guild representation, (3) Access to technology, (4) Association building, and (5) Services from the Solutions for Enterprise Development (SDEs).

Among the main barriers, the lack of access to financing was consistently cited as a major concern. 57% of the interviewees had attempted to subscribe to formal credit and 50% of them were denied. The in-depth interviews, which accompanied the research, showed that the SMEs complained about the banks´ lack of interest in helping them, their lack of trust, and the fact that they were always asked for personal guaranties for their business operations.

With respect to the secondary barriers, SMEs' complaints (75%) were basically about the lack of State support and they argued that the fact that more that 63% never worked for the public sector was self-explanatory. The smallest firms pointed at the high labour costs, which turn out to be a perverse stimulus towards informality because of their low profits margin.

With respect to the internal barriers, SMEs denounced that the existent guilds did not represent their interests, that collaborative work through cooperatives and associations was not a common practice, only 18% had been part of associations, and that the SDEs were based on wrong consulting approaches, 60% considered it ineffective, for their specific characteristics.

Overall, it can be said that at micro and macro levels, the existence of distrust was always present between and among key actors in the McSEs' field.

While 49% of the microenterprises belong to the commercial sector, 40% to the services sector, and 11% to the industrial sector, the small and medium sized enterprises (not including micro) are more in the services sector (59.2%) than in the commercial (21.5%) or industrial sector (19.3%). The growing informal economy is mainly made up of microenterprises (Echeverry *et al.*, 2011; DNP, 2007a; Caro and Pinto, 2007; Portafolio, 2005).

Essentially, the characteristics of microenterprises in Colombia are (DNP, 2007a):

- Low educational and technological level,

- Limited access to financing opportunities: the majority gets loans from friends and family members,

- Narrow markets for their products,

- Low level of association building, and

- High level of informality:

    - In a survey in 2003, among 28,871 microenterprises, 53.5% did not pay taxes, 42% did not generate accounting reports, and 45% did not hold a commercial registry (Rodriguez, 2003), and

    - In 2005 it was found that 58% of men and 60% of women had informal employment (Cárdenas and Mejía, 2007).

The characteristics of the small and medium sized enterprises are similar except that medium sized businesses have better negotiation power in selling their products; and informality and access to the financial system may still be predicaments in the smaller enterprises (DNP, 2007a) (Table 1-2). Because of their shared characteristics, this research concentrated on McSEs.

4.2.4   Leather industry

Figure 4-4 Relative size distribution of the Colombian tanning sector

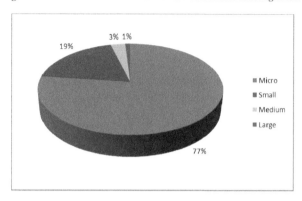

Source: DNP, 2007b

Within the scenario presented above, the leather industry in Colombia is constituted by 77% micro, 19% small, 3% medium, and 1% large industries (DNP, 2007b).

Table 4-1 Number and size of the tanneries in Colombia in 2003.

Place	Number of tanneries	Size				Production (skins/months)	
		Mc	S	M	L	Average	Maximum
Cundinamarca (Villapinzón, Chocontá & Cogua)	190	124	66			70.000	120.000
Antioquia (Medellín, Guarne & Sonsón)	7	-	2	1	4	62.000	74.000
Bogotá (San Benito, San Carlos)	350	298	42	10		33.000	140.000
Valle del Cauca	22	10	8	4		40.900	92.150
Atlántico	2				2	21.000	
Nariño	64	64				19.000	38.000
Quindío	27	16	10	1		12.000	50.000
Bolívar	1			1		10.000	
Risaralda	1			1		9.000	12.000
Santander	4	No information				No information	
Huila	1	No information				No information	
Tolima	8	No information				No information	
TOTAL	677					271.000	

Source: "Proyecto Gestión Ambiental en la Industria de Curtiembre en Colombia: Diagnóstico y Estrategias", Centro Nacional de Producción Más Limpia. 2004.

Figure 4-5 shows that most of the tanneries, 81% are located in the Cundinamarca and Bogotá regions put together, 10% from Nariño in the far south, 4% from the known coffee region, 3% from Valle del Cauca, 1% from Antioquia, and the last 1% from the rest of the country (DNP, 2007b).

Figure 4-5 Regional distribution of the tanneries in Colombia.

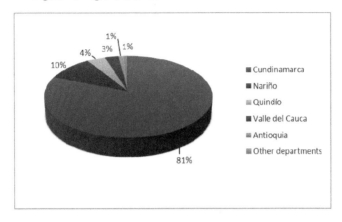

Source: DNP, 2007b

The tanning industry is considered a high impact industry because of its environmental effects, and is characterized by a low technological level, high informality, difficulties ensuring high leather quality in the production chain, and the difficulties of competing with leather substitute imports from China[29] (DNP, 2007b). From August 2011 to August 2012 DANE (2012c) reports a 20.5% decrease in leather production. The latest numbers show that a pair of shoes from China is sold for USD 1 while it takes a Colombian producer USD 7 to 8 to produce the same pair (El Espectador, 2012a).

International markets for Colombian finished leather products, which demand high quality are hard to find. At the bottom of the leather production chain, livestock are not carefully handled and the slaughterhouses use obsolete technology, which damages the skins. The latter is especially relevant because raw material and initial steps are responsible for 60% of the leather production costs (ICEX, 2004). The main leather exports are to USA, Italy and Venezuela. This sector represents 2% of the total Colombian exports and the wet blue constituted 20.7% of the leather exports (DANE, 2012a; ICEX, 2004).

In order to face the industry's threats, the DNP pointed at solving the gaps in the production chain, to stimulate association building, capacity building and CP programs. Fighting the industry's rampant informality through stronger and effective sanctions is seen as a crucial and main issue (DNP, 2007b). This last statement is backed-up by the in-depth interview withthe executive director of Conservation international (15) in Table 3-1.

---

[29] The national production of leather is dropping despite the fact that the shoe production grew by 15.4% in the first trimester of 2012 because the shoe industry is using Chinese leather substitutes.

4.2.5   Water - environment data

There are five main river basins in Colombia: Caribbean, Pacific, Orinoquía, Amazon and Catatumbo. Colombia's water resources are not uniformly distributed due to geographic, climatic, demographic and socio-economic factors; regional and seasonal flow variations and environmental vulnerability differ considerably among river basins (Rodríguez, 2009).

Colombia is considered one of the four most bio-diverse countries in the world (UNDP, 2007) having a run-off of 63 l/km^2 which is six times the world average (10 l/km^2) and three times the Latin American average (21 l/km^2) (ENA, 2010). The renewable water resources in 2000 were 50.600 /cap.year and in 2005 were 47.470 m^3/cap.year (WWAP, 2006; 2003). On the other hand, the environmental degradation has reached 3.7% of Colombia's GDP in 2004 and water degradation alone accounts for 1% of GDP. The latter is significant compared to the mere 0.31% of the GDP representing the annual participation of the environmental sector in the national budget (Canal and Rodríguez, 2008; Sánchez-Triana et al., 2007).

Carrizosa (2008) and an in-depth interview (12) in Table 3-1 states that more than a biodiverse country, Colombia is a highly complex one and that Colombians have historically disregarded such complexity. The huge complexity caused by highly humid areas within three tall mountain chains and a big variety of original indigenous groups have historically brought difficulties in exerting authority, causing a chaotic population distribution, and limited chances for social and economic mobility (Gaviria, 2011; Carrizosa, 2008).

The effects of environmental degradation mainly result in increased mortality, morbidity and decreased productivity (Sánchez-Triana et al., 2007). Over the past two decades, the situation has worsened due to climate change. The agricultural zones of several watersheds have been affected by alternating severe dry seasons in 2009 (because of the phenomenon of El Niño) causing water deficits, including the Bogotá River (Cundinamarca Department) with severe rainfall in the years 2010 and 2011 (because of the phenomenon of La Niña) causing the most dramatic floods Colombia ever experienced. As a result, La Niña water floods and landslides in 2010-2011 cost more than 2% of the country's GDP.

The estimated 1% of Colombian GDP, seen as the cost for water degradation could, hence, easily reach higher scores since there is only limited data on the cost of water-borne illnesses (Sánchez-Triana et al., 2007). Other costs, such as the potential impact of heavy metals and chemicals on health and recreational value, or the loss of biodiversity have not been estimated (Sánchez-Triana et al., 2007; Márquez, 1997). Despite monitoring and data collection difficulties, the overall trend for the country's environmental management reflected in the Environmental Performance Indicators report (EPI, 2012) acknowledges a declining situation mainly by the performance on forest and water issues. The effects on human health and ecosystems pose serious threats to water management (EPI, 2012; interviews 6, 8, 9, 10, 11, 12 in Table 3-1).

The most costly problems associated with environmental degradation are in decreasing order of magnitude: waterborne diseases, urban air pollution, natural disasters, land degradation, and indoor air pollution (Sánchez-Triana et al., 2007). The burden of these costs has been found to fall most heavily on vulnerable segments of the population, especially poor children under five years of age. Even though in recent decades progress has been made in addressing the water, forest and protected areas' agendas, 'the impact of environmental degradation on the most vulnerable groups suggests the need to increase emphasis on environmental health issues' (Sánchez-Triana et al., 2007: xlvi).

Water demand and water use

Water demand in Colombia in 2008 was 35,877 million cubic meters (Instituto de Hidrología, Meteorología y Estudios Ambientales, IDEAM, 2010). Of this amount, 54% was for agriculture, 19.4% for energy, 7.3% for household use, 7.2% for fisheries, 6.2% for livestock, 4.4% for industry and 1.5% for the services sector (Figure 4-6).

Figure 4-6 Water demand in Colombia 2010.

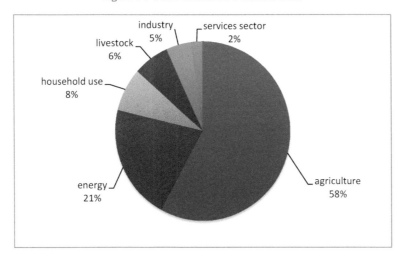

Source: ENA, 2010

It is important to mention that use efficiency is rather low especially in agricultural practices (Portafolio, 2010). 26% (9,351million cubic meters) of the water demand is extracted water not being used (ENA, 2010). IDEAM estimates that if water resources are not properly managed, by the year 2015, 66% of the municipalities will face a great risk of water shortages on a dry year; and by 2025, over 14 million inhabitants could suffer water shortages in the Andes and Caribbean regions of the country. In 2010, 44% of the population is already highly vulnerable to water shortages on dry years because of unsustainable uses of the big rivers, the aquifers and the wetlands (ENA, 2010).

The big rivers are not used for supplying drinking water due to pollution, economic reasons, corruption or uneven access to water (Guhl *et al.*, 2007; Sánchez-Triana *et al.*, 2007) (Box 4-1). The city of Quibdó in the Pacific coast does not have an adequate water utility whereas it has had budget to build it (interview 15, Table 3-1). The city experienced a water shortage in 2007, which affected 70% of the population while it is ironically located on the banks of the Atrato River, one of the largest rivers of Colombia and in an area having the highest precipitation rate in the world.

Box 4-1 Uneven access to water

In the Bogotá Savannah, while the big industrial sector does not consider water availability, price or quality to be limiting factors in its development and investment plans, small rural water users face scarcity every year during the dry months. It has been acknowledged that small users lack the allocation mechanisms and access to judicial appeal enjoyed by large water users (Merlano and Negret, 2006). At the same time, it is estimated that 70% of small users do not hold water permits and there is no information on their water consumption (Sánchez-Triana *et al.*, 2007). Finding long-term solutions to the problems of small water users has not been a priority for local, regional or national authorities in Colombia. Obtaining water permits or concessions is a burdensome and expensive process for small and local users and public participation in water allocation policy for them is non-existent (Sánchez-Triana *et al.*, 2007).

Over 30% of Colombia's freshwater resources are contained in aquifers, which makes groundwater an important potential supply source. Forthy percent of municipalities rely exclusively on groundwater for their drinking water (IDEAM, 2004). Although systematic information is lacking regarding groundwater, it is used inefficiently and in some areas, studies have shown overexploitation (La Guajira, Cordoba and the Bogotá Savannah) and pollution (The Bogotá Savannah) (JICA, 2002; Interview 14 in Table 3-1).

With respect to the wetlands, Colombia has over 2.5 million hectares of wetlands and almost 2 million hectares of floodplains. These serve an important role in recharging and discharging aquifers, controlling floods, retaining nutrients and sediments, producing biomass, facilitating aquatic transport, stabilizing coastlines and microclimates, filtering contaminants, maintaining fauna and flora and providing recreation and tourist attractions. Due to poor management of these wetlands, those environmental benefits have decreased. There has been drainage of marshes and contamination. Construction of infrastructure, effluent discharges from agriculture, fishing, and domestic uses have contributed to this degradation (Sánchez-Triana *et al.*, 2007).

Main water problems

The main water problems in Colombia are related to pollution, safe drinking water and sanitation, land use practices, and water flooding and landslides. 1450 to 1820 children die every year from diarrheal illness related to this (Sánchez-Triana *et al.*, 2007).

Water pollution

The main water pollution problems are associated with toxic substances, pathogens, and other hazardous substances, but there are very few policies addressing these problems. Furthermore, they omit nonpoint discharges, which are more damaging to the environment than point discharges of organic matters (Table 4-4); agriculture and livestock represents 84% of the BOD production (Sánchez-Triana *et al.*, 2007).

Table 4-2 Daily BOD production in Colombia per sector in 2007

Sector	Tons/day	Percentage of total production
Agriculture and Livestock	7100	84
Domestic wastewater	800	10
Industrial wastewater	520	6
Total	**8420**	**100**

Source: Sánchez-Triana *et al.* (2007).

Toxic pollution

Most toxic waste discharges in Colombia come from the improper disposal of solid wastes and hospital waste and from runoff from fertilizers, pesticides, and industrial waste coming especially from petroleum refineries, the chemical industry, and leather tanneries (ENA, 2010; Sánchez-Triana *et al.*, 2007). The use of polluting technologies in the manufacturing industry such as those traditionally used in tanneries, electroplating, metal smelting, and palm oil extraction promote the waste of raw materials and the high production of residual waste in uncontrolled conditions. Cleaner Production implementation has shown only isolated successful cases (Blackman *et al.*, 2012; Sánchez-Triana *et al.*, 2007; Van Hoof and Herrera, 2007). In Bogotá's Doña Juana landfill, for example, high concentrations of toxic compounds have been registered since 1995 (DNP, 1996). Already in 1982, research showed a level of Mercury in milk from the cows living around the Bogotá River that was more than two times the accepted standard limit for potable water (EPAM, 1993). Wastes can pose severe health risks to anyone consuming the water directly or through agricultural products irrigated with it. Key causal factors in groundwater pollution include agricultural run-off, septic tanks, and landfills. In 2002, the Ministry reported that in 66% of the cities they had studied, no industries treated wastewater as mandated by decree 1594 of 1984 and only 3.1% of the industries treated 100% of their wastewater (Sánchez-Triana *et al.*, 2007; interview 13, 15 of Table 3-1).

The National Water Report (ENA, 2010) shows for the first time the impact caused by large amounts of polluting chemicals from the use of pesticides in the agricultural sector by the illegal crops like coca and the fight against them, supported by the USA, through chemical pesticides. Estimates backed by the DNP data (1995) suggest that the application of such pesticides in all crops in Colombia were two to three times greater than the amounts recommended by manufacturers, implying that the problems of runoff pollution and agricultural percolation were critical already in 1995 in the 1.2 million hectares dedicated to agriculture. Crude oil spills from acts of sabotage by terrorists have become another cause of water pollution in the country. The most affected ecosystems have been slow moving bodies such as wetlands and low-flowing creeks. In the cities, because of deficient management and disposal of substances used for the operation and maintenance of vehicles, 250,000 barrels of motor oil, for example, ended up in the Bogotá River in 1990 (Sánchez-Triana *et al.*, 2007; Latorre, 1996).

Organic pollution

Organic material wastes from agricultural, industrial, and domestic sources reduce dissolved oxygen. It is estimated that the three largest contributors to biochemical oxygen demand in 1994 were non-point sources with 84% accounting for agriculture and livestock, and point sources with 16% including domestic wastewater from larger cities with 10% and industrial sources with 6%. Among point sources, 76% came from urban wastewater and 26% from industry (Sánchez-Triana *et al.*, 2007) (

Table 4-2).

Lacking Integrated Water Resources Management (IWRM)

The director of Conservation International (15) in Table 3-1 reports that having for years preferred the construction of drinking water plants, Colombia is lagging 50 years behind in terms of a framework for dealing with the consequences of the increased water supply demand, *i.e.* the generation of huge amounts of wastewater.

About 80% of the municipalities discharge their wastewater into creeks, lakes and rivers and between 8% (ENA, 2010; Contraloría de la República, 2003) and less than 1% (ENA, 2010; Sánchez-Triana *et al.*, 2007) of the wastewater is treated. The situation is so critical that a significant percentage of wastewater is not collected because many households are not served by municipal sewerage systems and when they are, it is generally poorly or not treated because many of the existing wastewater treatment plants operate inefficiently or not at all (personal communication, 2012-09/02[30]; UNIANDES, 2002a).

Another urban wastewater treatment issue is its high cost, estimated to be Col$5250 billion (US $2.5 billion), or 72% of the total cost of all water infrastructures. An important contributor to this is the prevalence of high-cost, high-technology conventional treatment plants and the limited use of low-technology, low-cost solutions, including lagoons, anaerobic processes and filters, and seasonal stabilization reservoirs for agricultural reuse (Sánchez-Triana *et al.*, 2007; UNIANDES, 2002a). The so-called conventional wastewater management approach uses end-of-pipe and centralised treatment technologies with effluents discharged into rivers. As a result, eutrophication and frequent outbreaks of water borne diseases are reported (ENA, 2010; Nhapi, 2004; Uniandes, 2002a).

This may change in the coming years since Colombia is in the process of implementing a new policy on Integrated Water Management, which received support from the Embassy of The Netherlands (MAVDT, 2010) and has recently named the water directorate under the heading of IWRM (Figure 4-9).

Appropriate monitoring

IDEAM (2004) has recognized that water quality information gathering is insufficient. Although monitoring stations exist for certain rivers and aquifers, coverage is limited and data collection and management are not standardized. Assessment of the state of water quality at a national level is incomplete while water pollution is evident in several lakes and rivers like Bogotá, Cali, Combeima, Medellín, de Oro, Otún, Pamplonita, and Pasto. Lately, due to climate change policy, monitoring has received more support and IDEAM's last Water Report claims significant improvement (ENA, 2010).

Safe drinking and sanitation

In the urban areas, 91% of the country's inhabitants have access to safe drinking water and 80% to sanitation services (MHETA, 2003). However, those figures are misleading as it is known that 35% of water utility companies do not have water treatment plants or that technical, operative and maintenance failures exist (interview 15 in Table 3-1). The most significant issue regarding drinking water is the enduring discrepancy between the rural and the urban areas. The rural sector accounts for

---

[30] Comment made by the National Comptroller at a meeting presenting the results on the control on the Bogotá River

23% of the population and its drinking water coverage is 66% and its sanitation coverage is only 57% (Guhl *et al.*, 2007).

In the Bogotá region, for example, where the AR in this thesis takes place, although there is sufficient water up to 2025, and important results have been achieved in cutting down water consumption from 160 to 120 l/cap/day (UNIANDES, 2002a), comprehensive water management is not considered a common practice (interviews 8, 16 in Table 3-1) and water pollution is overwhelming.

Land use practices

Regarding land practices, deforestation for agricultural purposes causes a decrease in the water retention of the soil. The latter is especially important in coca growing areas (ENA, 2010).

Floods

The problem related to the floods shows figures as significant as having had 400,000 people affected by floods in 1998 (Alfaro *et al.*, 2000) and more than two million people in 2010 (El Espectador, 2010).

4.3   National Water-Environmental policies

4.3.1   Introduction

This Section presents the history of water policy and law (4.3.2), the current water policy and law (4.3.3), the organization of water policy and law (4.3.4), governance instruments (4.3.5), the budget and SINA's finances (4.3.6), and an assessment of water policy and law (4.3.7).

4.3.2   History of water policy and law

The present Chapter would be incomplete without presenting the history of environmental policy in Colombia, as water policy is strongly tied to it.

The evolution of Colombia's environmental institutions over the past 50 years has led to a unique decentralized Environmental Management System (Sistema Nacional Ambiental, SINA) (

Figure 4-7). This framework evolved gradually and achieved a number of milestones from the years 1952-1974, 1975-1993, and 1994 - to date (Table 4-5).

Figure 4-7 Organizational framework of environmental policy and law: The SINA system

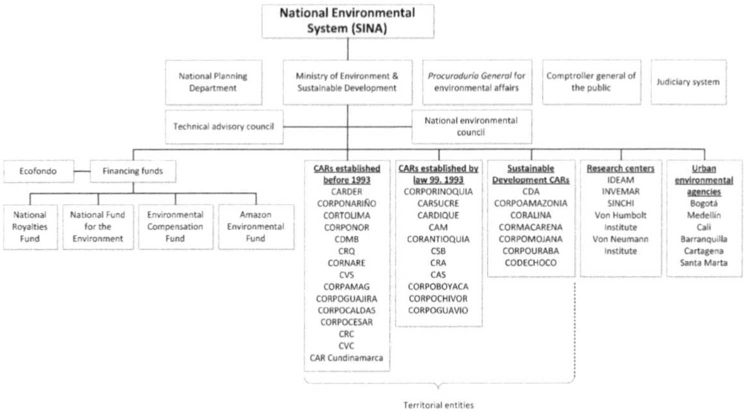

Source: Adapted from Sánchez-Triana *et al.*, 2007

Table 4-3 History of Water Policy and Law in Colombia

1500 – 1700 (Colonial)	Laws inspired from Castilla, Spain. Water: a public use. The king allowed private ownership of water to loyal servants.
1887 (Independ ence-year)	The Civil Code by Law 57 for the new Republic of Colombia. Public goods -inalienable, unprescribable and unseizable- such as water belong to the Nation.
1888-1928	1928: the first water concessions for 50 years. Two possibilities, valid today, provided for private ownership of water
1952-1974	1952: The Division of Renewable Natural Resources within the Ministry of Agriculture and the (CARs) agencies promoting regional and economic development. 1961: a CAR (CVM) dealing with the main artery in Colombia: The Magdalena River. The CVM became the National Institute of Renewable Natural Resources (INDERENA) and fuses with the Division of Renewable Natural Resources in the Ministry of Agriculture. 1969: The forestry Law defined. 1974: National Code for Renewable Natural Resources and Environmental Protection[31] (CNRN)- the first code in Latin America created. Decree-Law 2811- basic legal structure for water regulation

---

[31] The code has 340 articles covering **water,** air, solid and hazardous waste, soil, flora, and fauna and it was one of the first environmental protection laws in the world to incorporate pollution fees and environmental impact assessments.

1975-1993	1977: CARs attached to the National Planning Department (*Departamento Nacional de Planeación*, DNP).
	1978: Decree 1541- regulations related to ownership and use of all water resources, the framework for the national system of water concessions[32].
	1979-1984: Law 09 of 1979 and Decree 1594 of 1984 provides details of standards and requirements regarding environmental quality, treatment of wastewater, and compliance.
	1987: CARs' main priority is building infrastructure for water resources and the profits from these projects support the Ministry of Agriculture.
	1993: Law 99 aims at decentralized, democratic, and participatory environmental management, creating (not modifying regulations for the water sector or the management of water resources):
	• The National Environmental System (*Sistema Nacional Ambiental*, SINA
	• The Ministry of Environment fusing regulation, control, and coordination[33]
	• The CARs implementing the national environmental policy promote creation of new and Autonomous Sustainable Development Corporations, (*Corporaciones Autónomas de Desarrollo Sostenible*, CDSs) for indigenous, and Urban Environmental Authorities (*Autoridades Ambientales Urbanas*, AAUs) for cities with over 1 million inhabitants.
1994-to date	Strengthening the environmental agencies and consolidating institutions within SINA.
	2002: New ministry grouping the environmental sector with the housing and regional development sectors- The Ministry of Environment, Housing and Use Land Development (*Ministerio de Ambiente, Vivienda y Desarrollo Territorial*, MAVDT).
	Decree 1729 of 2002 Watershed Administration and Management Plans for CARs (*Planes de Ordenamiento y Manejo de Cuencas Hidrográficas*, POMCAs).
	2005-2006: a proposal to a unified water law presented to Congress and not accepted
	2006: Vice-ministry of Water and Sanitation
	2011: the MAVDT re-decomposed into separate ministries again: The Ministry of Environment and Sustainable Development (MADS) and the Ministry of Housing.

4.3.3 Current water policy and law

The National Environmental System (Sistema Nacional Ambiental, SINA) was conceived by Law 99 as a set of orientations, norms, activities, resources, programs and organizations that follow seven guiding principles: a) economic and social development will be led by the goal of sustainable development laid out in the 1992 Rio Conference; b) biodiversity must be protected and should be exploited only in a sustainable manner; c) human consumption is the highest priority for water use; d)

---

[32] Water concessions can be granted up to 10 years and maybe transferred to another beneficiary only if authorized by the authority and without changes to the original conditions.

[33] Once the ministry was created, INDERENA was subsequently phased out.

environmental policy will be based on the best available scientific evidence, and action should be taken to prevent possible serious irreparable damage even if that evidence is incomplete; e) environmental costs will be incorporated into policies and markets to help conserve the environment and renewable natural resources; f) environmental protection is a joint task for the state, communities, NGOs, and the private sector, and in order to promote this vision, the state will support the development of environmental NGOs and may delegate some government functions to them; and g) environmental impact studies will be the basic instrument for deciding whether to engage in activities that might affect the environment.   SINA was designed to be a decentralized, democratic and participatory entity (

Figure 4-7; interview 1, Table 3-1).

Colombia does not have a unified water law (An attempt to work on a water law resulted in a draft that was presented to Congress in 2005-2006. The Head of Senate herself explained that the draft faced strong opposition with arguments regarding the risk of turning water into a commercial good (mc1 in Table 3-2). The Law was presented again after some modifications for the period 2008-2009 but was not adopted.

Two policies speak about water related issues: The CP (Cleaner Production) policy (1997) and the IWRM policy, *Política de Manejo Integral del Recurso Hídrico* (2010).

Table 4-4). The legal framework has been developed out of the large compendium of norms and rules from each sector. Basically, the group of laws, decrees and *resoluciones* with respect to water are from the most recent to the oldest (MADS, 2012):

An attempt to work on a water law resulted in a draft that was presented to Congress in 2005-2006. The Head of Senate herself explained that the draft faced strong opposition with arguments regarding the risk of turning water into a commercial good (mc1 in Table 3-2). The Law was presented again after some modifications for the period 2008-2009 but was not adopted.

Two policies speak about water related issues: The CP (Cleaner Production) policy (1997) and the IWRM policy, *Política de Manejo Integral del Recurso Hídrico* (2010).

Table 4-4 Colombian legislation concerning water

Year	Legislation	Number	Summary
2011	Resolución	075	A new reporting format is based on the discharge limits to the public sewers
2010	Decree	4728	Modifies partially Decree 3939 from 2010
2010	Decree	3930	Modifies partially Law 2811 from 1974 with respect to water uses and liquid wastes
2007	Resolución	2115	Dictates characteristics and instruments for command-and-control, and for water quality for human consumption
2007	Decree	1575	Establishes the system for protection and control of water quality for human consumption
2007	Law	1151	In order to accomplish the National Development Plan 2006-2010, modifies articles 42, 44, 46, 111 from Law 99 from 1993
2007	Decree	1323	Creates the Information System for Water Resources (SIRH in Spanish)
2007	Decree	1324	Creates the Water Resources Users' Registry
2007	Decree	1480	Watersheds are considered by their importance
2006	Resolución	872	Establishes the methodology needed to measure the water scarcity index for underground waters
2006	Decree	1900	Modifies articles 43 from Law 99 from 1993
2006	Decree	2570	Sums Decree 1600 from 1994
2005	Resolución	2145	Modifies partially *Resolución* 1433 from 2004 on sanitation and management of discharges

2005	Decree	4742	Modifies article 43 from Law 99 from 1993 on water use fees
2004	*Resolución*	865	Adopts the methodology to calculate the scarcity index for running waters
2004	*Resolución*	240	Sets the reference basis to calculate depreciation and the minimum tariff for water use
2004	Decree	1443	Sets partially Decree-Law 2811, and Laws 253 and 430 with respect to pesticides, and dangerous wastes
2004	Decree	155	Amends article 43 from Law 99 from 1993 on fees on water use
2004	Decree	3440	Amends Decree 3100 from 2003 with respect to pollution fees
2003	Decree	3100	Sets pollution fees on point discharges
2002	Decree	1604	Amends article 33 from Law 99 from 1993 on joint commissions
2002	Decree	1729	Amends Decree 2811 on watersheds
1997	Law	373	Program on efficient water use
1994	Decree	1933	Amends article 45 Law 99 1993 on electric energy
1994	Decree	1600	Sets SINA with respect to the National Systems for Research
1993	Law	99	Creates SINA and the Ministry of Environment
1984	Decree	1594	Rules water uses and liquid discharges. Partially in force
1979	Decree	1875	Rules on marine pollution prevention
1978	Decree	1541	Rules on fresh water allocation
1978	Law	10	Rules on continental platform, territorial sea, etc.
1977	Decree	1449	Rules on riverbanks. Partially in force
1974	Decree- Law	2811	Dictates the CNRN
1973	Law	23	Sets the need to protect the environment, determines discharge limits and sanctions.

Source: MADS, 2012

4.3.4   Organization of water policy and law

The Presidency is directly responsible for the regulation, implementation and coordination of policies and plans with respect to the different water related sectors for the water market (

Figure 4-8). Besides the Ministry of Environment and Sustainable Development, and the Ministry of Social Protection, other Ministries end-up dealing with the Colombian water market. The Ministry of Housing, City and Territory rules on the territorial urban development, and regional authorities. The Ministry of Finance deals with the economic and fiscal policies and rules on the financing institution for territorial development (FINDETER). The Ministry of Transport is in charge of all related matters by sea, road, river or air. The Ministry of Agriculture deals with the rural policies on agriculture and rural fisheries (AnteaGroup, 2012).

Figure 4-8 Ministries dealing with the water market in Colombia

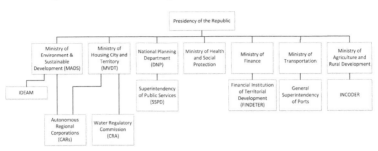

Source: Antea Group, 2012

The Ministry of Environment and Sustainable Development (

Figure 4-8 and Figure 4-9) was intended to consolidate the key environmental functions, and with regard to the water domain it is the supreme regulatory authority for water resources. Its responsibilities include formulating, managing and coordinating policies, regulations, and programs on drinking water, water resources management, wastewater discharges, and sanitation. In alliance with the Ministry of Foreign affairs it is also responsible for international cooperation on water related issues. Jointly with the Ministries of Social Protection and Housing it is responsible for defining the acceptable quality standards and uses for water and for developing a resource classification plan that includes existing uses, projections of water use needs, quality simulation models, quality criteria, and discharge procedures.

Figure 4-9 Organizational framework of Ministry of Environment and Sustainable Development

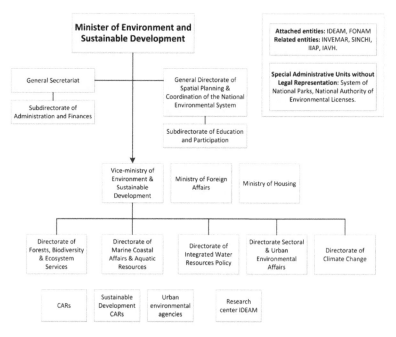

Source: MADS, 2012

As coordinator of the SINA system, the Ministry of Environment and Sustainable Development (Ministerio de Ambiente y Desarrollo Sostenible, MADS) is responsible for approving legal regulations adopted by regional autonomous authorities (CARs) and Urban Environmental Authorities. The Ministry of Social Protection alone is responsible for approving water treatment, storage, and transportation when water is for human consumption. The National Department of Planning (Departamento Nacional de Planeación, DNP) deals with the long-term plans (AnteaGroup, 2012) (

Figure 4-7).

Even though the drinking water framework is not a target in this thesis it is worth pointing out that the tariff structure does not include the costs for sustainability, or the concept of social equality. The public utility service operators target high water demand and have no motivation for water saving patterns to be adopted by society (interview 3 in Table 3-1).

The main entities (

Figure 4-7 and Figure 4-9) responsible for water resources management are regional authorities, including departments, municipalities, Autonomous Regional Corporations (Corporaciones Autónomas Regionales, CARs), and Urban Environmental Authorities (Autoridades Ambientales Urbanas, AAUs) which can formulate regulations that are more restrictive than those required under national law, but not less. The CARs and AAUs are responsible for monitoring and enforcing national water policies, granting concessions for water rights, and reviewing environmental impact assessments. Departmental and municipal governments help CARs to monitor and enforce water pollution regulations and to develop sanitation and wastewater treatment infrastructure. The municipalities are the owners of the wastewater treatment plants and must comply with the effluent discharge criteria established by law. Their relationships with authorities are not always easy. At the Bogotá River, for example, the fact that CARs and AAUs have built municipal wastewater treatment plants and are still operating them after quite a while has led municipalities to avoid compliance with the regulations (interviews 15, 16 from Table 3-1; mc3 from Table 3-2).

Apart from the above-mentioned organizations within the organizational framework for the SINA in Colombia, it is important to mention the National Environmental Council and the National Technical Advisory Council (

Figure 4-7). The first one is a consultative group attached to the MADS. Its role is to provide a forum to give both private and public stakeholders a voice in the design of important national environmental policies. The National Technical Advisory Council was created to advice MADS on scientific and technical issues related to policy issues. It gives advice on decrees that establish regulations subject to approval by the President. It has five to eight permanent members in contrast with the National Environmental Council that has none. This body has representatives from universities and the private sector. Elected representatives from the big industry, agriculture, mining, and the petroleum industry represent the private sector.

Apart from the last two consulting bodies, Law 99 established two types of institutes: Four that have primarily research responsibilities: The von Humboldt Institute, the Institute of Marine and Coastal Research, the Amazonian Institute for Scientific Research, and the Environmental Research Institute of the Pacific; and one that primarily has data collection responsibilities: The Institute of Hydrology, Metereology and Environmental Studies, IDEAM. They are said to focus primarily on basic sciences and not on sustainability and interdisciplinary approaches facing the impending challenges on poverty and environmental degradation (Guhl, 2008).

Within SINA, the control organizations are the Procuraduría Delegada para Asuntos Ambientales y Agrarios (Inspector General for Environmental and Agricultural Issues) that is responsible for supervision of all environmental authorities in Colombia; the Contraloría (Comptroller's office; and the Fiscalía (Attorney General), which deals with the highest penal sanctions.

The National Department of Planning (Departamento Nacional de Planeación, DNP) is considered part of SINA (

Figure 4-7). The DNP reviews and approves Colombia's national investment budget and evaluates the outcome of that spending. All ministries submit their budget requests for approval. The DNP coordinates the writing of the National Development Plan (PND) required for each presidential administration. Additionally, the DNP serves as the technical secretary of the National Council of Economic and Social Policy (Consejo Nacional de Política Económica y Social, CONPES). Chaired by the president, this governance body includes ministers and private sector representatives. DNP also exerts power by distributing national funds to municipalities, and by having the authority to negotiate and approve international loans to all public sector agencies. Until February of 2004, the Environmental Policy Office within the DNP handled environmental matters. The government phased out this office and its functions spread among other DNP offices. One of the main DNP concerns in its relationship with the CARs is a system for monitoring and evaluating them. This task today presents serious difficulties as the Environmental Policy Office does not exist anymore and CARs lack adequate indicators of environmental quality. The performance indicators of the CARs usually reflect regulatory processes rather than impacts (Blackman *et al.*, 2007a, 35). Even though policies prioritizing watersheds in 2007 were issued, systematic periodic planning exercises establishing priorities across environmental programs and subsectors are still vague (section 4.3.5). Planning is done sector by sector and seriously challenges the implementation of IWRM (Sánchez-Triana *et al.*, 2007; Guhl *et al.*, 2007).

Since 2010, licensing was transferred to a ministry-dependent body chaired by the Minister of Environment. It is called ANLA by its name in Spanish (Autoridad Nacional de Licencias Ambientales).

4.3.5   Governance instruments of water policy and law

The instruments of water policy and law can be for planning and coordination but can also be regulatory, economic, persuasive and participatory.

Instruments for planning and coordination

The National Development Plan

The 1991 Constitution envisioned a government with wide-ranging planning responsibilities, including those related to environmental issues. It created a National System of Planning and required the president to draft a National Development Plan (*Plan Nacional de Desarrollo*, PND) and to present it to Congress within six months after taking office. The National Environmental Plan from each government is taken from the PND and was introduced since 1993. As a basic statement, the 1991 Constitution ruled that public resources could only be used for public programs. Only recently in January 2012, under Public-Private Alliances (APP) (Law 1508, 2012) some public resources could be invested for general interest programs or under social responsibility endeavours.

The National Planning Department, DNP coordinates the National Development Plan proposed by the Executive.

The National Development Plan for 2006-2010 targeted a development that entailed not just economic growth but specifically social fairness and quality of life improvement. The plan defined the role of the government centred in the planning, regulation, and control aiming at a balance between the market and public interests. Water is managed by a two-pronged approach in this plan: 1) Improvement of the drinking and sanitation services by stimulating the involvement of the private sector; 2) Improvement of the regulatory framework for water conservation and protective land acquisition, and for the improvement of the economic instruments' efficiency such as with regards to effluent discharge fees. Nevertheless, the plan did not address other matters ensuing from the hydrological cycle such as the social dimensions of water and its use (Guhl *et al.*, 2007).

Regional environmental plans (PGAR, PAT, POIA, POMCA)

Subsequently, CARs were designed to draft a Ten-Year Regional Environmental Management Plan (PGAR), a Three-Year Action Plan (PAT) and Annual Investment Operating Plans (POIAs). All these plans were required to be aligned with the National Development Plan. From 2002, the Watershed Administration and Management Plans, POMCAs, made the watershed the unit for planning and management. They are considered to prevail over the rest of land planning development plans (Guhl *et al.*, 2007; Alvarez, 2003; Amaya, 2003).

Reality can be different with respect to planning and coordination

The MADS is not capable of exerting sufficient control on the planning and functioning of CARs as it is limited by insufficient information and limited human and technical capacity. In some cases, the general directors of some CARs are extremely powerful and have strong linkages with Congressional representatives, which in turn can exert influence on MADS (El Tiempo, 2010; Blackman *et al.*, 2007a; interview 1, 2, 7 in Table 3-1).

In general, limited evidence suggests that few CARs abide by the plans laid out in their PATs. It also suggests that many CARs operational expenses far exceed 30%, despite the requirements of law 617 that limits to a maximum of 30% of the total funds to be used for salary and administrative costs. The fact that CARs had been considered autonomous had perverted effects (personal communication, 2013-04/26[34]; Canal and Rodríguez, 2008; mc1 in Table 3-2). CAR Cundinamarca (Bogotá region), for instance seems to be one such CAR, that does not appear to have invested in a number of projects set out in its 2001 PAT. In this regard, spending on a project-by-project basis instead of category-by-category basis could help to improve priority setting (Sánchez-Triana *et al.*, 2007).

Efforts aiming at adopting IWRM are being developed in Colombia and it may be too early to draw conclusions on its implementation. In March 2007, a panel of experts worked on a guide to IWRM. It addressed issues related to the availability, use, impacts, treatment, and the instruments to improve its management. It took into account the strengths, weaknesses and priorities for Colombia and the proposals from the PND (Guhl *et al.*, 2007). At the end of 2010, the ministry elaborated the policy on IWRM, which was supported by the World Bank and the Embassy of The Netherlands (MAVDT, 2010). The Ministry is supporting two pilot IWRM cases in Colombia in order to test it at the regional and local levels (personal communication, 2013[35]).

---

[34] Comment made by the Head of IWRM at the ministry at a focus group for UNESCO-IHE at the Dutch Embassy
[35] *Ibid*

Regulatory

The regulatory entities have been presented in section 4.3.3, and adopt legal norms such as environmental laws, presidential decrees, enforcement actions, discharge limits to rivers or to public sewers, the water quality standards for potable water, agricultural use, recreational use or industrial use, or property rights. Under Law 1333 of July 2009, the Sanction Code was created and establishes the penal fines with respect to the environmental impacts caused. The licensing process regulates also. To obtain a license, projects are required to carry out an environmental impact assessment and hold a public hearing. Research activities also need special permits, which are reported to create barriers to scientists (personal communication, 2012-1[36]). The licensing process, however, does not encompass the degradation of bodies due to non-project related activities such as urban expansion, rural and urban run-off or the use of fertilizers and pesticides from unsustainable agricultural practices (Sánchez-Triana *et al.,* 2007).

Economic

The economic instruments are use and effluent discharge fees. Some CARs charge fees for water use based on article 159 of the CNRN, but since they do not provide specific criteria, the CARs decide who to charge and how to set the fees (interview 4s in Table 3-1). In 2003 the Colombian Farmers Society sued over the fees that some CARs had been charging. The Constitutional Court declared that specific regulations needed to be established in order to charge water fees. New regulations were developed following meetings with powerful stakeholders. Quantitative studies of the social value of water were not taken into account and new fees were set so low that the CARs have doubts about the financial justifications for even collecting them (Alvarez, 2003; Interview 11 in Table 3-1.

Effluent discharge fees have been used since the late 1970's in Colombia. As envisioned in Law 99, discharge fees were applied only to those discharges surpassing the established standards (interview 3, 4s in Table 3-1). In this regard, the Constitutional Court declared that not all discharges should be paying effluent fees. Very small discharges and the ones not surpassing the limits because of dilution, which should be punished, are exempted (Sánchez-Triana *et al.,* 2007 citing Alvarez, 2007). The effluent charge system addresses organic discharges from point sources to surface waters but hardly the impacts of other substances as priority pollutants (pathogens and other hazardous wastes), the contributions of the largest polluters, and the contamination of aquifers and marine waters. Neither does it address the impacts of the individual sewerage bills of the poorest segments of the population. Nonpoint sources such as agricultural runoff contribute an estimated 9 times as much BOD as municipal wastewater (UNIANDES, 2002a). An amendment of Decree 1594 of 1984 to control discharges from nonpoint sources and priority pollutants is urgently needed (Sánchez-Triana *et al.,* 2007; interview 16 in Table 3-1).

Persuasive

The persuasive instruments are environmental guides and voluntary agreements, *Convenios* in Spanish, and are presented in CP policy on 4.5.5.

---

[36] Comment made by a professor of UNAL on biodiversity ant the Environmental Symposium, June, 2012

Participatory

As part of the governance issue in environmental management, public participation has been considered an essential component by the Constitution of 1991. In addition to having a right to a healthy environment, citizens have an express duty to protect natural resources and the environment. The Constitution enables them to do so in some ways whereas the most frequently used are: 1) by filing a public interest claim to protect the collective right to a clean environment (*Acción Popular*); 2) by filing a compliance action to ensure that laws are upheld (*Acción de Cumplimiento*); and c) by filing a suit requesting injunctive relief to prevent the violation of fundamental rights (*Acción de Tutela*), including the right to a clean environment where environmental degradation threatens human health (Londoño, 2008; Sánchez-Triana *et al.*, 2007; Guhl *et al.*, 2007). Aside from these mechanisms of participation that depend on access to courts, the Constitution states that an essential purpose of government is to facilitate such participation by providing sufficient understanding about environmental challenges. The adoption of statutes guaranteeing community participation has been specifically required in decisions that affect the environment through the *right to petition* public authorities, *public hearings, open meetings, referendums* and *standard participation* in elections. In the case of the *Contraloría,* it requires the adoption of laws that create systems to allow citizens to monitor public fiscal management at all levels (Table 4-5).

Table 4-5 Mechanisms for public participation

Kind of mechanism	Instrument	Implementation
To obtain informative approval	Right to petition	Disrespected
	Environmental newsletter	Web pages lack complete information
To discuss environmental policies and production processes	Popular initiative	Only example was the Water Referendum. Not popular initiative
	Participation on directive boards of CARs	It is manipulated by big interests
	Audiences in the Congress	Still lacks more effectiveness
Political participation	Referendum	Almost non-existent
	Citizen oversight	They exist but supported by the control bodies. They lack independence
	Open meeting	Not used
To support administrative decision-making	Consensus building / Networking for social building / Constitutional tools to protect the environment	The right to petition is the most frequently used but underestimated by officials. It is not meant yet for individuals, which could be very useful.
	Environmental administrative intervention	Ignored
	Environmental public audiences	Seldom used. Institutions do not like them

	Consultation with Indigenous and Afro-Colombian communities	The territorial rights are disregarded
	(Accion de Tutela) Written for the protection of constitutional rights	Very frequently used from 1992-1998. Less used because of the *Accion Popular*
	(Accion Popular) Popular Action	Most used for water protection
	Accion de cumplimiento Civil Compliance Action	Limited by public expenditure. Too formal
To participate on judicial actions	*(Accion de Inconstitucionalidad)* Action against the inconstitutionality of laws	Not known. Violations to the Constitution are frequent and should be useful
	(Accion de nulidad) Nullity action	Limited
	(Accion penal) Criminal proceedings	Not used
	(Accion de responsabilidad civil) Civil responsibility action	Not used

Source: Londoño (2008: 534)

In general, it has been acknowledged that a number of requirements to promote public participation in environmental management and more so in the water sector, are still absent in Colombia (Londoño, 2008). In practice, there is evidence regarding the effectiveness of access to both justice and impartial administrative, and alternative settings for the resolution of conflicts — affordable and timely legal services, and active education by government on the participation and environmental rights – (Access Initiative, 2009/2008/2005; Londoño, 2008; Merlano and Negret, 2006). The problem is that, for the most part, the Constitution does not specify with any precision how these mechanisms are to be implemented (Interview 5, 7 in Table 3-1). In 2001 it became evident that relatively few Colombians were aware of the right to a healthy environment (Sánchez-Triana *et al.*, 2007). Apparent support for public participation is nevertheless found in the literature. Former CAR Cundinamarca's director states, for example, that IWRM should be a priority to be developed as a result of consensus with all the stakeholders in order to ensure legitimacy, long-term support and avoid coercive instruments (Alvarez, 2003). The mechanism that has been considered the most positive by people is the *Acción de Tutela*, but only when it has to do with the Constitutional Court and not through the normal jurisdictional channels (Sánchez-Triana *et al.*, 2007 citing Seligson, 2001). In 2006, 5008 *Acción de Tutela* were presented by people with respect to the environment; almost 20% of them were related to water and 12% to territorial development. The *Derechos de petición*, potentially a useful instrument for people to be heard, has been despised by officials (Londoño, 2008).

In general, people distrust participation (see 4.2.3). In a poll on this matter conducted in 2003 on 2031 interviewees in five municipalities including the three biggest cities in Colombia, the perception was that politicians were not interested in stimulating participatory processes (47.1%), that participative approaches were not understood (24.6%), and that there was no budget allocation for such methodologies (24.6%). When asked about the main barriers to participation, people considered that 68% was due to lack of information, 31% to lack of political will, and 24.4 % to political conflicts. People perceived nevertheless that participation had the potential to solve their own problems (31.9%), conflicts (12.7%) or to simply serve their own community (33.6%) (Velásquez and González, 2003).

4.3.6   Budget and SINA's finances

The total budget during 1995-2003 for operation and investment by MAVDT (former ministry including housing and territorial development), the CARs, the Unit for National Parks, and IDEAM was estimated at USD 1.98 billion or an average of USD 220 million per year (est 2003). This budget expenditure represented 0.31% of Colombian GDP. During the same time period this expenditure was exceeded by Mexico, Costa Rica and Chile (Rodriguez, 2008; Sánchez-Triana *et al.*, 2007).

The CARs accounted for 81% of the total estimated environmental expenditures; the MAVDT 10% (including some allocations to CARs), IDEAM 5%, the Unit for National Parks 2%, and four research institutions accounted for the remaining 2%. During that period, 71% of SINA funding for operational expenditures and 85% of its funding for investments was allocated to the CARs (Sánchez-Triana *et al.*, 2007).

Some 78% of funding for operational expenditures and 86% of funding for investments was self-generated by the CARs from national allocations. The CARs derive 35% on revenues from property taxes, 34% on profits on investments, 10% on electricity taxes, 10% on inter-institutional agreements and sale of goods and services combined, and less than 2% on effluent fees. It is worth mentioning that out of the property taxes, Article 111 of Law 99 has determined that 1% should go to water protection projects. Former Minister of Environment Rodríguez (2008) considers that all authorities have neglected this 1%. The CARs environmental investments in the period 1995-2003 accounted for 57.4% on water bodies' conservation, conventional wastewater treatment and water supply, and only 2.8% to the promotion of CP (Sánchez-Triana *et al.*, 2007). The wastewater infrastructure being just one category represented 30% by itself with a very small number of projects, but a high average cost of USD 4 million in 2001. Leaving aside wastewater infrastructure, industrial pollution control, solid and hazardous waste, clean technologies accounted for only 5% of total funds and 12% of the total number of investment projects in 2001 and also in 2009 (CAR, 2009; Sánchez-Triana *et al.* 2007 citing Canal, 2004)).

There are significant differences among CARs as 82% of the resources are concentrated in only 8 of the 33 CARs. The CARs' ability to generate revenues relies critically on the size of their populations and even though mechanisms exist to even out differences among CARs, the criteria for assigning funds to them by the national government are unclear (interview 7 in Table 3-1). Despite large budget allocations, some of the largest CARs receive further contributions from the national government. The Environmental Compensation Fund, the National Environmental Fund, FONAM and the National Royalty Fund were meant to finance investments in poor CARs but have not been effective in this regard (personal communication, 2011-10/12[37]; Sánchez-Triana *et al.*, 2007).

Out of the national funds assigned to the research institutes, three quarters go to IDEAM. The research institutes have money-raising difficulty for operational expenses because international funding sources often place severe restrictions on financing resources for that purpose. It has been identified that as CARs also face this problem, operational expenses are veiled under certain supposedly investment projects (Sánchez-Triana *et al.*, 2007).

---

[37] Comment made by director of Inderena at the Seminar on Floods causing emergency in October 2011 in Bogotá

4.3.7   Assessment of water policy and law

The normative spread in the water domain generates problems when it comes to applying the rules as these are not of recent origin coming from the Civil Code and have been partially modified (Guhl *et al.*, 2007). The law proposal from 2006 basically addressed issues related to the allocation of water rights, to clarifying the roles of agencies, to the promotion of public participation in water resources management, to the establishment of a users' registry, to regulations on management of water runoff and urban drainage, to the control of hazardous wastewater discharges, to defining the ecological flow rate, and to redesigning economic instruments, such as water pollution fees (Gutierrez, 2006; Gutierrez, 2005). New regulation is not targeting not even in 2010 (personal communication 2010-12/17)[38], non-point pollution or setting phased approaches and implementations that take into account the specific characteristics of the regions and the high level of informality. Subsequent norms have worked on the establishment of a users' registry, to regulations on management of water runoff and urban drainage, to the control of hazardous wastewater discharges, and to defining the ecological flow rate.

The Colombian water policy and law do not usually present specific criteria supporting enforcement and leaves it up to government officials to have their own interpretations. In such a state of affairs, the most vulnerable people are prone to be abused. Some examples are:

- According to Decree 1541 of 1978, water allocation should be undertaken according to the following order of priority: Human consumption, preservation of flora and fauna, agriculture, animal husbandry, recreation, industry and transportation. The Decree did not address the practical implications posed by setting priorities whereas the economic value of water, social preferences, and water allocation priorities vary enormously from one region to the other in such a diversified country (Carrizosa, 2008; Sánchez-Triana *et al.*, 2007). The Decree did not specify how to use the order of priorities and how to resolve conflicts. As a result, often the most powerful local stakeholders drive local water policy and the small are left with limited opportunity for administrative or legal appeal. This Decree also established unwieldy procedures for obtaining water concessions. In fact, the process can take several years (personal communication, 2012-08[39]; Sánchez-Triana *et al.*, 2007).

- The charge fees face similar problems. The CNRN and subsequent decrees did not specify who and how to charge. Quantitative studies of the social value of water were not taken into consideration and new fees were set so low that they represent only between 1.5 and 6.5% of the marginal contribution of water to the production of potatoes and peas in the area of Bogotá for example. Besides, the CARs charge water fees to just legal users. This fact challenges the role of CARs since it has been well known that usually 70% or more of the water users do not hold water permits (Sánchez-Triana *et al.*, 2007).

- Regarding the environmental licensing process, Decree 1729 did not encompass the degradation of bodies of non-project related activities.

---

[38] Comment made by Head of Senate at the SWITCH meeting on Participation in Water issues in Zaragoza Spain in December 2010
[39] Comment made by Minister of Environment at Universidad Piloto

On the one hand, most policies do not address the country's main problems. The pollution policies do not rule on toxic substances, pathogens and most hazardous wastes; policies do not rule on non-point discharges or even consider groundwater and seawater; policies on wastewater management set discharge standards extremely high and beyond the country's reach, and result in a strategy of non-action that has caused significant water degradation and health impacts in the country; no specific policies exist that address the so-called new wastewater management approach based on prevention and decentralised treatment technologies in spite of the new policy on IWRM (Sánchez-Triana *et al.,* 2007).

On the other hand, other needed policies take a long time to show a real input because of implementation weaknesses. For example, Decree 1200 of 2004 on environmental indicators relies heavily on the improvement and standardization of the environmental information system that is in progress. The IDEA, or Institute for Environmental Studies from the National University in Colombia, has worked on setting those environmental indicators. It acknowledges that municipalities are not undertaking follow-ups in order to update these indicators on a continuous basis (Personal communication, 2010-10/14[40]).

Regarding voluntary agreements aiming at CP implementation, the policy needs to improve on issues addressed in section 4.5 and has been long thought to be a way to evade the law (Rodriguez, 2008).

With respect to public participation, the Constitutional framework states that the government has the duty to facilitate public participation by providing reliable information and instruments to assure that people have a voice in the decisions that affect them. None of these provisions are currently working effectively. The state has developed instruments to promote public participation, but it is not clear how to promote and support public participation and how to face social conflicts (personal communication, 2012-09/02)[41]. It has been acknowledged that in the best of cases, these are used simply to comply with formality, but there is no real opportunity for people's comments to be taken into consideration (Londoño, 2008; Blackman *et al.*, 2007). In general, Londoño (2008) argues that powerful business sectors evade real participatory processes, that participation is used for political manipulative purposes, and that the government hardly recognizes the voices from minorities such as ethnic groups or NGOs. Traditionally, the state has only two approaches towards protecting water resources: planning and designing policies, and enforcing judicial actions (Amaya, 2003).

With respect to planning instruments, they were expected to provide a possible mechanism for establishing priorities. The reality reveals a different state of affairs. The POMCAS were intended to force authorities to set the planning at the watershed level but it is not clear how to avoid conflicts whenever territorial plans have been set previously to POMCAs since the POMCAs were set from a decree (1729) that in theory, cannot jump over the mandates from the Constitution (Alvarez, 2003). Although CARs are required by law to create POMCAs, available information suggests that only a few watersheds have them and that there is no monitoring and enforcement of those plans (Sánchez-Triana *et al.*, 2007). Since it has been acknowledged that a sound water information system is far from being attained, the water planning exercise turns out to be a simple diagnosis of water availability and of the water status (ENA, 2010; Guhl *et al.*, 2007; Sánchez-Triana *et al.*, 2007; Alvarez, 2003). The only one that is considered to work accurately is the one related to the hydrometeorology system (Guhl

---

[40] Comment made by professor Burgos at IDEA institute in October 14 2010.
[41] Comment made by Minister of Environment at a meeting in Villapinzón on Spatial planning September 2 2012.

et al., 2007). Besides the above, since the planning is done sector-by-sector, implementing IWRM becomes a difficult task (Sánchez-Triana *et al.*, 2007; Guhl *et al.*, 2007).

With respect to the organizational framework, various reports have shown that coordination between the institutes and other SINA entities is far from optimal (El Espectador, 2011; El Tiempo, 2010; Contraloría General de la República, 2003):

- One reason for that lack of coordination is the fact that the institutes tend to specialize in research that is academic and not especially relevant to policy making. It is worth mentioning that none of these institutes focuses on industrial pollution control and human health – even though more than 70% of the population lives in urban areas and is exposed to significant health risks (Sánchez-Triana *et al.*, 2007).

- The situation is especially urgent regarding IDEAM as it is not receptive to requests for specific data, partly because it has no resources dedicated to facilitating coordination and seems to place priority on research over that of data collection. In a country with such high environmental degradation and lacking accurate monitoring, IDEAM could be responsible for certifying that data forthcoming from the CARs are accurate (Sánchez-Triana *et al.*, 2007).

- CARs are also responsible for the poor coordination prevalent among environmental entities. They have been slow in developing a consistent set of environmental indicators; maybe because those indicators could be used later on to evaluate their performance by the control entities themselves (Sánchez-Triana *et al.*, 2007). In order to overcome this problem, the government issued Decree 1200 of 2004 to establish the framework for environmental indicators.

Apart from the poor coordination between the above organisms, problems exist at the level of the National Environmental Council, which does not appear to play its intended role and has become simply a formality (Sánchez-Triana *et al.*, 2007); and at the level of the National Technical Advisory Council which although it includes representatives from universities and the private sector, seems to be dominated by the large private sector (Sánchez-Triana *et al.*, 2007).

With respect to SINA's finances and budget allocation, funding varies greatly from region to region and relies on the large private sector or/and on foreign aid. Traditional approaches to water management are still clearly prioritized. It is then worth recalling that during the beginning of this century, the CARs' environmental investments accounted for 57.4% on water bodies' conservation, wastewater treatment and water supply, and only 2.8% for the promotion of CP for example (CAR, 2009; Sánchez-Triana *et al.*, 2007). The operational expenses of CARs are veiled and are potentially a source of corruption findings in the near future as the Santos' government has announced on-going investigations in this regard (El Tiempo, 2010; Semana, 2010; El Tiempo, 2008).

Finally, with respect to the control bodies, the *Procuraduría* with its nine lawyers and three technical staff members is seriously limited in its human and technical capacity. Each year, it chooses a specific area on which to focus in order to be more effective (Sánchez-Triana *et al.*, 2007). In 2012, its institutional weakness has been sensed as its director has been running for his reelection and personal conflicts of interests have been denounced (El Espectador, 2012). Due to lack of appropriate environmental indicators, the control offices use administrative indicators of performance rather than indicators based on environmental quality improvement (Sánchez-Triana *et al.*, 2007).

4.4 National SME policy

4.4.1 Introduction

This section presents the history of SME[42] policy and law (4.4.2), current SME policy and law (4.4.3), organization of SME policy and law (4.4.4), instruments for SME policy and law (4.4.5), and assessment for SME policy and law (4.4.6).

4.4.2 History of SME policy and law

The history of SME policy in Colombia runs parallel to the historical approach to informality in Latin America.

- Even though bodies dealing with SMEs exist since the 50's, it was not before 1988 that formal policies and laws were adopted. Until the mid 80s, the informal sector was viewed as an expression of economic marginality that was bound to lead to unfair competition, and that should be eliminated (Caro and Pinto, 2007).

- In the decade of the 90s, statements from economists like Hernando De Soto in Peru delved into the phenomenon of informality. Initiatives were developed in order to lessen the huge requirements and paper work SMEs were faced with. The problem of the transaction costs was nevertheless not confronted (Caro and Pinto, 2007).

- Over the past ten years, more generalized policies on '*ventanillas únicas*[43]', on simplified tax models, micro-financing programs, improved information technologies in public service organizations and policies on entrepreneurship have been developed but have not been sufficient (Caro and Pinto, 2007).

- New paradigms towards a more positive and less repressive approach regarding SMEs in Colombia started with the replacement of law 78 of 1988 with law 590 of year 2000 or the so called first SME law. There, for the first time, it was stated that there should be a clear difference between informal and illegal activities.

In Colombia, different degrees of economic activities have co-existed in terms of their relationship with the institutional and legal framework. There is legal economic activity in full compliance with the law; a formal economic activity with traits of avoidance and evasion; a twofold activity that stands for compliance with the law on the one hand, while on the other hand, fulfilling the law only partially or not at all; an activity that has no regulatory framework; an activity where the law is not clear and is open to interpretations and illegal or underground activity typically involving delinquent and criminal elements of society. Experts consider that the denial of this quandary involving the presence of the different economic activities ends up supporting powerful groups interested in preserving a *statu quo* that supports exclusion, corrodes the social fabric and causes technological and legal segregation (Caro and Pinto, 2007). Recognizing the damaging results of this denial may lead to opportunities to understand the phenomena and be able to design the right initiatives towards more effective policies to address it.

---

[42] Since there are no specific policies for micro and small enterprises, this Chapter presents the traditionally targeted policy on small and medium enterprises.

[43] Could be translated as unique "windows" or entities created to simplify the registry of new enterprises.

4.4.3   Current SME policy and law

The policy

At present, the policy is framed first by the *Agenda Interna* or Internal Agenda, a proposal of 10 strategies for the Micro, Small and Medium Sized Enterprises that was presented in May 2006. This document basically focuses on developing an information system on SMEs; on stimulating a comprehensive policy towards exports based on association building; on concentrating policy on strategic sectors such as the industrial process of leather and its sub-products; on working on developing a banking system with entities such as Fondo Nacional de Garantías, Findeter, Bancoldex, Finagro and Fonade; on developing performance indicators; and on offering programs for the implementation of Cleaner Production processes and techniques. The agenda was then included in the Plan Nacional de Desarrollo, PND 2006-2010 and became a central part of the National Policy CONPES 3484 on SMEs, defined by the DNP. It is worth pointing out that this initiative was also incorporated in the strategic plan of the Ministry of Commerce, Industry and Tourism (MCIT).

The national system for SMEs was created to support the policy and is described in section 4.4.4. The policy is integrated through two components: A financial and a non-financial one. The financial one aims at increasing the capital of the SMEs and the non-financial one on empowerment and capacity building. From 2006-2010, the budget of the financial component was: 150,000 million Colombian pesos (COP), USD 88 million from the IFI (Industrial Promotion Institute) for 100 days, and from the commercial banking system for micro-credits - 300,000 million COP ($176 million US dollars[44] per year).

A linkage between the MCIT agenda and the MADS agenda does exist but evidence shows that in the past three years, the ministries did not meet for the above purpose (mc2 in Table 3-2). The agenda is expected to aim at sustainable development with special emphasis on SMEs. It was established in 2005 and its specific actions are:

1. To identify the strategic sectors. Among them, leather was included;

2. To develop green markets;

3. To expand CP with agreements, self-management, regional empowerment;

4. To work on sectors affected by climate change; and

5. To stimulate the creation of clusters with special emphasis on association building, and on the recycling of sub-products as environmental priorities.

The law

The current SME law is law 905 from 2004 that replaced law 590 from the year 2000. It basically developed a set of incentives for enterprises committed to employment policies. It offers the

---

[44] Exchange rate of 1700 colombian pesos- USD 1 in June 2008.

possibility for chambers of commerce to offer credit to SMEs. It establishes a major emphasis towards differentiating between microenterprises and the so-called small and medium sized ones.

4.4.4 Organization of SME policy and law

The organisational framework for SMEs is basically operated by the Ministry of Commerce, Industry and Tourism (MCIT) that formulates, adopts, leads and coordinates the general policies on economic and social development. These policies are inspired by the principles of competitiveness, integration and development of the different sectors from the industry, micro, small and medium sized enterprises, the promotion of foreign investment, the trade in goods and services, and technology and tourism. For 2010, the aim was to reach an important impact on the level of economic growth by doubling exports, multiplying by three the number of international tourists and reducing by 50% the level of economic informality (MCIT, 2008). The Ministry is seen as the leader with respect to competitiveness, sustainability and a major added value with respect to the manufacturing processes.

In relation to SMEs, the ministry has two consulting bodies: The superior council for the micro enterprise and the superior council for the small and medium enterprises. Among the vice-ministries, the one in charge of the SMEs is the vice-ministry of Enterprise Development (Figure 4-10).

Figure 4-10 Organizational framework of SME policy and law

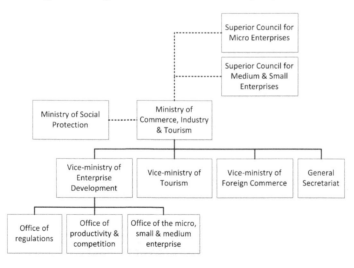

Source: MCIT, 2012

Entities assigned or linked to the Ministry are the ones responsible for executing the policies (MCIT, 2008). Those entities can be either public or private or both.

The *Servicio Nacional de Aprendizaje*, National Learning Service, SENA is a body linked to the Ministry of Social Protection but working closely also with the MCIT. It is a public body dedicated to offering free capacity building and training on intermediate careers. It has an autonomous financing system based on the tax contributions from businesses.

Colciencias is the Institute for the Development of Science and Technology. It is also a public body at a national level which was assigned to the DNP and that has been given the status of a ministry. Both SENA and Colciencias had signed cooperation agreements on applied research proposed in the National Development Plan since 2003.

In relation to SMEs, ACOPI, created in 1951, is basically the private body that groups the formal microenterprises, and the SMEs from the different production sectors (ACOPI, 2007). It offers its affiliates (6000 in 2007), representation towards public opinion, private organizations, government authorities, and international organizations as well as a big array of services such as programs on Cleaner Production. ACOPI can sign inter-institutional agreements with public institutions. The chambers of commerce deal with SMEs as well as the big enterprises. They offer the CAEs, centres for enterprise attention, which deal with the registry process for the microenterprises (Santamaria and Rozo, 2009; MCIT, 2008). All the different guilds (like ACOPI) come together at the *'Consejo Gremial Nacional'*, the National Sector Council, where 16 macro guilds represent a total of 36 production sectors.

All industrial sectors gather at the ANDI - the National Industrial Association. In terms of the tannery and leather sector, the ANDI has a leather office that is focused mainly on the medium and big sized industries. Two additional bodies represent the leather industrial interests: Colombian Association of Industries of shoes, leather and manufactures (ACICAM) - a political body and Technological Centre for the industries of shoes, leather and related (CEINNOVA) - a body dedicated to providing technological assistance on the quality and design of leather products.

A set of financing bodies was also defined. Bancoldex, known as the development bank for enterprises and external commerce, operates as a second floor bank and is supported by the Fondo Nacional de Garantías, or National Guarantee Fund. These organizations are meant to work mainly for SMEs and both are assigned to the MCIT. Other second floor banks like Findeter work for territorial development, Finagro mainly for the farmers, and Fonade for projects with big impacts outlined in the National Development Plan (Stephanou and Rodriguez, 2008).

4.4.5   Instruments of SME policy and law

The instruments for the SME policy and law are basically of an economic and persuasive nature. They target tax reductions and exemptions under the creation of special regimes; and towards organising municipal committees for microenterprises, which promote sectoral and local policies based on association building and productive chains. The municipal committees rely on the mayors' leadership; mayors are expected to buy government supplies from the McSEs.

The instruments privilege enterprises created after 2000. Those enterprises are exempted from paying the *parafiscales* or non-wage costs[45] by 75% in the first year, 50% in the second year, and 25% in the third year (Mipymes, 2012; WB, 2010).

With respect to financing, support exists through the following programs: Young entrepreneurs-Bancoldex, Multipurpose- Bancoldex, Bogotá- Bancoldex, Bancoldex- Colciencias, and Microcredit.

---

[45] They represent to the employer 4% to supporting abandoned youth, 2% to oral health, and 2% to the technical institute SENA.

All of the latter are basically targeting newly created enterprises with no negative credit history and entail cumbersome formal requirements (Mipymes, 2012). Even though the Chambers of Commerce can offer special microcredit, the credits are targeted to very small financing that are not always sufficient for the microenterprise (personal communication, 2012-06/04[46]).

Lately, in the commercial area, programs focusing on technological development for microenterprises have been created and are supported by microenterprises' incubator companies that can be of a private, official and/or academic nature. The chambers of commerce have created the CAEs, which constitute centres for enterprise attention and are meant to advise new entrepreneurs on business registry requirements (Santamaria and Rozo, 2009).

With respect to the participatory instruments, the judicial one like *Acción de Tutela* or the community participatory one like *the Derecho de petición* are the most commonly used. As stated in section 4.3.7 the reality is that the powerful interests who do not always share the common people's interests are the ones benefitting from these instruments.

4.4.6   Assessment of SME policy and law

The SME policy in Colombia needs to evolve towards more effective strategies. Countries like China and India have pushed over 450 million people above the poverty line. Understanding the phenomenon of informality and its transitional nature has allowed China to improve incomes for millions of people. A progressive approach to informality in India has meant significant improvements in the fight against poverty (Caro and Pinto, 2007).

In Colombia there still persists a lack of awareness, real participation and efficiency in terms of the design and implementation of policies to address informality. Enterprises find measures oppressive and overregulated. There is a lack of coherence in the regulatory frameworks among the national, regional and local settings; there is a lack of consulting bodies to solve contradictions and problems over the transition to formality. There is a lack of a long-term policy approach (Caro and Pinto, 2007).

There are still many barriers towards achieving economic formality for SMEs particularly for micro and small enterprises (McSEs). There are many inappropriate strategies of public entities struggling against informality, which deepen inequity, exclusion and in some cases obstruct the productive activities of vast regions (WB, 2010; Caro and Pinto, 2007).

Experts have proposed the implementation of the following actions (Caro and Pinto, 2007):

1) Lowering the transaction costs to stimulate access and permanence in the formal world: An example would be to use the mercantile registry in a progressive form as a unique requirement, solidarity programs among chambers of commerce, and establishing a set of benefits such as training on and funding of CP;

2) Favouring the transition to formality: information system through association building practices, and planning based on regional and local priorities;

3) Establishing a system of opportunity banking using also non-financial institutions; and

---

[46] Comment from researcher at DNP at an informal meeting in June, 2012

4) Establishing a digital agenda.

The effort cannot just be restricted to promoting firm formalization but also geared towards enhancing worker welfare through stronger human capital and by strengthening government institutions (WB, 2010).

Since Perry *et al.* (2007) has proposed that informality is a function of both, exclusion and exit, for those being voluntarily informal, the CAEs could basically lead not only to registry but to counselling for growth in order to join formality and to lessen bankruptcy costs. Microenterprises could well be seen as a natural business stage towards larger firms in that scenario (Santamaria and Rozo, 2009). The challenge remains within the excluded, the most vulnerable communities that are low skilled people, that are rarely entrepreneurs and that entail higher comprehensive and complex solutions (Mel *et al.*, 2008). For those groups, Colombian labour market rigidities related to the minimum wage and the non-wage costs are excessive, and the transition to formality represents an overwhelming national issue (Mondragón-Velez *et al.*, 2010).

4.5   National Cleaner Production Policy

4.5.1   Introduction

This section presents the history of CP policy and law (4.5.2), the current CP policy and law (4.5.3), the organization of CP policy and law (4.5.4), the instruments of CP policy and law (4.5.5), and the assessment of CP policy and law (4.5.6).

4.5.2   History of CP policy and law

Blackman *et al.* (2007) consider that voluntary regulation on CP has been often used in Colombia since the passage of Law 99 from 1993. It was inspired by international practices from countries such as the Netherlands rooted in a strong enforcement of command-and-control regulations. CP was included for the first time in the objectives of the National Development Plan (PND) from 1994-1998. In 1993, the Colombian Business Council for Sustainable Development (CECODES) was established. In 1995, the first voluntary agreements were drawn, and in 1997 the National CP policy was created. In 1998, the National Centre for Cleaner Production was set up. In 2004, the National Council on Tanneries was defined (Van Hoof and Herrera, 2007).

Despite the institutional efforts presented above, at the beginning of the   90s in Colombia, the regulation based on end-of-pipe solutions did not stimulate technological innovation at the industrial level as prevention solutions would have. The work was focused more on formulating environmental norms than on determining the institutional capacity that would ensure efficiency upon its implementation (MAVDT, 1997a).

4.5.3   Current CP policy and law

The current CP policy from 1997 is based on four principles:

1) To work in a comprehensive manner,

2) To build consensus with the stakeholders,

3) To internalize environmental costs into the normal operating costs, and

4) To recognize the importance of the phased approaches in order to reach compliance with the law at the end of pollution abatement processes according to the specificities of the local settings.

4.5.4   Organization of CP policy and law

The MADS (MAVDT before year 2012) is responsible for designing and defining the CP policy. Entities belonging to SINA such as the municipalities, CARs and NGOs execute the policy. They include private bodies such as CECODES, The National Centre of CP, and the different Chambers of Commerce support the industries' CP implementation through the CARs' supervision. Joint agendas among ministries were also thought to supervise the implementation of CP (Figure 4-11).

Figure 4-11 Organizational framework of CP policy and law.

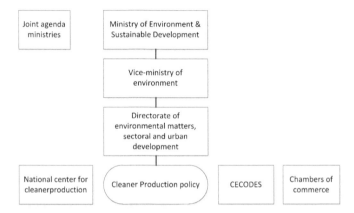

4.5.5   Instruments of CP policy and law

In Colombia, CP is seen through the lens of persuasive instruments. Two types of voluntary regulations are commonly applied: 1) CP agreements, and 2) the issuance of environmental guidelines.

1. The purpose of the CP agreements is to motivate towards compliance of the law by "building consensus" among polluters on the need for it and giving them guidance on how to comply (Blackman *et al.*, 2007b). Voluntary agreements aim at capacity building and to sidestep institutional weakness than for environmental management (Blackman *et al.*, 2012).

2. The guidelines fill a gap in Colombian regulation, which is the lack of technical guidance on how emission standards are to be met. They clarify the process of obtaining licenses to sell products. They also facilitate the international certification for the international marketplace.

Regulatory instruments such as signing *Resoluciones de obligaciones* - Resolutions of Obligations – can be used to tie-up CP to command-and-control by CARs.

Economic and financial instruments such as tax exemptions and a credit line from the Institute for Industrial Promotion also exist but are restricted to formal enterprises (personal communication, 2007-03[47]).

Participatory instruments are based on the fact that CP implementation needs building consensus among common interests. A variety of methods exists for this implementation but is hardly prioritized by CARs or by NGOs (mc1 in Table 3-2).

4.5.6   Assessment of CP policy and law

Isolated successful results are reported on CP implementation in large industries in Colombia in sectors such as: agro-industries (flowers, sugar cane, coffee and pigswill), mining, hotels & restaurants, or in industrial corridors.

In the industrial corridor close to the second city in Colombia, Medellín, water consumption reduced by approximately 25% between 2000 and 2002. Despite the potential benefits of these results, data suggest that overall industries have not been successful in improving their environmental performance (Blackman *et al.*, 2012; Blackman *et al.*, 2007a; Van Hoof and Herrera, 2007):

1) During the grace period that was given to polluters it has been observed that they do not make any significant improvement.
    - Regulators have no means of assessing environmental performance because those agreements do not include indicators or establish a baseline,
    - The agreements did not identify sources of financing for costly pollution abatements and prevention investments, and
    - The legal status of the agreements was unclear.

2) The guidelines have been found
    - To not link up with existing command-and-control regulations,
    - To provide limited technological alternatives for pollution prevention and control, and
    - To be appropriate for large firms, but not for the McSEs that dominate many sectors (Blackman *et al.*, 2007a).

In general, it has been identified that the initial analysis in any CP implementation process lacks appropriate institutional capacity analysis and deep knowledge regarding the strengths and weaknesses of the target community or enterprise (Sánchez-Triana *et al.*, 2007).

The CP policy does not target SMEs/McSEs. SMEs/McSEs are simply mentioned with respect to the fact that it is compulsory that 20% of a credit line from the Inter-American Development Bank be used to educate McSEs on CP. Experts consider it essential to emphasize the need to reach SMEs/McSEs in an effective manner; to encompass environmental management and enterprise development (Van Hoof and Herrera, 2007).

---

[47] Comments made by tanners in Cerrito, in Villapinzón and by the Head of CP in Cerrito March,2007

4.6   Inferences

Having presented the concern on the McSEs in Colombia in terms of the general and environmental national context, a SWOT analysis can be undertaken in order to support the understanding of the national framework and the construction of pathways towards the elaboration of the local tannery case studies.

A SWOT analysis on McSEs

Overall, McSEs have more weaknesses and threats than strengths and opportunities. Their weaknesses are related to their own specificities: causing substantial environmental impact, having low education, no access to technology and /or credits, and lack of monitoring systems. Since they have been living through a culture of survival, no forecasting exists and they are informal. Turning such weaknesses into strengths asks for internal changes that are difficult to lead. Their external world offers overwhelming threats for them and beyond their possibilities to interfere in their traditional settings. However, the literature emphasizes that they can potentially innovate and adapt to new markets; work through clusters provided they get appropriate support on integral capacity building programs based on CP implementation. The State has to adopt a new vision towards McSEs and through political will, transform their threats into opportunities: creating specific policies based on inclusion for CP and for competitiveness, supporting prevention and affordable technologies, supporting IWRM, improving control agencies and instruments, being flexible on phased approaches and access to credit, supporting real participation and related indicators and creating accurate sustainable indicators for performance.

All the above does not seem possible without agents of change that can catalyse such endeavours.

Table 4-6 A SWOT analysis on McSEs in developing countries (Colombia)

	Strengths	*Weaknesses*
Internal	Small scale enterprises have potential to adapt to new market trends (Blackman, 2011; UNCTAD, 2008; UNIDO, 2009; 2005; Ashton et al., 2002) Potential to innovate (Blackman, 2011; UNCTAD, 2008; UNIDO, 2009; 2005; Ashton et al., 2002) Existence of clusters (Shi *et al.*, 2010)	Cause substantial env impact - 70% industrial pollution (Soni, 2006, Le van Khoa, 2006). Low education, no monitoring systems, no access to financing and technology, culture of survival (no forecasting), informal, narrow markets, low level of association building (DNP, 2007) Difficulties in building symbiotic relationships, information sharing and awareness, low financial benefits, poor organizational structure and legal and regulatory frameworks (Sakr *et al.*, 2011: 1168) Distrust is high (Rodriguez, 2003)
	Opportunities	*Threats*
External	Raising consciousness on water crisis. IWRM being implemented (MAVDT, 2010) CP can potentially handle harmful effects; existence of phased approaches; and liability agreements (Sanchez-Triana *et al.*, 2007; Van Hoof and Herrera, 2007). Customer-oriented policies, collective, cumulative improvements may be possible (Sverisson and Van Dijk, 2000) Role on the fight against poverty (Echeverry *et al.*, 2011; Caro and Pinto, 2007; Rocha, 2007; ACOPI, 2004; Ocampo, 2003; Stiglitz, 2003) National government prioritizes land issues for excluded people in social conflicts areas (Law on land restitution, 2011). Creating networks with public sector may be possible (Walker and Preuss, 2008) Decentralization is reinforced (PND, 2010). Agendas between MADS and MCIT exist. A subdivision for micro enterprises exists at the vice-ministry of Enterprise Development. A council on McSEs exists at the Ministry of Industry (MCIT, 2008). The SME policy gives more importance to McSEs (MCIT, 2004). Public Private Alliances are created based on social responsibility (Law 1508, 2012)	Growing environmental concern (Soni, 2006). Water shortages in localities Inefficient command-and-control (Alvarez, 2003). Water use and policies privilege large industry (Sanchez-Triana *et al.*, 2007) Official organizations have low capacity building and exclusive end-of-pipe focus (Sanchez-Triana *et al.*, 2007). Policies towards formalization are not flexible and target only administrative issues. Wrong consultation for SMEs (Rodriguez, 2003) Lack of appropriate credit (Personal communication, 2011, Stephanou and Rodriguez, 2008; Rodriguez, 2003) Guilds not representing their interests (Personal communication, 2011; Rodriguez, 2003) Growing competitiveness at global level, resilient corruption at national and local levels (AnteaGroup, 2012) Regions and localities are weak on control and evaluation instruments. Punitive approaches are most commonly used for handling pollution (Amaya, 2003) Lack of culture of participation. Information is a privilege (Londoño, 2008, Sánchez-Triana *et al.,* 2007) Rural life conditions are the lowest (Hoyos, 2012) McSEs, socially excluded, no environmental policy address them (Mondragón-Vélez *et al.*, 2010; Rocha, 2007) Distrust is high (Rodriguez, 2003) Weak judicial systems and no alternative dispute resolution channels (Merlano and Negret, 2006; Herrero and Henderson, 2003) Public agencies cannot invest on McSEs because they are private (1991 Constitution). No CP policy on McSEs (MAVDT, 1997a).

4.7   Conclusions

Although at the national level, the Government is making an effort towards addressing inequality, effective policy compliance at regional and local levels is still far from being a reality, especially in the environmental sector.

Colombia has historically failed to understand that complexity is the main characteristic of McSEs. If Colombians take into account the local characteristics, needs and interests for policy design and implementations, and manage to prioritize the common good over the individual good, the country can eventually succeed in facing its main challenges.

Colombia has made significant efforts at the institutional level establishing a decentralized and unified environmental system, but is a long way from implementing policies that address those most vulnerable like industrial McSEs. The Colombian Water Agenda has to catch up with the following change in priorities:

-   Governmental support to IWRM and hence to pollution prevention technical options and policies besides traditional pollution control; and command-and-control policies that can be more appropriate for McSEs (Canal, 2004).

-   A new emphasis placed on the most vulnerable groups demanding new mechanisms for public participation that offer alternative venues for conflict resolution, and for the effective involvement of a broad range of stakeholders or effective financial policies for their specific needs (Londoño, 2008).

-   Improved monitoring, like the design of follow-up indicators on participatory processes, and the dissemination of information with regard to outcomes; and the assignment of accountability for environmental actions and outcomes (ENA, 2010; Londoño, 2008; Sánchez-Triana *et al.,* 2007).

5   Comparative case studies: Cerrito and San Benito

5.1   Introduction

This chapter deals with the second sub-research question, which is to identify the perceived mechanisms supporting CP and how they operate in the context of McSEs in Colombia. Its purpose is to present the two comparative case studies of the micro-tanneries of Cerrito and San Benito in Colombia in which the involvement of the researcher was relatively limited.

The structure of this chapter is as follows: Cerrito (5.2), San Benito (5.3), Cerrito and San Benito compared (5.4), and conclusions (5.5).

5.2   Cerrito

5.2.1   Introduction

Cerrito is a village of 14,962 inhabitants (DANE, 2005) one hour's drive from Cali, the third largest city in Colombia. The village residents sustain themselves to some extent from the activity of 21 tanneries, among which at least 16 are micro firms. For decades, the tannery industrial activity had caused a substantial collective impact on the Cerrito River, a tributary of the Cauca River, the second largest in Colombia.

This section presents the problem in Cerrito: Pre-case study (5.2.2), and outcome of the case study (5.2.3).

5.2.2   The problem in Cerrito: Pre-case study

Relevant socio-economic data

The whole metropolitan area of Cerrito has a population of 38100 inhabitants. Indigenous people that were reluctant to accept the Spanish settlers originally occupied the area of the Cerrito Village in the 17th century. It became afterwards the site for one of the major Afro-Colombian communities that were brought as slaves to work on the sugar cane fields at the nearby major sugar industries, Providencia and Manuelita (CVC, 2007).

The economy of the area is primarily based on the monoculture of sugar cane, which (a) offers low salaries compared to other activities in the country. In fact, 80% of the people of working age earn less than a minimum monthly income[48] (USD 205) and only 11% earn more than one minimum monthly income salary (CVC, 2007); and (b) cannot offer employment to a growing population between 20-40 years of age since it is also becoming less and less labor intensive as it goes through technological change.

Besides underemployment and unemployment, tensions with this growing population is also caused by the rising demand for housing facilities, and by the fact that land ownership is in the hands of few

---

[48] The official minimum monthly salary for year 2009 in Colombia is $ 497,000 Colombian pesos (Decreto # 4868-December 30 2008) (An equivalent of USD 205 -taking as per April 3 2009, the Representative Exchange Rate of $ 2413 pesos/dollar).
http://www.finanzasydinero.com/blog/analisis-dolar-en-colombia-abril-32009/

sugarcane landlords while 57 small properties of less than 1 ha represent 0.54% of the whole area; 2 properties alone represent 29.81% of the total basin area of 12,612 Ha (CVC, 2007). Within this scenario, the municipality of Cerrito offers limited educational opportunities; 35% of the population represents those between 5-24 years of age and does not receive an adequate education (CVC, 2007) or health care. Two major concerns are reported to the health care centers covering the area: (a) Intestinal infectious diseases due to unsustainable water practices and difficulties in accessing safe water, and (b) acute respiratory diseases caused by the cultural practice of *Pavesa* – burning of the sugar cane residues (CVC, 2007).

As a result, the Cerrito poverty indicator NBI (Unsatisfied Basic Needs, *Necesidades Básicas Insatisfechas)* of 18.7% is higher than the regional average for the Valle del Cauca, which is approximately 15.6%. (The NBI indicator is defined in terms of 4 parameters: (1) The quality of the house in terms of the number of people living in it; (2) The availability of convenient public services; (3) The high economic dependency on one member of the family; and (4) The number of young children not attending school). Contrary to what happens in the rest of the country, the NBI turns out to be higher in the urban centers than in the rural areas in the area of Cerrito (CVC, 2007).

Within a context of limited labor opportunities, the tanneries offer 380 direct and around 1200 indirect job opportunities in the area, distributed as 63.2% in the two medium sized industries and as 26.3% and 10.5% in the small and micro sized enterprises respectively (Jaramillo AD *et al.,* 2003).

Relevant geographical, environmental, technical and institutional data

The village of Cerrito is located to the east of the department of Valle del Cauca in Colombia on the western flank of the Central Mountain range (*Cordillera Central)* (Fig. 4.1). The village is located on the Cerrito River basin that has an area of some 126 km^2. The Cerrito River runs along 39 km from its source at 3250 meter above sea level (masl), flows into the Cauca River[49] at 935 masl and has a flow of 0.8m^3/s (CVC, 2007).

The main aquifer in the southwest of Colombia is found in the geographic valley of the Cauca River. The water reservoirs, which have been identified as three hydrogeological units, total 1600 million m^3. The annual recharge to these aquifers is in the order of 3500 million m^3. The latter entails a privileged flow of 110 m^3/s (*Ibid*).

The whole drainage area of the Cerrito River has an average rainfall of 1151 mm per year. During the months of April and October, the rainfall can reach 2000 mm per year. The temperature ranges from 24^0 to 10^0 Celsius. This drainage zone can be divided into three main sectors. (1) The high zone that goes from 3850 masl to 1285 masl, (2) the Honda zone from 1285 masl to around 935 masl where the Honda stream flows into the Cerrito river, and (3) the flat low zone that has literally no slope and constitutes the most fertile soil area where the Cerrito River flows into the Cauca River (*Ibid*).

In the flat zone area, 95.5% of the original wetlands and the dry wood ecosystem have been replaced by the sugar cane crops, which already account for 56% of the basin itself. As the demand for biofuels

---

[49] The Cauca River is the second main water artery of Colombia. It is born at the Colombian Massif in the south at 4200 masl and runs 1350 km parallel to the Pacific coast forming an intertectonic depression about 200 km long between the western and the central mountain ranges. It goes into the Magdalena River, the main river of Colombia. The Cauca River has a flow between 400-1000m^3/s depending on the season.

is rising, traditional agriculture like coffee and even livestock activities are being pushed aside. The urban center with its tanneries is located also in this third sector (*Ibid*).

Despite the natural availability of water, the drainage area of Cerrito experienced water shortage for human consumption mainly because water use has traditionally privileged the water demands from the sugarcane agro-industrial activities and because of water polluting practices. Two strategies were adopted in order to face those problems: (a) the Amaime River (tributary of Cerrito River) has been moved to the flat low zone of the Cerrito's river basin, and (b) a comprehensive strategy was been implemented on CP at the tanneries (Jaramillo and Vásquez, 2003).

The tanneries' impact

For decades, the tannery industrial activity caused a substantial collective impact by discharging effluents straight into the sewer, which discharged into the river without treatment. As of 2009 a treatment plant has been installed. The two largest industries still discharge directly into the Cerrito River without any treatment. These industrial effluents with high levels of BOD_5 and TSS (Total Suspended Solids), Chromium and Sulphates had harmed the sewer system itself and the water quality of the surface and underground waters.

Since the industrial raw waters were mixed with domestic waters, designing an appropriate wastewater treatment was very expensive. Since the beginning of the 1990s until the year 2002, the regional environmental authority CVC (*Corporación Autónoma Regional del Valle del Cauca*) ruling the area was trying to regulate the tannery discharges without success. The environmental authority had identified that the tanneries needed to implement CP to make their discharges suitable for the domestic wastewater collective system being designed by the regional university.

Besides the pollution problems, the spatial planning policy determined that there was need to relocate the tanneries out of the residential area and out of the 30m protection zone of the Cerrito River.

5.2.3   Outcome of the case study

History of activities

In spite of the difficult social and economic problems in Cerrito, the authorities decided that opportunities to overcome them could arise from stimulating economic activities derived from tourism, commerce or industry such as the tanneries.

The tanneries had shown a tendency to hire more people and to compensate for the layoffs caused by the sugarcane industry. Facing the challenge of regulating the tannery industrial activity could entail a win-win situation in the region because a healthy tannery activity could become the seed for prosperity in the area.

By 2003, the CVC decided to create an independent CP body (CRPML, *Centro Regional de Producción Más Limpia*) to support CP adoption in order to make the discharges suitable for a collective and affordable wastewater treatment system for the municipality (Jaramillo AD *et al.*, 2005; Jaramillo and Vásquez, 2003; mc4 in Annex 5.2).

Besides this new body, the CVC decided to innovate in terms of creating a policy, the *resolución* 00028 of 2004, for CP diffusion bound to command-and-control rules and not to voluntary agreements. The latter was very much against the general wisdom of CP theory that highlights its voluntary character (mc4 in Annex 5.2). This policy was meant basically to comply with water quality standards from the water and sewerage utility company for the proposed collective waste water treatment system and the discharge limits were set individually for each industry in terms of its maximum installed production capacity. Such limits were determined in terms of load and not in terms of concentrations in order to stimulate prevention and to reduce resource use and waste generation (CRPML, 2006).

The industries went into individual negotiations and signed mandatory policies, *resolución de obligaciones (resolution of obligaciones)*, with the authority to meet a timeline complying with the corresponding CP activities and technologies needed to be implemented. The CP body developed a participative project where the tanners were making their own decisions and receiving training along the process.

As a result of this effort, in 2005 the industries were willing to adopt the technological changes and all the tanners had attended the CP workshops (Jaramillo AD, 2005; mc4 in Annex 5.2).

A strategy aiming at CP implementation was being supported by CVC, the regional body dealing with the leather chain (*CDP del cuero*), the most reputable public university of the region, Universidad del Valle along with its water institute CINARA, and the newly created CP body (CRPML) (mc4 in Annex 5.2).

Training and education were not only focused on the tanners. Simultaneously, the population was also encouraged to participate in associations for environmental education called CIDEA (*Centro de Educación Ambiental*) supported jointly by the Ministry of Education and the Ministry of Environment. The population was entrusted to co-develop the Municipal Plans for Environmental Education, which were expected to be integrated into the Development Plan of the Municipality itself (CVC, 2007). The latter gave legitimacy to the environmental plans in the Municipality, and enhanced public participation and decision-making (mc4 in Annex 5.2).

With respect to the spatial planning issue on the riverbank, the CVC has recognized the need for the relocation of most of the tanneries either because they are located in the riverbank-protected area, or are in residential areas. Nevertheless, stressing the fact that the tanners had committed to CP adoption, and that relocation was very expensive, the CVC postponed spatial planning decisions (interviewsTable 3-4 and mc4 in Annex 5.2).

Despite all the efforts, in 2006, the micro-enterprises were not meeting the time lines. The owners of such industries asked for a new policy to replace the *resolución* 00028 in order to set new deadlines to comply with the CP implementation activities and the technological changes: *resolución* 000196 of 2006 resulted from this process (mc4 in Annex 5.2).

The quality of the river water was rated in 2006 as of *Very Bad Quality* by the quality indicator CETESB[50] used for human consumption at the last monitoring station before reaching the Cauca River and after passing the tanneries. With respect to the quality of the river water for irrigation purposes,

---

[50] Indicator built by the environmental authority of the Sao Paulo State in Brazil based on the National Sanitation Foundation and adapted for tropical rivers.

the quality indicator DINIUS[51] in 2006 had considered the water to be usable only after treatment. With respect to the oxygen content, the quality was at 0 mg/l (CVC, 2007).

In 2007, the quality of the river water had been considered to be of *Bad Quality* by the quality indicator CETESB[52] at the last monitoring station before reaching the Cauca River and after passing the tanneries and hence, showed a slight improvement from 2006 onwards. The same was true with respect to the quality for irrigation purposes by the quality indicator DINIUS[53] that qualified the water to be used for most of the crops. In terms of the oxygen content, the quality of the water remained at 0 mg/l (*Ibid*). It is important to point out that none of those indicators measure the heavy metals content.

Preliminary results suggested reductions of the total discharge loads of 47% on BOD_5, 64% on TSS and 42% on water use (CRPML, 2008; CRPML, 2007). An innovative solar energy installation to heat water was established at a tannery (Jaramillo AF *et al.*, 2006). The CP body, CRPML received the Green Apple prize in London for the effort in CP implementation by the industries from Cerrito (CRPML, 2007) and the tanners were contacting international markets for their products for the first time (La República, 2009).

As of 2009, the end-of-pipe water treatment system was about to start functioning and most of the tanners were ready to be connected to it. Only the smallest tanners were facing difficulties meeting the discharge limits because of limited financing opportunities (interview Rg in Table 3-4 and mc4 in Annex 5.2).

Impact of policies and policy procedures on tanners

The implementation of the above policy was brought about through sound academic support from the *Universidad del Valle* and its water institute CINARA and from the regional trade body dealing with the leather chain (*CDP del cuero*). The CP body-CRPML played the dual role of facilitator between the authority's command-and-control and the industries' interests, and additionally that of a capacity building office. The CP policy was tied to command-and-control and not to voluntary agreements. Its discharge limits were set on loads. CP implementation was understood as a process and the timeline was negotiated individually with CVC. Since the solutions were developed with the community, the policy was successfully set in accordance with the local needs (interview Rg in Table 3-4mc4 in Annex 5.2).

The perceived mechanisms for CP implementation

In 2004, at the *initial exploratory in-depth interviews* (aiming at responding to the which question, seeTable 3-3), the interviewees mentioned the importance (all scoring 2) of the following kinds of mechanisms supporting CP adoption: (a) a *regulatory framework* that recognizes their needs, (b) access to *technology*, (c) *economic incentives* and *financial support*, and (d) access to sound *information* on all related issues like legal and technical options (see Fig 5.1) and Annex 5.1.

In fact, for each mechanism:

---

[51] Indicator used in Mexico built by Dinius in 1987.

[52] Indicator developed by the environmental authority of the Sao Paulo State in Brazil based on the National Sanitation Foundation and adapted for tropical rivers.

[53] Indicator used in Mexico developed by Dinius in 1987.

- There was a common sense of pride and hope regarding their new policy based on measuring loads and on negotiating the timing for the CP implementation phases. People were anxious about the challenges that CP implementation would entail for them.

- Stakeholders pointed out the importance of targeting incentives, market opportunities, scale economies and reliable information that suit the needs of the McSEs.

The attitudinal and organizational factors mentioned in the theoretical background of this dissertation (section 2.6.8) were cited in Cerrito only by the head of the CP body. She emphasised a *high degree of participative* and *trial-and-error oriented process* (score: 1) (

Figure **5-1**) and Annex 5.1.

Figure 5-1 Mechanisms supporting CP in Cerrito in 2004

From the years 2005 to 2008, the program went through difficult times. Tense relationships between the tanners, the authorities and the head of the CP program were observed at the national leather board committees. The CP project was on the verge of being cancelled and disputes were taking place (interviewees Nt, HdCP, Tech, Rg, Lc, and tanners in Table 3-4).

The authorities blamed the tanners for being too slow in implementing the agreed upon CP options (interviews Rg and Lc in Table 3-4).

The CVC (regional authority from Valle del Cauca) threatened once to stop the support from the CP body and to call for legal action; the tanners in Cerrito renegotiated the timelines for implementations and 76% worked actively on arriving at solutions. The CP body started working in parallel on a complementary project aiming at stimulating competitiveness in the leather chain through the *Universidad del Valle*.

Even though the whole process had been slow, the same officers from CVC and from the CP body remained involved and proactive throughout that difficult period.

As of 2009 in Cerrito, even though the industries delayed meeting the discharge limits by more than a year, 76% of the tanneries were on the verge of being connected to the end-of-pipe wastewater treatment system because they were complying with the law.

In 2009, the *explanatory in-depth interviews* (aiming at responding to the 'how' question) showed that success was a consensual issue among all the stakeholders. Even though in 2004 only the head of CP was conscious about attitudinal mechanisms, by 2009 all the interviewees supported and complemented her vision in order to explain the reasons for success through the following mechanisms (score: 2) (Figure 5-2):

1) Dealing with McSEs' social and economic impending issues, which built trust among all the stakeholders involved;

2) Working through a high degree of participation and trial-and-error in order to build solutions and to sort out CP problems generating a long-term commitment;

3) Creating consensus on the perceived definitions of the problems each community faces in order to foster unified action, even before the technical implementations were taken into account; and

4) In Cerrito, the CP policy was considered successful. It was set as part of command-and-control, and not as part of voluntary agreements where the discharge limits were based on load (stimulating prevention) and not on concentrations. Because the tanners set their own timetables, they felt committed to honour them.

Besides the above, all the interviewees considered the following as essential (score: 2):

5) The regional authority's leadership and commitment to CP. The CVC had prioritized developing a sustainable solution towards cleaning the river overlapping with spatial planning issues and was effectively enforcing the law. The tanneries on the riverbank were not asked to relocate in the short-term provided they met the discharge limits. This issue was especially important for the two largest industries in the area;

6) The role of an expert played by the head of the CP body in helping and facilitating processes and in the choice of technology;

7) The support from tannery experts chosen and approved by the community; and

8) Long-term[54] support to break down barriers.

Consensus was found among all the CVC officers. The head of the National Tannery Committee, of the CP body and of the regional officer considered that support from academia was essential for them (score: 2). In contrast, not all the tanners mentioned the role of academia as essential (score:1). The regional officer, with an engineering background, mentioned that she had gone through training on facilitation at the university once she realized the importance of learning how to manage social issues.

---

[54] Interviewees consider long-term support to be over 5 years.

The head of the CP body had just finished a master's degree at the university and was aware of the importance of supporting a high degree of participation and on fostering self-reliance.

The relevance of green champions and/or of the collective (building association) work was not mentioned here (score: 0).

Figure 5-2 Mechanisms supporting CP in Cerrito in 2009 per group of actors

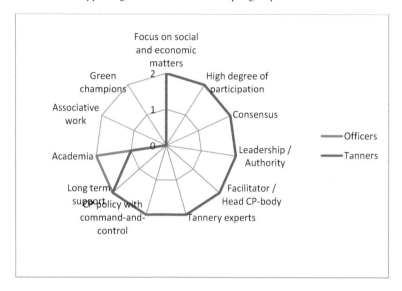

The *focus group* at Cerrito deduced several observations and lessons learnt from their experience:

(a) Having an environmental authority that has a sense of self-criticism is inspiring for all actors. They called attention to the fact that from the beginning of the process in 2004, the authority itself started recognizing its own weaknesses, as historically pollution had not been handled successfully.

(b) Positive conditions prevailed since the big tanner behaved as a role model for the smallest tanneries. The big industry was ready to help the smallest firms in their transition to formality and to share information with the community.

(c) The head of the CP body pointed out that it is important to keep in mind that it was not until 5 years following the implementation of the project that savings due to CP were really noticed. Only in 2009, as the big tanner was investing in the secondary treatment, he realized that he had saved more than 800 million pesos (50% of original investment without CP implementation) ($270.000 Euros of 2009). The lesson was not to rely only on quick financial win-win situations.

(d) The tanners mentioned that maybe if they had focused from the beginning in working through collective or associative projects, their results could have been better.

(e) When they were asked to try to condense the process in a sentence they declared: "Before 2004, nobody knew anything even if we pretended the opposite. Today, we are all knowledgeable in our own contexts."

5.3   San Benito

5.3.1   Introduction

San Benito is a neighborhood of Bogota, the capital of Colombia, where 365 tanneries heavily impact the water quality of the Tunjuelito River, a tributary of the Bogota River.

This section presents the problem in San Benito: Pre-case study (5.3.2), and outcome of the case study (5.3.3).

5.3.2   The problem in San Benito: Pre-case study

Relevant economic and social data

San Benito is a neighbourhood that reflects some of the acute problems of the capital city of Colombia, which relates to this thesis: high informality, low competitiveness, lack of long-term plans for IWRM and severe problems regarding wastewater management (Villegas, 2012). In fact, it has been determined that even though Bogotá represents 25% of the national GDP, it is expected that in 2025, its performance will drop to less than 20%. The city is not adequately dealing with urban inequality. While cities like Lima have increased the importance of its middle class, Bogotá is even increasing the size of the marginalized neighbourhoods. With respect to wastewater management coverage, Bogotá is below the Latin American standard, which is 80%. It has not been able to surpass a mere 25% (Villegas, 2012).

San Benito was originally a rural region close to the Tunjuelito River in the outskirts of Bogotá that in the 50s seemed appealing for tanners from Villapinzón and Chocontá, coming from the north part of the Sabana of Bogotá region (CNPML, 2004). Ever since that time, the use of the land moved from agricultural to residential and industrial. The spatial planning framework determined a mainly residential purpose and conflicts between the tanners' community and residents arose because of harmful production processes (Vásquez, 2012). In theory, 39% of lands are residential, 28% industrial, 24% commercial and 2% are for the services sector (*Ibid*). In practice, a recent survey has found that out of 621 properties that are considered for residential use, 69% are misused. The situation gets even more complex because many tanners live in their own industries. Out of 356 tanneries, 127 are also used as residences *(Ibid)*. The survey also reported 10 industries illegally located on the 30 m riverbank.

The population in San Benito is mostly between 27-44 years of age and there are almost no families with babies. The major health problems are due to suboptimal waste management. Diarrhea and acute respiratory illness are hence reported among the population (*Ibid*).

The tannery industrial activity in this area has historically developed mainly on properties smaller than 500m², which do not even have room to store the products. 85% of the tanneries hire less than 10 employees (Poveda and Sánchez, 2009). Because of their high informality and small sizes, it has been determined that these industries can open or close depending on issues like commercial opportunities, law enforcement or institutional support (CNPML, 2004).

Relevant geographical, environmental, technical and institutional data

San Benito is located southwest of Bogotá. Its natural southern border is the Tunjuelito River. The Tunjuelito River is the most important river after the Bogotá River in the Sabana of Bogotá. It flows for 53 km and covers the south of the capital city. The Tunjuelito basin is home to 2.5 million people (SDA, 2006).

Once the Bogotá River crosses Bogotá with a flow of $24m^3/s$ its Chemical Oxygen Demand (COD) and its Biochemical Oxygen Demand (BOD_5) increase by more than 100mg/l. Once the Bogota River receives the discharges from the Tunjuelito River, its COD reaches the highest limit of 350 mg/l and its BOD_5 of 250 mg/l. In that area the river is considered technically dead since its dissolved oxygen (DO) is 0.07 mg/l (ACUEDUCTO-UNAL, 2010). The Tunjuelito River is considered responsible for 22.2% of the domestic pollution and 7.8% of the industrial pollution of the Bogotá River. The Tunjuelito River drains the poorest areas of Bogotá, which can be flooded in the rainy seasons. This river is responsible for 46% of the solid waste thrown into the Bogotá River (UNIANDES, 2002b). The last bio-assays in that area of the river, given in toxic units (TU), showed a peak of 37 TU (second worst after the previous discharges from the other tributary, the Fucha River) (ACUEDUCTO-UNAL, 2010).

San Benito is ruled by the urban environmental authority of the capital of Colombia, SDA from its Spanish name Secretaría Distrital de Ambiente (District Environment Secretary Office). As such, the urban environmental authorities are not autonomous and they depend on the District's directives (Figs 4.5 and 4.7). Since the early 2000s, the SDA has been trying to find solutions to the problem of the tanneries, which along with the brick industry are the major polluters of the Tunjuelito River. Spatial planning and land ownership conflicts have blocked long-term solutions. The tanners had neither the financial support nor the space to build an industrial park where they could more effectively control the pollution they cause by separating industrial processes and constructing the right infrastructure. Relocation has been considered an option that is too expensive (CNPML, 2004; El Tiempo, 2004a; El Tiempo, 1994). Since 2003, SDA was signing agreements with the Chamber of Commerce to run CP programs with the tanneries through the office called *Ventanilla Ambiental (environmental window-ACERCAR strategy)*, a mainly private initiative. Even though a CP project was implemented in the 1990s by an NGO regarding CP called PROPEL, diffusion of CP was not taking place. Since the nineties, forced closures had been occurring without solving the pollution problems (El Tiempo, 1993).

5.3.3   Outcome of the case study

History of activities

In 2004, a court order on the recovery of the Bogotá River and a national governmental plan brought the issue of its recovery to the highest level. For the first time in Colombian history the problems with respect to the pollution of the Bogota River were on the national agenda (Court order, 2004; CONPES 3320, 2004).

The Bogota River

The Bogotá River is the main river of the region covering $4600km^2$ of the plain of Bogotá from its source at 3400 masl in Villapinzón in the north to the Tequendama Falls of 1925m in the south. The river runs 370 km southwest and flows into the Magdalena River at 280 masl. Along its path, the river drains an area of 599,561 ha whereas 71.8% constitutes the upper basin and 28.2% the middle (urban area of Bogotá) and low basin (EPAM, 1993).

The river receives 11 tributaries, whereas 3 are considered urban rivers crossing Bogotá from east to west and known as Juan Amarillo, Fucha and Tunjuelito, and which are mainly responsible for the extreme pollution of the river. The organic pollution at the end, after the Tunjuelito River, showed already in 1993 values for medium flows of 94 mg/l representing loads of 264 Ton O2/d with Total Coliforms of 28 million/100ml with maximum values of 79 million (EPAM, 1993). Besides, the organic pollution it was calculated that the chemical and the physical load discharges were 79 kg of Chromium, 79 kg of Lead, 20.4 tons of Iron, 5.2 tons of detergents and 1473 tons of suspended solids. It was estimated that if no definitive and effective treatment was implemented, the pollution at the very end after the Tunjuelito River would be 1.5 times the pollution cited above by 2020 (*Ibid*).

The hydrology of the river has been modified in the upper basin since the fifties for the generation of hydropower, flow control, potable water for Bogotá and crop irrigation (see 6.2). The regional authority CAR in the upper basin built wastewater treatment plants, but those aren't working either because of high monthly operative costs or because they are already obsolete (El Espectador, 2012c). As the river reached Bogotá, it was reported in 1993 that its flow was 12m^3/s and that through its path it received 15m^3/s of wastewater from the city without effective treatment. The domestic wastewaters are just partially treated at the tributary Juan Amarillo at the only wastewater treatment plant. The EPAM study from 1993 estimated that in 2015 the flow of wastewater would be 25 m^3/s.

The concern with respect to the region's wastewater management was not really taken into account until 1974 with the Camp Dresser and Mckee-CEI study and later on in 1985 with Hidroestudios-Black and Veatch (EPAM, 1993). As a result, from 24 different alternatives, the ones proposing the end-of-pipe treatment in the lower basin were discarded (EAAB, 2000b; EPAM, 1993). In 2006, the alternative of building a second treatment plant at the very end of the plain of Bogotá was defined (Semana, 2006). In 2012, besides the construction in the capital city of a collector meant to bring the wastewater to the future second treatment plant, no solutions have been implemented. As Bogotá has grown to have 8 million people, the Bogotá River has been suffering continuous severe degradation and has become a sewer (El Espectador, 2012c). The poorest inhabitants live along its path and suffer the consequences from the extreme conditions.

Among the effects of the severe pollution it is worth mentioning (EPAM, 1993):

–   Contaminated milk and crops for the city with microorganisms and heavy metals such as Chromium and Mercury above limits;

–   High incidence of waterborne diseases among communities living along the river. In the lower basin, the village of Apulo has 28.86% of these illnesses contrasting with a mere 2.5% in areas far away from the river.

–   High concentrations of heavy metals in the sediments of the beds of rivers;

–   Absence of river fauna;

–   The Muña reservoir in the far south becoming an oxidation pond; and

–   High maintenance costs of the generators and general equipment of the hydropowerplants because of the corrosive waters.

A cost of USD 6,27 million per year for the consequences of water degradation (US$ from 1993) was estimated without taking into account costs such as the loss of biodiversity or of well-being etc. The

Bogotá River has been known as the river that is practically dead at its source because it receives in the upper basin within its first 10 km, chemicals from the potato crops and the raw waters from more than one hundred rural tanneries (Semana, 2006; EPAM, 1993).

The tanners'community in Bogotá

In 2004, the tanners were seeking to move to the outskirts of Bogotá or to the upper basin to the rural area of Villapinzón because of the urgent need to relocate, and the fact that the district considered it could not invest in their end-of-pipe solutions because they belonged to the private sector. The regional and district environmental authorities worried about the risk of transferring pollution from Bogotá to the source of the Bogotá river where pollution problems and social conflicts already existed (SDA, 2006).

The biggest tanners were buying expensive nearby district properties and installing their own secondary (end-of-pipe) treatment. The smallest tanneries could not afford to build their own. With respect to CP, SDA was continuously organizing workshops on CP that were attended by the majority of tanners (interviews HdCP, Tech, Reg and Lc in Annex 5.1).

By the end of 2008, 50% of the tanners were going through forced closures, 30% (80 tanners) were investing on an industrial park for the polluting phases and the other 20% were implementing different and individual technical options basically focused on end-of-pipe solutions. Because of inadequate results, the head of the CP body at SDA had just resigned and new officers had been hired (mc5 in Annex 5.2).

As of 2009, San Benito registered only isolated successful CP implementation cases. Only 20% of the industries had improved technically and no solid waste valuation was developed (PAHO; 2010). 89% of the workers' conditions did not meet the minimum requirements set by the law (Poveda and Sánchez, 2009). The SDA highlights that 80 tanners had proposed working towards developing a collective strategy for the de-hairing and tanning processes that are considered the most polluting steps of the whole industrial process (mc5 in Annex 5.2). The CP consulting company was no longer in charge of the project and the CP body had basically established an alliance with the body responsible for economic development at SDA itself. A new vision from the head of the CP body towards improving the leather productive chain included a holistic approach to the problem: The SDA was considering that the problem should be seen in a broader perspective and not limited, as in the past, to the analysis of the technical aspects (mc5 in Annex 5.2).

Impact of policies and policy procedures on tanners

The local CP policy was basically the same CP policy at the national level: based on voluntary agreements. The command-and-control decree 1594 did not stimulate prevention measures (mc5 in Annex 5.2).

Mechanisms supporting CP

In 2004, the interviewees (Table 3.3) in the in-depth exploratory interviews - mentioned the importance (all scoring 2) of the following kinds of mechanisms supporting CP adoption: (a) a *regulatory framework* that recognizes their needs, (b) access to *technology*, (c) *economic incentives* and *financial support*, and (d) access to sound *information* on all related issues like legal and technical options (Fig 5.3).

In fact, for each mechanism:

- The tanners expressed that the discharge limits were not realistic and did not meet the local circumstances of the people;

- People were anxious about the challenges that CP implementation would entail for them, and pointed out the importance of targeting incentives, market opportunities, scale economies and reliable information suiting the needs of the McSEs; and

- The tanners argued about the need to have clear rules and *coordination* among institutions (score: 1) (Figure 5-3).

Figure 5-3 Mechanisms supporting CP in San Benito in 2004

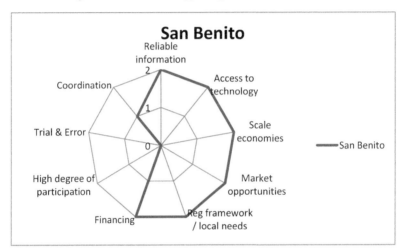

Between 2005 and 2008, the program went through difficult times. Tense relationships between the tanners, the authorities and the head of the CP program were observed at the national leather board committees. The CP project was on the verge of being cancelled and disputes were taking place (mc5 in annex 5.2).

The authorities blamed the tanners for being too slow in implementing the agreed upon CP options. The need to relocate tanners without clear rules on spatial planning like not defining the appropriate relocation area in the outskirts of Bogotá was a permanent issue creating uncertainty and high-risk perception in terms of investment (mc5 in Annex 5.2).

In 2009, with respect to the final explanatory in-depth interviews (responding to the how question), the results showed a lack of a unified vision among all the stakeholders involved (score: 1) (see Figure 5-4). This finding asked for deeper analysis regarding the diversity of the answers found among the groups of stakeholders themselves.

The answers were organized and put into a new radar graph into two groups: SDA officers and tanners. The kinds of stakeholders targeted in the interviews per group of officers and tanners are explained in the table below (

Table 5-1) where, (a) each officer represents a responsible person in the CP implementation process from the national to the local levels, and (b) the tanners were selected for (1) being part of CP projects,

(2) being micro-tanners, (3) being at different stages in the implementation of CP, and (4) belonging to specific interest groups like trade associations.

Table 5-1 Types and number of stakeholders interviewed in San Benito

Type of stakeholder	Description	Number
Officer	Head of National Tannery Committee	1
	Head of CP body	1
	Technician in charge	1
	Head of CP program at regional environmental authority	1
	Head of local environmental authority	1
Tanner	Tanner belonging to the group of 80 tanners	3
	Tanner belonging to COESA (trade association)	2
	Tanner belonging to 2nd association	4

Among the officers, consensus was present with respect to three mechanisms: associative work, focusing on social and economic issues before handling the technical ones, and of establishing leadership from the authority (score: 2). The technician in charge (Tech) did not mention the importance of participation or of building consensus. These results can be explained by the fact that he was not involved in the social work that the others were implementing. The officers in charge of the CP project (HdCP) and of the National Tannery Committee (Nt) highlighted the importance of promoting green champions.

Among the tanners, no consensus was identified with respect to any of the perceived mechanisms. Some tanners mentioned only certain mechanisms (score: 1). The three tanners belonging to the group of 80 tanners that are focusing on an associative strategy were optimistic but cautious about receiving support in the long-term. Two other isolated CP implementation cases felt they could not commit to CP in the long-term because they considered that support was missing. The four tanners left, which had serious financial and land problems did not sense support and had no perceived reasons for success.

A further analysis of the answers found that these interviewees recognized the importance of acknowledging the characteristics of the McSEs and of (1) adapting the programs to the social needs, (2) working in a clearly participative manner stimulating trial-and-error and the choice of technology, group efforts and collaborative work, (3) stimulating a unified vision, before any technical issues are considered, and (4) building a strong leadership on the part of the urban environmental authority.

Figure 5-4 Mechanisms supporting CP in San Benito in 2009 per group of actors.

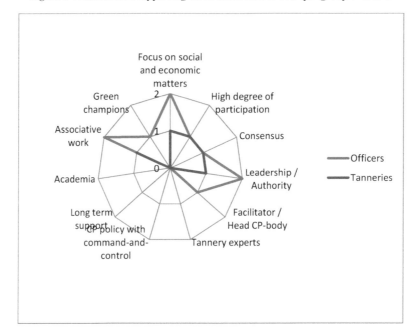

Focus groups on the lessons learnt pointed out that:

a) The leadership on CP from the authorities is essential. As long as SDA was not completely in charge of the tannery case, results were slow.

b) Once a holistic approach focusing not only at the environmental challenge but also on the market was adopted, tanners felt more self-assured about implementing CP.

c) Effective law enforcement is important but should be complemented by removing all the barriers towards CP adoption and a comprehensive approach.

d) Tanners did not wish to take action while others were free riding: 'If we are not the only polluters of the Tunjuelito River, why should we bother if others are not acting'? There was need to change this attitude.

5.4   Cerrito and San Benito Compared

5.4.1 Introduction

This section compares the case study results from Cerrito and San Benito by discussing the pre-case study situations (5.4.2), and the outcomes of the case studies (5.4.3).

5.4.2   Cerrito and San Benito: Comparison pre-case study

Stakeholder analysis

An initial insight into the situation of the tanners allowed the researcher (i.e. myself) to understand the basic principles, steps and targets of each CP project being implemented, as well as the interests and needs of the stakeholders involved. Table 5-2 characterizes both cases and was developed in 2004:

Table 5-2 Relevant issues for Cerrito and San Benito in 2004

Issues	Cerrito	San Benito
Industries	21 tanneries, 77% micro	265 tanneries, 90% micro
Culture	Individualistic, no experience with collaborative work	Individualistic, no experience with collaborative work
Environmental focus	Curative	Curative
Educational level	Low secondary school	Low primary school
Legal status	Critical, threatened with closures in early 2000, needing to meet limits for a joint wastewater treatment plant with municipality, sense of living through a crisis, attendance at meetings on CP 80-90%	Critical, threatened with closures in early 2000, sense of living through a crisis. Attendance at meetings on CP 80-90%
Authority	CVC: autonomous and relies on academia for technical advise	SDA: district dependent and relies on private branch organizations for technical advise
CP body	Legally independent from CVC but financially dependent Supported by academia	Originates from an agreement between the SDA and a branch organization
Relationship among stakeholders	Mistrust	Mistrust
Legal framework	Industrial control set on loads and not just on concentrations implemented in 2004	Discharge limits set on concentrations
CP implementation dead lines	Negotiated agreement between tanners and authority to be enforced	Negotiated agreement between tanners and authority Voluntary nature
CP approach	Holistic, aiming at increasing competitiveness	Only technical
Interlinked land issues	Sustainable solutions for McSEs and recovery of river prioritized	Environmental conflicts compounded by spatial planning issues

The above table presents common elements with respect to the local culture (both individualistic and with no experience of collaborative work), to the curative end-of-pipe approach of both environmental authorities, to the lack of trust between stakeholders, and the legal status of the tanners (facing both penalties and threats of closures).

The stakeholder analysis also revealed differences with respect to the endogenous circumstances of the specific McSEs such as the size of the community, the level of education; and exogenous circumstances focusing on the way the CP policy was being set, how the legal framework operates, how the interlinked land issues are dealt with, and how the CP body operates on the basis of its focus and characteristics.

Following up on how the implementation of CP was being handled, starting from common elements could help provide insights regarding how the mechanisms supporting CP operate over a period of time long enough to show either effective change or the lack of it.

5.4.3   Outcomes of the compared case study

Mechanisms supporting CP

While the perceived mechanisms supporting CP implementation were basically the same as those cited by the stakeholders from both case studies in 2004, how those perceived mechanisms supported CP implementation showed differences and similarities between contexts (Cerrito or San Benito) in 2009.

In 2004, basically all the stakeholders mentioned (responding to the exploratory questions) the following mechanisms supporting CP (all scoring 2) (

Figure 5-5):

- a regulatory framework that recognizes their needs,

- access to technology,

- economic incentives and financial support, and

- access to sound information on all related issues like legal and technical options.

Only the head of the CP body in Cerrito pointed out a high degree of participation and trial and error as the preferred way to teach and handle McSEs. Only some tanners mentioned the importance of coordination.

Figure 5-5 Mechanisms supporting CP in Cerrito and San Benito in 2004

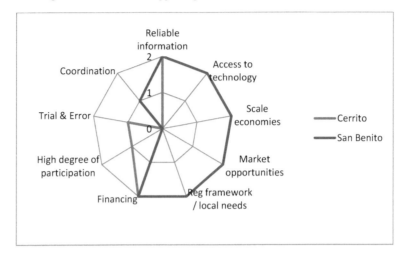

In 2009, even though there are differences with respect to the degree of CP implementation between cases, 5 mechanisms explaining a successful CP implementation are common to both (

Figure 5-4):
- focus on social and economic issues first (specially including inter-related land issues);
- emphasis on participative and trial-and-error processes;
- building consensus;
- regional authority's leadership and commitment to CP; and
- an expert on facilitating processes and the choice on technology – the head of the CP body.

Stakeholders in Cerrito added three explanatory mechanisms (scoring 2):

- CP policy should be promoted in parallel with the command-and-control's measuring loads and not just concentrations,
- long term support, and
- technical advice from tannery experts.

Besides the above mentioned mechanisms, the CVC officers from Cerrito brought-up *support from academia* as an essential one. Academia had inspired them about participatory approaches based on trial-and-error from the very beginning of this process.

Figure 5-6 Mechanisms supporting CP in Cerrito and San Benito in 2009

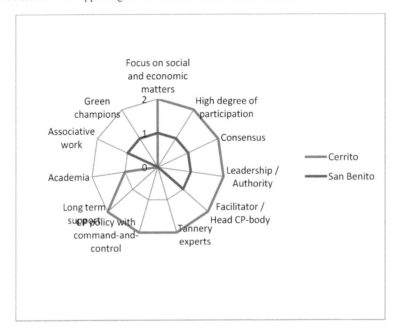

5.5   Conclusions

The case studies suggest that exclusive top-down methods from the classical consulting approach used in developing countries (San Benito) in the context of McSEs are less successful in the long term than participatory, holistic learning methods (Cerrito).

The case of Cerrito illustrates the power of having the political will and leadership from authorities prioritizing (a) the McSEs' agenda, (b) approaches based on prevention such as CP implementation, (c) stakeholder participation aiming towards social inclusion, (d) building sustainable solutions for communities and the environment over straight spatial planning concerns.

The case of Cerrito brings up the fact that academia can play a role in supporting sustainability because it can successfully bring knowledge to and from relevant real challenges in society. Solving the San Benito case has reached a dead end because the solutions seem to be caught in vicious cycles from the legal framework, lack of leadership, spatial planning restrictions, or lack of financing programmes for the smallest industries.

From the comparative analysis of both cases it can be concluded that successful CP implementation programs for McSEs:

- Focus on attitudinal and organizational mechanisms: have long-term supportive, participative and multidisciplinary capacity building programs based on building consensus; learning through trial-and-error supported by academia; tackling first the social and economic issues (specific contexts); choosing technology between tanners, researchers, and tannery experts (the authority stays as an observer); and operate with policy entrepreneurs who take on a leadership role;

- Run parallel to appropriate command-and-control policies; and are not hampered by overlapping jurisdiction of spatial planning generating uncertainty and a risk for the micro-entrepreneurs' investments.

Through this analysis, it became clear that the mainstream consulting approach has not been appropriate for the McSEs' needs.

The comparative analysis proposes that the initial perceived benefits from CP implementation couldn't be the quick savings derived from CP implementation; and that CP policies operating in parallel to command-and-control are more effective as well as CP programs taking into account interrelated land problems in the context of McSEs.

6 Action Research in Villapinzón

6.1 Introduction

This chapter deals with the third sub-research question, which is to identify how AR can be developed and tested to help McSEs implement cleaner production in the context of Colombia. Its purpose is to present the case of the micro-tanneries of Villapinzón in which my involvement as the ARer (action researcher) was intense.

The structure of this chapter is as follows: the problem in Villapinzón- pre AR (6.2), AR in Villapinzón (6.3) and conclusions (6.4).

6.2 The problem in Villapinzón pre action research

6.2.1 Relevant socio-economic data

Villapinzón is a village of 8000 inhabitants (DANE, 2005) one hour's drive from Bogotá, and is located in the upper basin of the Bogotá River valley. The Bogotá River valley is home to almost 20% of Colombia's total population of 44 million people (DANE, 2007) and represents 26% of the GDP (DNP, 2007). One third of the Bogotá River's water in the upper basin is used for the water supply ($4m^3/s$) of Bogotá's 8 million inhabitants, for hydropower generation, for hydrology control with 7 reservoirs, and for crop irrigation (see section 5.3.3).

The village residents of Villapinzón sustain themselves mainly through the activity of 120 tanneries[55], among which at least 90% are micro enterprises. In the late nineties, the regional authority reported that there were 132 tanneries classified (Table 6-1).

Table 6-1 Tanneries in Villapinzón

Micro	Small	Medium	Definitive Closure	Total
95	23	2	12	132

Source: CAR, 1999

They were classified with respect to the number of treated skins produced per month: micro tanneries process up to 500 skins, small tanneries up to 1500, and medium up to 10,000. This small-scale tanning industry offers 700 direct jobs and 4000 indirect jobs and represents the main commercial and industrial activity in the area (SINA II, 2005).

The tanneries community is one of the first, from the beginning of 1800s, to be in charge of producing the leather goods for horseback riding at a strategic area where most of the fights for independence took place (Ojeda, 2004). The industries are spread over an area of 7 km along the river and south of the village of Villapinzón. 51 of these micro-tanneries have been for decades within 30m of the riverbank, a zone that since 1977 has been considered 'for preservation and protection use only' (INDERENA, Decreto 1449, 1977). The latter has made the tannery activity illegal in that specific area.

---

[55] Because of the high informality, the real number increases or decreases depending on the stringency of law enforcement, institutional support, or market opportunities.

The tanners' community, with a native Indian background, has only a primary level of education, is mainly informal, lives on a subsistence economy and uses obsolete technology. Descendants from the original families that settled down such as López, Fernández, Contreras, Bernal and Lizarazo are still part of the community (Ojeda, 2004).

6.2.2   Relevant geographical, environmental, technical and institutional data

Because Villapinzón is located at no more than 10 km from the source of the Bogotá River, a national *resolución* 076 from 1977 from the Ministry of Agriculture has considered the area a forest reserve since the seventies and was asked to declare an exclusion area with a restricted industrial zone for the village and the tanneries. The source of the river is considered a strategic and fragile ecosystem called *páramo*[56], which is a water producing area with water retention plants such as *frailejones*, moss and lichens.

Before the arrival of the Spanish conquerors, the native population called Muiscas worshipped the *páramos* because of their water-producing ability to the point that the lakes at high altitude were sacred places[57] (EAAB, 2000a). Such was the case of the Guacheneque *páramo* where the Bogotá River is born.

In the 20th century, few individuals became landlords of big properties and the poorest people were pushed to live and work in the only available land at the *páramos* (Procuraduría, 2008). The *páramos* hosted smallholdings for more than 50 years. The area has suffered from the unsustainable practices of the potato crops with organophosphate pesticides, livestock presence and coal mining. Severe degradation of these fragile ecosystems hence caused 53% of the 68 species of *frailejones* to become endangered; between 1970 and 1990 45% of the *páramo* surface has disappeared. As the green surface area for water retention has diminished, the river flows have decreased (*Ibid*). In the last decade, at the Guacheneque *páramo*, the environmental authorities have been buying as many smallholdings as possible in order to transform them into protected areas and have designated an environmentally motivated native person as the ranger of the area (*Ibid*).

Even though the study area is called Villapinzón in this thesis, it also includes the nearby town of Chocontá that has the same kind of industrial activity but without riverbank occupation. Villapinzón is then located at 95km from Bogotá, and Chocontá at 75km at altitudes of 2715 and 2655 masl respectively. The average temperature is 13°C at noon for the whole year and the rainfall is 1200mm/year (Ojeda, 2004).

The regional autonomous corporation CAR governs the area and has had historically tense relationships with the capital district authorities. The regional authorities have always complained that the capital city uses their resources and does not remunerate enough for that service.

---

[56] In this area, it is a strategic ecosystem above 3000 masl. Only 7 countries in the world have it namely, Costa Rica, Panamá, Venezuela, Colombia, Ecuador and Perú. Colombia has the biggest páramo area.

[57] This region was the place for the ceremony of El Dorado, which stimulated the gold search by the Spanish conquerors. The Chief of the native indians called Muiscas used to cover his body in gold powder and swim in a lake on full moon.

6.2.3   The tanneries' impact

For decades, the tannery industrial activity has caused a substantial collective impact on the Bogotá River. Since the tanners are located close to the source of the Bogotá River, this *river was known* 'as the only river in the world that is born and dies in the same place'.

Natural tanning agents were used until 1984 when the ruling Regional Authority CAR (not yet an Environmental Agency), taught the use of synthetic tanning agents and then was absent for more than 10 years (CAR, 1994). During that period, salesmen of chemical products and representatives of companies providing end-of-pipe technologies were the only technical advisors of these enterprises. In the late 1990's CAR started to blame the tanners for the use of those chemicals. In 2000, at the request of CAR many tanners presented Environmental Management Plans (EMP) but never got an answer regarding their approval until 2004 when the CAR said that the plans were of poor quality. The technicians who knocked on the doors of the tanners after they got the request from CAR, and the salesmen of chemicals that visited them developed these initial plans.

Today, tanning entails two basic processes that impact upon the environment: the classical dehairing with sodium sulfate and the tanning process itself using chromium sulfate.

The effluents of these industries were discharged into the Bogotá River with disastrous consequences for river water quality. According to the CAR ruling decree 043 of 2006, the discharges made it impossible to meet the water quality parameter limits of the Bogotá River determined for this part of the river for the year 2020, *I.e.* 7 mg/l for BOD_5 and 10 mg/L for TSS. On leaving Villapinzón, which did not even have a domestic wastewater treatment plant, the river had a COD of 102mg/l in 2004 and high levels of Chromium Sulfate ($Cr2(SO4)3$) that at 0.3mg/l were three times the safe limit for agricultural and domestic use. The discharges of Chromium Sulfate from the micro-tanneries were found to be between five and nine times higher than what is allowed by Decree 1594 on industrial discharges (CAR, 2004).

For over 20 years, the Regional Environmental Authority considered that they had tried to solve the environmental problems of the community of Villapinzón. However, 67 potential solution proposals remained on the shelves. Since the agency has always had a focus on end-of-pipe solutions without a CP branch, only one of the presented proposals was directed towards the prevention of the polluting flow (CAR, 1998). None had been implemented and the tanners were blamed for the lack of results. The conflict had not benefitted the economic livelihood of the tanneries, whose owners all have been sued by the authority, banned from credit and faced fines that they were unable to pay (Ojeda, 2004). Lawyers had made a living from their conflict manipulating the tanners without presenting effective solutions to the authority and letting the legal terms expire. Closures were executed by CAR since the late 1990s.

Looking back at judicial court proceedings, newspaper articles, and statements from the tanners, CAR officials, and the Public Prosecutor, it becomes clear that interrelated land issues had placed the tanners in even more conflict with the authorities. Their industrial activity was not being formally recognized in the local zone planning and the tanners from the riverbank were considered invaders without property rights. CAR was not even sure whether the protection zone should be wider for this part of the river (El Tiempo 2004c).

The tanners' association did not enjoy the support of the community and their leader was a lawyer that was making a living out of the conflict, stimulated fights and disputes without a problem solving orientation. The owner of the biggest industry wanted to organize an industrial park, supported by the mayor, where the owners of the micro-tanneries would work for him.

Realizing that no environmental rehabilitation project was being implemented, in February 2005 the Regional Environmental Authority closed 58 tanneries (El Tiempo, 2005b), which in 2006 rose to 80. The authority considered that the tanners had wasted all the opportunities that had been offered to them.

A story like this in Villapinzón has, as a backdrop, the fact that Colombia has one of the highest inequality rates in South America (GINI of 0.56 in 2012) and that as a consequence the government efforts have not been geared to battle inequalities but at strategies that stimulate overall national growth. In the water sector the situation is especially difficult for the marginalized groups (Chapter 4). Table 6-2 characterizes the case in Villapinzón. It was developed in 2004 from the initial insights:

Table 6-2 Relevant issues for Villapinzón

Issues	Villapinzón
# Industries	120 tanneries 92% micro
Culture	Individualistic - no experience with collaborative work
Environmental focus	Curative
Educational level	Low primary school
Legal status	Critical- threatened with closures in early 2000
	Sense of living through a crisis. Attendance at meetings 90%
Authority	CAR regional autonomous corporation relies on private branch organizations for technical advisory
CP body	Beginning 2004 inexistent (one person)
Relationship among stakeholders	Mistrust
Legal framework	Discharge limits set on concentrations
CP implementation dead lines	Inexistent for tanneries. For other industries based on voluntary agreements
CP approach	Only technical
Interlinked land issues	Environmental conflicts compounded by spatial planning issues

6.3   Action Research in Villapinzón

6.3.1   History of activities

This investigation was conducted for a period of six years (2004-2010). Table 3-5 presented the six steps of the SASI approach, the aim of each, the theory and the elements needed for each to be reached, and the expected results. Table 3-7 summarized the opportunities for observations and interventions at Villapinzón. Table 3-8 presented the time line for the AR. Table 3-9 presented the stakeholders interviewed in 2004.

In Figure 6-1 the steps of the SASI approach as applied in Villapnizón are presented and analysed.

Figure 6-1 SASI steps in Villapinzón

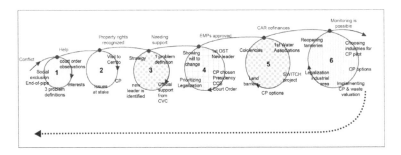

Step 1.Preparation

By the year 2003, all the judicial claims from civil society on the pollution problems on the Bogotá River were being grouped under one big court order called *Sentencia sobre el Rio Bogotá* (Court Order on the Bogotá River). The court proceedings constituted a unique opportunity for the researcher (i.e. myself) to get acquainted with the multiple forces operating at the local, regional, and national levels regarding the context of the McSEs impacting on the Bogotá River. The purpose was to make a follow-up to such a process with a low key inquiry step aiming at assessing the risk of intervening (Schein, 1996) and the health of the system from the analysis of the Hs indicators (Table 3-5) in the in-depth interviews. As stated by change theory (Burnes, 2004; Schein, 1996; Lewin, 1946) and the nature of AR, this initial intervention should be a low-key inquiry-oriented diagnostic intervention, designed to have a minimal impact on the processes inquired about.

The court proceedings then followed once a week during a year. Two important issues were identified in relation to the tanners in Villapinzón:

- Renowned lawyers always represented the medium-sized tanner, whereas the small tanners had only occasional legal support. When small tanners spoke up in court this often turned out into aggressive claims against CAR and the government leading to clashes with the judge. This was particularly the case with the smallest tanneries located within the 30 meters preservation zone of the river.

- The discussion focused on potential solutions that had been offered in the past and that had not been adopted by the small tanners, which, according to them, was due to lack of resources. These suggestions all focused on end-of-pipe treatment and did not explore pollution prevention through cleaner production (CP).

These initial observations brought the analysis towards identifying two basic issues associated with the tanners' situation:

- Facing social exclusion, and

- Dealing with mainly an end-of-pipe focus on the part of the environmental authorities.

After six months of the follow-up of the dynamics of the court order (March 2004) as an observer, the researcher asked to have an informal meeting with the judge aiming at providing technical advice based on prevention, which could be more affordable for the micro-tanners. The researcher established also a first contact with the tanners' leader, who presented himself also as their legal representative.

The researcher visited Villapinzón and CAR at the national and regional levels, and conducted initial observations on the group dynamics (a general meeting was taking place regarding the conflict with the mayor, political actors, and tanners) and in-depth interviews (Table 3-7). The meeting was claim oriented; the mayor told the tanners that they were stupid for not accepting the solution the medium sized tannery owner was offering. The meeting ended with aggressive attitudes and clashes. The tanners' leader at that time had told the tanners that the policy on the riverbank did not exist and that CAR was ignoring their rights. The initial experience with the tanners confirmed the findings from the observations during the court proceedings: There was a need to overcome the two issues at stake presented above.

During the first in-depth interviews, the small tanners asked to be helped to resolve their conflict. They stated that they wanted to change. The researcher accepted this request, as it became clear to her that it was necessary to dig deeper into the causes and consequences of their problems and to empower this underprivileged group. Based on the Change theory, they were ready to go to the second step on trust building (Figure 6-1).

The fact that the researcher interviewed them about their interests had caught their attention. The first experience with the tanners showed that they were vulnerable, that had gone through similar cycles of psychological anxiety and had not been able to overcome their barriers towards positive change. Previous 'helpers' had manipulated them and had taken advantage of their conflicting situation.

Together with the tanners, once the in-depth interviews were conducted, a number of issues of relevance to the situation were identified (Annexes 6.1 and 6.2). The tanners started to realize that if united as a group they would have a better chance of overcoming the problems at hand.

Five issues were identified as most important. The issues were identified first with the McSEs in the preparation step (from the SASI process), and then with all the stakeholders in subsequent steps aiming at building common ground. These five issues later served to inspire the strategy for the McSEs:

1. To work towards empowered and united micro-tanners,

2. To legalize the micro-tanners with respect to environmental issues (This was a specific request constantly argued by the regional authority),

3. To work on the inter-related land problems as the tanners understood the spatial planning policy was real and could threaten their rights,

4. To implement a CP project supported by a well known academic institution, and

5. To develop business plans for the tanners based on association building and aiming at competitiveness.

The 12 in-depth-interviews with representative stakeholders and observations on the real and actual social interactions targeted also the understanding of the power games, discovered the long-term relationships, determined the health of the system through the Hs indicators, and unveiled the different problem definitions perceived by the actors in order to understand the complexity involved (Table 3-9 and Annex 6.2). From the in-depth interviews, micro-cases were developed (mc6 and mc7 in Annex 6.1 and mc2 in Annex 4.1).

Initial observations of group dynamics in four regional and local meetings in Villapinzón in 2004 showed that at least 90% of the tanneries were micro & small industries, that the big tanner was a negative leader, that the meetings were claim oriented and not solution oriented, that there was an absence of appreciative attitudes, and that there was mutual aggressiveness. .

Table 6-3 shows a social system where the perceived definitions and causes of the conflict were manifold in May 2004. Three different definitions expressed three different ways of looking at the problem.

Table 6-3 Three initially perceived definitions of the problem

Control authorities Mayor and Big Tanner	The small tanners have consistently rejected the options offered to them.
Small tanners' leader	The small tanners need the money to build their end-of-pipe treatment plant.
Small tanners	The authorities do not listen to us. We want to solve the environmental problem while keeping our identity.

The lack of consensus regarding the original definition of the problem at the preparation step suggested that the different stakeholders had limited access to the relevant information and/or that the interests were opposed to each other. The results and analysis of the three indicators assessing the health of the social system showed:

- An Hs_1 of 0: the micro-tanners said that they had no opportunity to be heard.

- An Hs_2 of 0: all the stakeholders perceived that their relationships were aggressive, even among the group of micro-tanners.

- An Hs_3 of 0: consensus was non-existent regarding the perceived causes of their problem.

The above three values showed a conflicting situation.

Additionally, the preparation step included (a) a stakeholder analysis with identification of as many actors as possible and their different interests and priorities in order to establish the possible strategic alliances in the process; (b) data collection on as much information on related cases and public documents to unveil the history of the conflict; (c) the identification of BATNAs (Best Alternatives to a Non Negotiation Agreement); and (d) a small tanner's profile.

a) Stakeholder analysis

Initially, 11 kinds of stakeholders were identified (from 2003 - April 2004) and were asked through semi-quantitative surveys to value from 10 to 100 the issues that represented their interests. This information was drawn from the initial 12 in-depth interviews, from the observations at the initial 4 meetings and from informal conversations with the actors. Table 6-4 shows the main stakeholder interests ranging above a value of 70.

Table 6-4 Stakeholders and their interests in 2004 in Villapinzón

Stakeholders	Interests	Values
CAR main office	To build efficiency image	100
	To clean the river at short term	80
	To abide judicial orders	100
CAR Regional Office	To support the Mayor	90
Public Prosecutor	To build efficiency image	100
	To reinforce command & control	80
Big Tanner	To get environmental status	100
	To invest in end-of-pipe plant	90
	To maintain their identity	80
Small tanners not from the River Bank	To be heard	100
	To clean their image	80
	To clean the river	80
Small tanners on the River Bank	To have their legal property rights	100
Mayor	To build efficiency image	80
	To support the interest of the big tanner	90
	To stimulate new coming tanners	100
Council Villapinzón	To support the Mayor	90
Governor	To support the Court Order	70
	To support sustainable development	90
Ministry	To get a consensus on the national tanners' politics	100
	To reach sustainable development	70
Court order Bogotá River	To recognize the tanners' legal rights	80
	To stimulate CP	100

The results in the Table 6-4 show that the control authorities (CAR and The Public Prosecutor) and the local and powerful stakeholders such as the mayor, the council and the big tanner considered building an efficiency image a priority together with short-term implementation and command-and-control approaches (values 100-80; and 100-90 respectively). The small tanners showed the will to change and solve their environmental conflict provided they were heard and were able to keep their identity as small enterprises (values 100-80). Stakeholders such as the Ministry, the Magistrate (only after getting to know about CP from the researcher) from the court Order on the Bogotá River and the Governor of the Province were interested in reaching sustainable development and CP implementation (values 100-70).

The analysis regarding the main stakeholders' interests suggested that:

- The conflicting situation was bound to get worse since the control entities were aiming only at quick and drastic measures to solve the environmental problem;

- All the stakeholders showed an interest in finding a solution to the industrial pollution;

- The tanners needed to make other stakeholders listen to them. They needed empowerment;

- The tanners were not a homogenous group but they all were interested in maintaining their identities as McSEs; and

- The Ministry, the Magistrate and the Governor could potentially become allies of the small tanners, even as indirect stakeholders.

b) Data collection in order to unveil the history of the conflict. For analysis of this, see section 6.2.

c) The BATNAs

The Best Alternative to a Non-negotiated Agreement (BATNA) was very bad for the small tanners: Definitive shut down of the tanneries. There was also a bad BATNA for the region and for the authority itself: by shutting down the industries, poverty and illegal businesses would flourish since people did not have any other choice than to undertake underground activities for their livelihoods. The above statements suggested that leading both communities and CAR through negotiations was the best choice for the community and the authority. They both needed to recognize that their relationship was not one of independence but instead of long-term mutual dependence.

d) A small tanner's profile

A small tanner's profile was defined based on a study of 60 tanneries in November 2004 (50% coverage) (Table 6-5).

Table 6-5 Average profile of tanners in Villapinzón in 2004

90% with low educational level and a humble personality
90% are afraid of CAR
60% live on subsistence economy
100% have a family oriented industry
80% considered the tanning chemical products salesmen to be their teachers
95% are working below their installed capacity
100% want to solve their environmental problem but distrust CAR.
90% have invested in individual end-of-pipe solutions, but have over-invested in such plant infrastructure.

Step 2. Building relationship

Helping the tanners on the riverbank to present their property documents to the Public Prosecutor (who considered them illegal settlers), built their trust in the researcher. Hence, this step included (a) building trust by providing help for urgent needs, (b) sharing information, and (c) trial and error approaches by the target community.

(a) The researcher, acting as a change agent, aimed at opening communication channels in order to build trust and long-term relationships. By using the Dialogue methodology, the tanners asked for help and the researcher became an empowerment agent in the beginning. First, direct access to the Public Prosecutor and the Ministry was worked on. Direct communication was established for the very first time between the authorities and the small tanners (in the past the authority contacted only the medium sized tannery owner).

(b) Sharing information took place for the first time. The tanners started to be heard. The land property legal papers were handed to the Public Prosecutor. CAR arranged meetings with the small tanners and realized that they seemed also interested in cleaning the river.

(c) The target group was actively seeking solutions to their problems and was undertaking trial-and-error approaches. The change agent became a source of containment and support for their 'psychological safety' (Schein, 1996).

The approach was based on appreciative attitudes and caused, on the one hand, that the tanners showed an interest in learning about CP. They chose the El Cerrito case as a legitimate example to follow. They visited El Cerrito. A first alliance was established and a new leader was proposed. The tanner that went through the initial experience in the preparation step with the researcher (myself) and that asked her for help was chosen to be their leader. On the other hand, the authority CAR distrusted the participatory process and the researcher (myself) in charge of it. Aggressiveness was observed from the side of the authority together with the mayor and the owner of the largest tannery. Negative rumours coming from their side attempted to discourage the tanners from trying to implement CP. However, the Public Prosecutor recognized the property rights on the riverbank and the tanners and other actors were ready to start redefining the problem (Figure 6-1).

Step 3. Redefinition of the problem

At this stage, new possibilities for the solution of the conflict had arisen.

(a) By using the Dialogue method, individual meetings and a first initial big meeting, a consensus was built over a new and only definition of the problem: 'The past options to solve the problem in Villapinzón did not take into account the small tanners' interests'.

(b) Elements of a strategy were developed based on the five initial issues identified.

The authority recognized that its relationship with the tanners is interdependent, on a long-term basis, and that it was also responsible for making the McSEs viable. Neither the tanners nor the authority had a good BATNA. If an agreement was not reached, both parties recognized that definitive closures would not solve the pollution problem. As the tanners did not know any other job, clandestine tanning activity would flourish and the pollution problems and social unrest would get worse.

The first 3 steps in the SASI process were mainly followed in small groups and lasted 3 months all together. Reliable information on the situation of the tanneries reached the Ministry of the Environment, the authority CAR, the Public Prosecutor and the local actors themselves. The property rights of the tanners on the riverbank were respected. As a result of a visit to the tannery community of tanners from Cerrito in the Colombian region of Valle del Cauca, and after receiving official support from the CVC's director, and advice from the regional officials in charge of Cerrito, individual tanners were eager to participate in a CP program (Annex 6.3.1). Trust was built with the researcher (myself) playing the role of a change agent. The tanners started to participate in the national leather committees at the Ministry of the Environment. However, actors like the former leader of the tanners, or the owner of the largest tannery tried to harm the relationship between CAR and the change agent (myself).

The strategy

At the end of step 3 of the SASI process (Figure 6-1), an intervention strategy was established based on the issues at stake identified in the initial steps. Table 6-6 presents the intervention strategy.

Table 6-6 Intervention Strategy

#	*Issue*	*Aiming at*	*Actions to be taken*
1	United tanners willing to change	Empowering tanners in order to give them negotiating power and improve the outcomes of a negotiation	- Initial 3 SASI steps - Broadcasting on Media -Building powerful alliances (Presidency, Senator)
		Social inclusion	Trust building
		Behavioural change	Reliable information
2	Process of Legalization	Social and economic inclusion: Financial support to become a possibility for CP implementation	Motivating the environmental authority CAR, control authorities (Public Prosecutor and National Comptroller) and the judge (magistrate) in charge of the court order on the Bogotá River to support the social inclusion of the tanners (Annex 6.3.2).
		Influencing the Court Order	Motivating a prestigious lawyers' firm to present a judicial action (not charging them) claiming the tanners' rights as socially excluded people (Annex 6.4.4)
		Getting approval from CAR for CP	Motivating the local judge to switch penal fines for environmental recovery work (Annex 6.4.11)
3	Inter-related land issues	Reaching sustainable solutions by integrating land and environmental issues	Motivating Ministry and regional environmental authorities to participate in OST, AI and Dialogue technologies and discuss such issues. Following 6 steps process Denunciations in newspapers
4	CP PROJECT SWITCH	Environmental sustainability and social inclusion through education and technology: Reaching a technical solution for McSEs in accordance with the requirements from the law, that supports prevention, and	Motivating academia, and environmental authorities to support the technological change chosen by the tanners. Following the 6 steps SASI process

		recognizes the interests of the tanners.	
		International and national support	
		Technical support	
		Political Support	
		Financial support	
5	Business Development program	Social inclusion through economic opportunities	Searching new markets
			Growth prospects
			Competitiveness through associative strategies on the leather chain
			International and national support
			Technical support
			Political Support
			Financial support

Step 4. Building common grounds among targeted group and among all the stakeholders

The process needed feedback and needed to be brought back to big group settings (big groups methodologies like Open Space technologies (OSTs) and Appreciative Inquiries (AIs) constantly in order to build common ground (steps 4-6) (Annex 6.4.1).

As a result of the first OST (June 2004),

1) CP was identified as the tanners' choice to solve their problems; the magistrate responsible for the court order on the Bogotá River supported that decision.

2) The woman tanner who had initially asked the researcher for help was officially chosen as the new micro-tanners' leader and the association grew stronger supported by 120 tanners.

3) The senator that was invited to participate since she was working on a proposal on a water law decided to support the conflict resolution process and opened channels of communications with the Office of the Presidency, the President himself and the Chamber of Commerce of Bogotá in order to support a legalizing process demanded by CAR that entailed the elaboration of Environmental Management Plans (EMPs), which included discharge and water use permits.

4) CAR stated that it needed to get proof of the tanners' willingness to change before considering evaluating that the EMPs could be based on CP.

In August 2004 following the participation of the magistrate in the SASI process, the Bogota River court order ruled on CP and pollution prevention implementation by the micro-tanneries. The tanners ought not just to implement CP, but they ought also to be supported by the environmental authority CAR, something that had been neglected for years. The magistrate had taken into consideration the specific characteristics and context of the micro-tanneries, which the researcher had presented to her in a document, used as supporting material during the court proceedings (Annex 6.3.2; 6.4.2). The magistrate designated the researcher as the only personal supervisor of the court order once it would be enforced (Court Order, 2004: 410, 459, 460) (Annex 6.4.2). She acknowledged this in a video conference with UNESCO-IHE Institute for Water Education in 2003 (Barrios and Siebel, 2003 DVD). As per 2013, the court order has had so many appeals from powerful stakeholders at the highest Court (*Consejo de Estado*) that it has not yet been enforced. The magistrates from this court asked the researcher to guide them on the solutions to be ruled for the tanners in August 2013 (Annex 6.4.3).

Following the strategy (Table 6-6) developed with the whole group of tanners, it took more than one meeting at the Office of the Presidency to reach solutions as initially CAR stated that CP was designed for clean industries becoming cleaner and not for dirty industries, and that CAR had no legal control over CP processes (personal communication, 2005[58]). As a way to show CAR their willingness to change, the tanners co-financed 15% of their own legalization process. Thanks to the support issuing from the Presidency, the CAR created a CP branch with the Chamber of Commerce (Annex 6.4.21).

In spite of all the efforts, the closures could not be avoided (6.4.5). The Supreme Court considered that the tanners had lost all the formal appealings (6.4.4). The CP Environmental Management Plans supported by the Chamber of Commerce were developed with even greater difficulties as the tanneries had been shut down (6.4.6). All the environmental plans based on CP (84)[59] ended up being technically approved by CAR for the first time in the history of the conflict, and the tanners grouped themselves in 7 water associations for water control use after a second OST was held (Annex 6.4.7 and 6.4.8). Despite the closures, the process was ready to go to the next step on agreements (Figure 6-1).

Steps 5 & 6. Creation of options and commitments, agreements, implementation, monitoring and evaluation

At the same time as a solution was negotiated with the authority, the negative actors like the initial tanners' leader and the owner of the medium sized tannery slowly started to get excluded from the process by losing power among their former supporters.

Reaching step 5, agreements on the legalization of the tanners and CP implementation plans, took 2.5 years while the first steps were reached in three months (Table 3-8) (Annex 6.4.8). Five AI and OST methodologies were implemented during that time and conducted visioning exercise for the years 2010 and 2020[60].

Once the tanners and the authorities were on speaking terms, international funding from the EU financed SWITCH project supported the needed implementation on CP in 2006 (Annexes 6.4.9 and 6.4.10). The National University of Colombia through the institute IDEA carried out the CP implementation and became an ally of the process (Annex 6.4.9). It was highly respected by the authority. The Colciencias institute and CAR decided to co-finance the activity (Annex 6.4.10). At this stage the last step started (Figure 6-1).

---

[58] Comment made at a public hearing at time of conflict by the CAR director in August 2005.

[59] This number was limited by the financing resources available to develop the EMPs.

[60] In 2005, through the visioning exercise the tanners expected to be open, implementing CP, in speaking terms with the authority, doing collective composting, and having a marketing plan for their products in the year 2010. In 2010, during another visioning exercise, the tanners were surprised to find out that their plans had been met.

The activities of the SWITCH project involved (responsibilities are presented in Annex 6.4.15):

1) Research on sustainable discharge limits from tanneries through understanding of the long-term effects of pollution,

2) Determining the best CP options for the micro-tanneries (Material & Energy balances),

3) Implementing CP options in six tanneries initially, and then with six more with the support from the environmental authority CAR,

4) Supporting training (30) on the best available technologies (BAT) but also on the good operational practices (GOP) and elaborating training material,

5) Monitoring the progress and impacts of implementation based on sustainability indicators (including the indicators assessing the initial situation Hs), and

6) Developing demonstrations on solid waste valuation such as composting derived from residues from the de-hairing and de-fleshing process. The tanners purchased a land plot for this purpose and are learning to work together.

The tanneries were not reopened until January 2008 (Annex 6.4.14). The formal policy on the riverbank issue was not put in place until the year 2010. The legalization of the industrial area, which implied discussions with large stakeholder groups, was finally established after six years in August 2011 (the tanners' leader support was strategic) and a study from UNAL confirmed the vulnerability of the Bogotá River and the importance of limiting the number of tanneries (Santos and Camacho, 2011).

6.3.2   Impact of policies and policy procedures on tanners

In the region of Villapinzón, the McSEs had been excluded (a) from governmental pro-poor policies because they were seen as private enterprises and (b) from opportunities offered to the private sector because they were considered illegal. The policy procedures used were exclusively based on punitive approaches from the judicial system.

The command-and-control policy on industrial discharges (decree 1594) did not stimulate prevention but pollution control measures entailing end-of-pipe technical approaches usually unaffordable for McSEs.

The environmental authority did not support CP for these kinds of industries because they were not considered clean industries. The CP policy is seen exclusively as part of voluntary agreements for industries willing to go further in their environmental endeavour. The only effort to implement CP in the past was disastrous. On the one hand, the private consulting company considered that the tanners were impossible to deal with; on the other hand the tanners argued that they were never asked to be part of a learning process.

The spatial planning policy created a non-action situation for environmental investments and implementations. The spatial planning policies entail a multiplicity of actors and interests that may interfere with environmental solutions (Annex 6.4.16).

6.3.3   What tanners need

In Villapinzón, the tanners needed to have their interests respected. They wanted to count on governmental support over the long-term through integral strategies on social, environmental and economic issues. They needed to be part of a learning process based on a high degree of transformative participation.

Their needs can be summarized by this tanner's statement at the time of the closures to the new head of the environmental authority CAR (2004): "Doctora Gloria Lucía, we have lived through a 30 years conflict which we realize we have not understood yet. Please, teach us how to disentangle our critical situation; we are people that did not have educational opportunities....".

6.3.4   Meeting the needs of tanners through policies and approaches

This AR worked towards meeting the tanners' needs with policies and approaches on marginalized groups. The court order on the Bogotá River (2004) supported the SASI approach based on transformative participation and integral solutions based on prevention; CP implementation took into account the characteristics of McSEs; CAR supported the process by financing six additional tanneries implementing CP, is transitioning towards phased approaches, and created a CP body run by the Chamber of Commerce; academia working on environmental interdisciplinary work, engaged in the process and was acknowledged by the environmental authority; international funding and support was offered by an international AR project called SWITCH which was designed specifically for the complexity of the situation; a credit line adapted to the tanners' needs was created by the governor (unfortunately, the local banks blocked this initiative); the tanners signed an agreement to work formally within the leather chain with suppliers and wholesalers; the tanners are part of the tannery committee at the Ministry; the tanners have a say on the meetings on spatial planning with support from the Governor (Annex 6.4.13); and a judicial order switching penal fines for environmental work was accepted by a local court (Annex 6.4.11). The ministry supported the judicial order (Annex 6.4.17). After seven years at the highest court in Colombia (*Consejo de* Estado), the Bogotá River court order (2013) ruled on the Villapinzón situation, acknowledging the efforts of this AR thesis and recognized the needs of the smallest tanners against powerful regional interests to build an industrial park, which was beyond the smallest tanners' reach. All the latter supported the integration of McSEs (Table 6-7).

6.3.5   Outcome of the process

Besides the above-mentioned change in policies and approaches, the 6 years' participative CP approach built affordable technical solutions for the McSEs, created collective solutions for solids valuation, and adapted the monitoring to more qualitative methods due to difficulties for an exclusive quantitative assessment.

- Substantial reductions of the negative impact from the industrial discharges were possible (Table 6-7). Between 2004 and 2009, reductions in pollution loads of 32-68% in Chromium, 60-72% in BOD_5, and savings of 24-68% on the effluent discharges to the river were measured in three of the twelve pilot tanneries. (Only in three tanneries measures could be consistently taken because other tanneries had been closed in 2004). UNAL plans to continue the monitoring scheme to more industries as 70 industries have been gradually implementing

CP as a result of the process with support from the Governor's office (personal communication, 2012-2)[61].

– Thanks to the mentioned reductions in liquid waste discharge, meeting the final legal standards will entail smaller and thus less costly wastewater treatment plants. A proposal to measure loads instead of just concentrations was being studied by CAR at the time of the completion of this thesis in order to adjust command-and-control to prevention strategies.

– The tanners are innovating their dehairing and production processes.

– A pilot project was developed with 30 tanneries on composting of fleshings and hair. Results are promising so far.

– When asked to evaluate SASI, 100% of tanners gave it a maximu score. The mayor (at the time of conflict) gave the lowest score (Annex 6.4.12). However, the board of directors of tanners wrote a letter of support (Annex 6.4.18).

Out of 120 tanneries:

1. 30 are actively implementing CP since they did not have land conflicts (12 of them were the SWITCH pilot industries),

2. 51 on the riverbank will have their lands purchased by the Governorship and will be relocated,

3. an additional 20 obtained their industrial land permits in July 2011 and are implementing CP, and

4. 19 were purchased by a highway consortium.

The above figures show that tanners are moving towards implementing prevention at slow motion not only because of difficulties related to education, culture or financing but because of inter-related land issues. With respect to the SASI process itself, Figure 6-1 highlights the key successes that allowed moving to the next steps. These successes confirmed the theories of N/CR and AR that inspired them (see Table 3-5). From step 1-2, asking for help was ideal; from step 2-3, finding the solution to an impending need like their property rights recognition allowed passing to the next step; from step 3-4, the need expressed by the tanners to build wide and open support from the entire community and later on by all the actors was essential: legalizing was a priority; from step 4-5, showing will to change, the fact that their EMPs were accepted for the first time despite the closures and that they were supported by high decision-makers and powerful actors allowed them to move to real agreements for their first time; from step 5-6, the SWITCH co-financing from Colciencias and CAR was a winning outcome that speeded-up the CP implementation. The empirical experience in Villapinzón confirmed the importance of thoroughly developing the preparation step; of working based on interests and not on positions; of applying the big group methodologies, which have surprising transformative powers; and of developing skills as observers on social dynamics for ARers.

---

[61] Comment made by the director of IDEA institute of UNAL at the Environmental symposium. June 2012. In 2013, a marketing plan is done with support from UNAL and the CCB.

Table 6-7 Results of intervention strategy in 2010 in Villapinzón

United tanners willing to change	Process of Legalization	Inter-related land issues	CP project SWITCH	Business Development program
The micro-tanner association (120 members) has become stronger: • participation at national committees • supporting the change process • visit to the Cerrito Tanneries to learn about CP. • a positive and new leader responsible for the association • consensus on the tanners' problems among all the direct stakeholders • cleaner production acknowledged as the right technical pathway for tanners • The Office of the Presidency of Colombia, the Public Prosecutor, the Governor and the Ministry of the Environment supporting the solutions built by the tanners. • positive reports in public media on tanners establishing principles based on conflict resolution as part of their organizational framework. learning to deal with high-level decision makers such as the National Comptroller, the Public Prosecutor, and the Magistrate • winning elections for mayor. • the leader of the tanners awarded as one of the best 35 leaders of Colombia (Semana, August 18 2011)	From 120 micro-tanneries: • 84 legalized in 2005 through their EMPs based on CP supporting 15% of the costs of this process • court order ruling that Cleaner Production should be implemen-ted in the area • a strategic alliance esta-blished after the first OST with a senator for the process building sup-port from Chamber of Commerce and the Office of Presidency • Pressure on CAR from the Office of the Presidency to allow CP and hence formali-zation. • Appealing on rights to work was denied by Supreme Court • A local judge allowing a pro-ject for envi-ronmental reco-very switching penal fines wor-king instead of paying the fines. February 2010 (first time in Colombia). • *Consejo de Estado* acknow-ledges SASI (2013)	Control authorities and political stakeholders supervising the process on the land conflicts • Governorship influencing CAR's board of directors to work by integration and proposing to buy lands on riverbank • A new major, sup-ported by the tanners, giving priority to the land issues • Setting the definitive spatial planning in 2009. Tanners beyond 30m from the riverbank allowed to work • The industrial use standards established finally in Dec. 2010	SWITCH-UNAL started in 2006: • Colciencias (Colombian Institute for Science and Technology) co-financing activities from 2007 • CAR itself co-financing training in CP, relying on support from academia • training leading to innovations on dehairing processes originating from the tanners themselves • between 2004-2009, there were reductions in pollution loads and water conservation in 3 of 12 pilot tanneries as the others were closed down. • future wastewater treatment plant costing less • solids valuation (composting) from the grease and hair residues • a land plot was bought for this common purpose and collective work stimulated • a monitoring tool created and based on sustainability indicators • public and academic media reporting positive change from the tanners in Villapinzón (6 newspapers articles, 3 magazine articles, 4 radio interviews, 1 TV note, 5 academic newspaper articles) • SWITCH project awarded sustainability prize at IWA in 2012	Colciencias-UNAL also focusing on competitive issues. The Technical Leather Body training on better quality products. UNAL training on business matters by the faculty of Business Administration The change agent working out access to loans from the Ministry of Commerce.

Inferences

The approach aimed at reducing social exclusion, and complementing end-of-pipe approaches with CP. Relative to social exclusion the research showed that:

- The tanners became united and chose a positive representative leader[62]. The tanners changed their polluting practices. Their negotiating power increased as they received support from the CRPML (CP body in charge of the Cerrito tanneries) and CVC, the Senator, the President's legal office, the magistrate,, the Governor, the Bogota Chamber of Commerce, and the National University. This led to press coverage and the Environmental Authority started to listen and negotiate with them. They gained political and organizational influence: A new mayor was elected, thanks in part to their support in 2010, and represented their interests for the first time in their lives. The tanners took the lead on spatial planning solutions in order to formalize the industrial area at the village council, which was delaying the technical solutions on CP implementation. They acquired a multi-scalar presence: The tanners are participating in the national tannery committee, which gathers formal representatives from all the tannery communities. The magistrate ruled in 2004 that the tanners were required to implement CP and that the authority was expected to support that initiative. The president of the tanners' association was designated as one of the best 35 Colombian leaders (Annexes 6.4.19, 6.4.20). The *Consejo de Estado* acknowledged all such efforts in 2013 and supported the interests of the underprivileged.

- The environmental authority CAR co-financed the CP implementation project and recognized that their past punitive approaches were inappropriate as (a) the pollution fines and lawyers' fees were unaffordable; (b) there were few positive incentives to encourage them to participate and implement policies effectively; (c) forced closures had not differentiated between the industries willing to change, provided they were offered support, from the ones that were not. CAR realized that their punitive approach stimulated clandestine industrial activities and corruption. Their participation in SASI helped them understand that since their relationship with the tanners was of a long term interdependent nature, they needed to help build solutions.

- The facilitator in this process was the researcher (i.e. myself) who committed on a long-term basis and whose role was backed by 'identified champions' (positive leading persons supporting commitment and change) within the different stakeholder groups. My independence was a key factor, which ensured that   the process did not lose momentum and assured persistence towards meeting the goals. My self-reflexivity allowed flexibility, improving processes, correcting mistakes and the capacity to go back on the cyclic processes.

Relative to the end-of-pipe approaches, the research showed that

- The AR led to the adoption of technical solutions based on CP and financed partially (15%) by the tanners once impending social issues were handled, mutual trust was built with the researcher and the tanners and the authorities were on speaking terms. The tanners were thus able to reduce their levels of pollution. They are doing solid waste valuation through composting the dehairing residues and have carried out innovations in their own production processes.

---

[62] This leader was appointed as one of the 35 best leaders of Colombia in August 2011 by the NGO Liderazgo y Democracia (leadership and democracy) among 300 nominees (Semana, 2011).

- The spatial planning solutions took three times as much time (compared with the technical discussions based on prevention) to be worked out by all the stakeholders and slowed down the CP implementation process itself.

Despite the success mentioned above, the SASI process presented weaknesses related (a) to theory, (b) methodology and (c) practice:

(a) SASI asks for long-term engagement and must, hence, deal with the continuous costs of engagement. As SASI's demands on real life engagement is high, dealing with every day's challenges hampers time for publishing. Since the power effects are a consequence of the engagement, diminishing the researcher's influence in the process as it becomes sustainable is advised, as well as involving multiple actors and "champions" in the process as early as possible, which can ensure legitimacy and ownership to the process itself.

(b) Since obtaining initial reliable quantitative data on CP implementation by McSEs is difficult, and the results on improving the environment have a time delay, it may be discouraging for academic institutions to support the on-going processes. It is hence advised to rely initially on systemic indicators (like organizational or the Hs built to assess the initial conflicting situation), which are more appropriate to evaluate the McSEs' initial transition towards sustainability.

(c) Three concerns are worth being recalled:

1. The intrinsic vulnerability of the under-privileged may slow down a SASI process. Despite the tanners' empowerment, their vulnerable nature due to their social and economic background makes them easy prey for political actors who want to manipulate them in order to obtain votes at the time of elections for example. In 2011, some tanners voted for a candidate for mayor that promised that they would not be bothered anymore with industrial environmental pollution abatement measures. Despite the slowdown, the most committed tanners since 2004 have been less vulnerable to manipulative promises and assured the survival of a long lasting process. The leadership of the president of the tanners' association, who received the national recognition as one of the 35 best Colombian leaders was diminished but not banned. The authority was able to regain control counting on the champions that had gone through a robust process of change. In 2012, as a result of the newly created APP (public, private alliances), the mayor has attempted to discredit the CP process promoting an industrial park supported by a multinational and by bringing big tanners from Bogotá. Strong opposition was found and the researcher was called to intervene. The empowered tanners of the process dismantled local corruption even at the CAR's regional office.

2. The lack of awareness of the financial institutions on the McSEs' particular situation makes it difficult to support the SASI processes. The researcher asked for support from the vice-minister of agriculture and from the minister of environment but at late stages in the process. Starting earlier may give more time for an awareness change to take place in this regard.

3. The interrelated land issues can severely hamper a SASI process in a country like Colombia if they are not taken into consideration right from the beginning. Property issues raise highly emotional cultural matters.

6.4   Conclusions

The AR, which helps to provide leadership in policy development and implementation fosters positive change, responds to an urgency to act in complex situations and provides effective sustainable results.

SASI offered an alternative to legal entanglements and judicial appeals, which entail expensive processes requiring the enlistment of legal services that are understandably costly for McSEs and not always willing to represent their cause faithfully. Also, the alternative opened opportunities for building relational capital and empowerment processes, easing the path towards formalization.

In particular, the application of the SASI method empowered vulnerable McSEs by

(a) turning the tanners into experts in their own field,

(b) helping the tanners pool their resources by establishing strategic alliances and raising their negotiation power,

(c) helping the tanners to overcome their exclusion from the policy process through a high degree of participation, and

(d) empowering the tanners to proactively participate in a change process adopting approaches towards prevention.

In addition, the SASI method fostered the transition towards prevention-oriented policies and approaches like CP implementation on the part of the governmental institutions which are not just tied to voluntary agreements, towards raising the awareness about the McSEs' needs, and towards supporting real participative approaches.

The SASI approach constitutes nevertheless a difficult endeavour that asks for long-term commitment, and faces high engagement costs, and delays in research and publishing. It must rely initially on obtaining data for change based on systemic indicators, which may suggest supporting groups and processes that may well be otherwise abandoned through traditional approaches based on obtaining just quantitative short-term data.

7   Conclusions and contributions

7.1   Introduction

This research focuses on the complex problems of micro- and small-sized enterprises (McSEs) impacting upon water bodies in the context of developing countries. It is estimated that this category of firms contributes significantly to employment *i.e.* in Colombia McSEs represent 67.9% of the work force (DNP, 2007a), but is at the same time responsible for 70% of industrial pollution (Alvarez and Durán, 2009; Le Van Khoa, 2006; Soni, 2006; Hillary, 1997).

Despite contextual differences related to where they operate, McSEs have several issues in common, such as not having more than 10 workers, limited access to technology, finance, capacity building, monitoring, markets, and social security coverage. Moreover, large numbers of McSEs operate in the informal economy (Alvarez and Durán, 2009). Research on McSEs is scarcely available in the Economics or Business Management literature (Audretsch, 2009), or in the environment-related fields (Blackman, 2010; Tokman, 2007; see also Chapter 1). McSEs in developing countries receive far less attention than medium-sized enterprises (Blackman, 2010; Caro and Pinto, 2007; Tokman, 2007; see also Chapter 4).

This research contributes to filling this gap by studying the situation of the micro tannery industry in Colombia. In particular, it studied the policy context relevant for the micro tannery industry on the basis of two case studies, Cerrito and San Benito, in addition to action research case study in Villapinzón. The study focused on the pollution caused by micro tanneries on water bodies, and the implementation of cleaner production (CP) measures.

This chapter aims to integrate findings from the preceding chapters, the literature review, the content analysis and the layered case study including the three sub-case studies. First the research question and its three subquestions are addressed, followed by how this contributes to science and recommendations for future research. Finally, the limitations of this research are noted.

7.2   Answering the research questions

7.2.1   Revisiting the questions

This research was driven by the question: How can micro- and small-size enterprises be effectively engaged to achieve national water and environmental policy goals in developing countries?
Literature reports that institutions usually privilege medium-sized production units (Alvarez and Durán, 2009) instead of smaller enterprises for two specific reasons. The first being that these small enterprises usually cannot afford the favoured end-of-pipe solutions, and secondly, that these enterprises are more present in disenfranchised communities that formal institutions find difficult to approach and communicate with, causing these smaller enterprises to be excluded from business and credit opportunities available to others (Thabrew *et al.*, 2009; Van Berkel, 2007; Montalvo, 2003; Ashton *et al.*, 2002). Against this background, this dissertation focused on Colombia as a single, but layered case study. Water polluting micro-tanneries were examined at the national level and in each of three sub-national locations. Specifically, three issues were investigated:

1) The institutional framework within which the firms operate: Sub-research question (1) - What is the current institutional framework in Colombia for integrating and supporting micro and small industries in their adoption of cleaner production? With this question we targeted the national context to determine whether the specific characteristics of McSE are recognized, see chapter 4;

2) The mechanisms for change that would result in CP implementation: Sub-research question (2) - Which perceived mechanisms support cleaner production and how do these influence the adoption of cleaner production by micro and small enterprises in Colombia? This answer to this question reveals the differences with respect to the local approaches regarding CP implementation in Colombia. It also provides valuable lessons for CP implementation in general and stakeholder participation, appropriate policy instruments in particular. Because approach and methodology turned out to be key for success, the insights from this research also contributes to social science methodologies for action research projects. See chapter 5 for the case studies of Cerrito and San Benito, and of Villapinzón in chapter 6;

3) Promoting social inclusion of disenfranchised tanner community through an action reseach approach: Sub-research question (3) - Can action research methodology be developed and tested to help McSEs implementing cleaner production and, if so, how? This question was addressed by the action research in Villapinzón during which a new method was developed to implement a more comprehensive and systemic approach of stakeholder participation that focuses on processes of social change, in which consensus building and decision-making were central. See chapter 6 on the case of Villapinzón.

7.2.2   The institutional framework

In order to understand how McSEs are included in the policy process, this thesis has specifically looked at a range of policies and laws that have relevance for this group, with particular attention to the water sector. The research revealed that even though there is considerable water legislation in Colombia, it is not comprehensive or fully integrated. It is not linked with land use planning or sectoral policies, nor does it address the entire population (Sánchez-Triana *et al.*, 2007). Accordingly, Colombia's institutional framework does not coordinate or address the needs of all the actors - and in particular neglects the McSE sector - and the multiple issues involved in the country's economic development.

This water governance framework is rooted in Colombia's socioeconomic inequality, historically exacerbated in rural areas. Coverage of basic needs is significantly greater in the country's larger urban centers (López, 2011; Echeverry *et al.*, 2011). The layered case study and the three sub-cases show that legislation, which takes place at the national level at the Ministry of Environment, is not always relevant to the local contexts. Water and environmental policy is administered at the regional department level - where most of the environmental budget is concentrated - and often fails to address the needs of the underprivileged groups, including the micro enterprises. Policy implementation at the regional level is often haphazard even when policies are appropriately designed. Private and political interests at the regional and local levels, where corruption and cronyism often hold sway, overcome the aims of national legislation that seeks to privilege the common good. In such a context, underprivileged groups have limited opportunities to be heard or plea their cause. The controlling authorities may either lack effective control instruments or can be prone to manipulation by the powerful local interests through corrupt practices. Accordingly, the country's institutional framework is poorly equipped to address urgent environmental, economic and social problems, including access to basic public services such as clean water and sanitation, health, education, justice and personal security, let alone financial or other support for a small business.

Not surprisingly, McSEs are barely taken into account for knowledge and capacity development programs. They are usually left on their own and their relationship with public agencies is largely limited to being at the receiving end of judicial sanctions emanating from their petty violations of environmental regulations. Challenges related to attending development needs of small firms, especially industries in rural areas, is left to the willingness of local politicians with varying capacity to deal comprehensively with the problems of their constituency.

Because water and environmental policies in Colombia do not take McSEs characteristics into account, or the informal economy context in which they operate, incentives for change are beyond the McSEs' reach. Many micro enterprises are relegated to a category identified as illegal or undesirable. Whatever business development programs public agencies may offer, these are tapped, if at all, by the larger group of SMEs, but not by the micro enterprises.

This thesis confirms the general characterisation of McSEs as provided in the rather scarce literature on the subject (Alvarez and Durán, 2009; Soni, 2006). These firms and their generally underprivileged owners tend to be excluded from (a) access to higher education because of its costs or because of low quality secondary schooling; (b) adopting cleaner technologies because of lack of information on affordable technical options or because of lack of financial opportunities; (c) public participation and decision-making processes, including those processes in which decision-makers are charged with addressing McSE issues; (d) associative or cooperative undertakings through trade organisations that exclude working with informal firms; (e) new market opportunities because of their lack of access to capital; and (f) programs that might assist them in improving product quality.

Accordingly, micro enterprises generally face a dire situation, with little or no recognition on the part of public agencies that indirectly discriminate against them. This thesis concludes that large numbers of families that make a living from micro enterprises comprise a disenfranchised and precarious group in Colombia's population and that the policy context contributes to this disenfranchisement.

7.2.3 Mechanisms of Change for CP implementation

This thesis confirms the importance of the classical mechanisms for CP implementation first presented in 1997 by Gunningham and Sinclair, namely: (a) the regulatory framework, (b) access to technology, (c) economic incentives, and (d) sound information. But, findings on the three sub-case studies by means of in-depth interviews, focus groups, and stakeholder analysis also contribute to enhancing the knowledge base on systemic (attitudinal and organizational) mechanisms that are considered by stakeholders as the backbone of successful CP implementation (Chapter 5), and by the relevant recent literature (Van Berkel, 2011; Mitchell, 2006; Montalvo, 2003).

Table 7-1 presents the identified mechanisms for successful CP implementation in the three sub-case studies from the standpoint of actor-oriented approaches. Mechanism (1) which aims at promoting systemic (attitudinal and organization) changes was mentioned in detail only after an initial CP implementation period; mechanisms ranging from (2 to 4) constitute the classical ones and were mentioned since the early stages of CP implementation. Underlying structural issues relating to unclear land ownership and use, specific to the Colombian context, fall under the regulatory mechanisms. Each change mechanism is represented by a dot, whenever there was consensus on its relevance for San Benito (SB), Cerrito (C) and Villapinzón (V).

Table 7-1 Comparing perceived mechanisms of change implemented in Cerrito, San Benito and Villapinzón.

1. Attitudinal & Organizational	SB	C	V
Focus on social and economic issues first		●	●
Emphasis on highly participative and trial-and-error processes		●	●
Consensus building		●	●
Regional authority's leadership and commitment to CP		●	●
Head of CP body: an expert on facilitating processes*		●	●
Long-term support		●	●
Academic engagement		●	●
Support from tannery experts		●	●
Collective work	●		●
2. Regulatory			
CP policy enforced with command-and-control regulation, measuring loads and concentrations		●	
Discharge limits according to local situations		●	
Lack of interference of land policies (unclear land tenure and spatial planning)		●	
3. Economic and persuasive instruments			
Financing on a long-term basis for projects and access to low interests rates **		●	
Market incentives	●	●	●
Access to information as a multilevel and continuous process		●	●
4. Choice of technology			
Decision-making developed by communities, researchers and experts. Authorities acted as observers		●	●

*In Villapinzon the researcher played the role of the Head of the CP body.

**Although policies to provide financing for CP exist (MADS, 2011), McSEs have major difficulties in accessing these finances because the financial markets consider McSEs high risk investments.

***Unclear land tenure and spatial planning create confusion and slow down CP implementation. In the 30 m riverbank protection area, legitimate property of the disenfranchised communitites, no industrial activity could be developed. Moreover, this land could not be sold to the state.

From Table 7-1 the following general tendencies/patterns become evident:

1) Actors' perceptions on the likelihood of successful CP implementation are tied to the existence of systemic (attitudinal & organizational) instruments.

   This is observed among all actors in the case study in Cerrito, in San Benito, only among the group of tanners that organized a common project for the wet blue stage in 2009 when the approach became more reflexive, and among all Villapinzón's actors in 2009. A major step towards problem resolution and systemic change occured once marginalized groups, who had been largely overlooked by institutions, were heard and respected and their impending issues and interests considered. This accelerated once these groups were supported by academics on a medium to long-term basis. This fact was clearly identified in Cerrito once the environmental agency designed its new strategy in 2003 (Chapter 5) and in Villapinzón once the tanners started to build trust with the researcher and the magistrate in charge of the court order of the Bogotá River. Through Open Space Technologies common ground can be targeted and positive change can be fostered because there is a common feeling of accomplishment among the people that participate (Chapter 6, see also Holman *et al.,* 2007; Holman, 2004; Raiffa, 2002; Owen, 1985). The AR component of this research illustrates that despite the fact that adjusting policies to local contexts and specific characteristics was not always possible at the start of CP implementation programs, long-term investment in time and effort helped to create an environment in which positive change was possible in terms of the institutional setting, the communities, and the environment. This finding seems to suggest that changes for McSEs may occur even in the absence of formal policy support if there is a third party willing to promote such change.

2) Implementation of distinct processes enables certain mechanisms of change to prevail over others. Each of the three cases provides evidence of this. In Cerrito, local authorities agreed to articulate and prioritize an agenda for McSEs; allowed, together with leadership from the academics involved in the project, for the launching of formal approaches for pollution prevention by means of a long-term CP implementation program; permitted a policy of CP implementation in parallel with command-and-control regulations; and allowed a CP approach based on social inclusion with transformative participation. The authorities in Cerrito decided to privilege the environmental over the spatial planning policies that had prevented industrial activity on the riverbank, provided that the tanners implement effective solutions that significantly reduced negative impacts. This meant that the tanners on the riverbank were allowed to continue working as long as they implemented their best available technical options and best practices under academic supervision. The experiences in Cerrito offered a lesson on how regional and local institutions with political will and social awareness can ensure long-lasting change and ease the transition to formalization of inclusive policies for marginalized communities. In fact, such a comprehensive process has produced an ongoing and continuous change process for almost ten years (2003-2013).

At the other extreme, in San Benito, actors from both the institutions and the community failed to invest in consensus-building. Decision-makers only consulted occasionally with the local actors. It was only in 2009, when a more comprehensive approach based on engagement of local stakeholders by the local environmental authority was implemented, that solutions were found. In fact, in that period of time, the urban environmental agency, amongst others, designed a new initiative supported by the leather trade body. Consequently, the San Benito experience suggests that mainstream instrumental approaches to CP implementation are not appropriate for McSEs insofar as they fail to respond to the needs of the community. Moreover, despite the engagement of some stakeholders in 2009, San Benito's interrelated spatial and environmental problems render the outlook for this community both confusing and vulnerable. San Benito was an exclusive industrial area at the beginning of the 20th century. Relocating the tanneries to a less vulnerable site was too expensive for the tanners since no institutional support was made available for that purpose. Today, environmental policies in the capital city do not allow tanneries in the San Benito neighbourhood because spatial planning policies are unclear. San Benito illustrates the lack of coordination of policies and of long-term systemic mechanisms entailing engagement and commitment with marginalized communities by the Colombian Government.

The community of Villapinzón was inspired by what transpired at Cerrito. Tanners in Villapinzón were facing a series of problems such as the lack of effective technical and financial assistance and the consequences of legal problems including judicial and penal fines. Because instrumental approaches were not working, intensive and long-term AR was carried out in conjunction with the community and the relevant institutions to understand and change the systemic (attitudinal and organizational) mechanisms that hindered progress. Activities included empowering tanners by building trust with the researcher, searching for strategic alliances at regional, national and international levels, and targeting effective technical and financial assistance through a new approach based on AR (SASI, Chapter 3). Although the process did not lead directly to appropriate regulation and the supply of credit by the state (Table 7-1), the AR in Villapinzón helped to empower the tanners to create active communication channels with the authority, and to achieve a feeling of self-accomplishment by community members committed to attain sustainable results.

The efforts to implement CP were widely broadcasted. Technicians that had been trying to implement technical solutions in the past testified to the change (SWITCH video, 2010). Between 2004 and 2009, reductions in pollution loads of 32-68% in Chromium and of 60-72% in BOD_5 were recorded. Savings in water use reduced liquid discharges into the river by some 24-68%. Their community leader was, with 35 others, recognized as belonging to the 35 best Colombian leaders in 2011. Moreover, the Court order recognized CP as the best solution to be supported by the institutions. Finally, the environmental agency, which only a few years earlier opposed CP implementation, became a financial supporter of the CP initiative. Last but not least, the tanners were able to sustain the change process despite the opposition of local political interests that had been supporting the construction of an industrial park to be managed by the big tanner in the area.

The Villapinzón case showed the need to design creative solutions for assessing CP implementation by means of semi-quantitative approaches. For example, UNAL assessed the extent of sustainability of the implemented solutions by considering conflict resolution indicators, and social and economic indicators in addition to the technical data. The former indicators could not be adequately assessed at the beginning of the process because of the closures that took place from 2005 to 2008 and the lack of monitoring tools. Initial efforts made to deploy exclusive quantitative methods proved impossible - only three out of twelve tanneries could be consistently assessed in the EU SWITCH project because of the closures when the project was starting. This confirmed the literature that argues that semi-quantitative methods may be more suitable for assessing McSEs (Del-Río-González, 2009; Howgrave-Graham and Van Berkel, 2007), at least in the initial steps.

Accordingly, this thesis suggests that long-term investment in reflexive and creative approaches can help to promote systemic change that allows (Cerrito & Villapinzón) effective CP implementation and, hence, less water pollution and violation of environmental laws. Disengaged short-term instrumental approaches are particularly unsuitable for promoting CP implementation in the specific context of the most vulnerable McSEs (San Benito). A correlation seems to exist between long-term reflexive approaches with stakeholder engagement and sustainable CP implementation in this context. The Villapinzón outcomes suggest that CP implementation need not necessarily work in parallel with command-and-control regulation in the initial stages. What seems essential is to provide communities with comprehensive support from organizations such as academic institutions, until such time as formal policies on discharge limits, for example, and strategies addressed to provide affordable credit or creative evaluation methods are adapted to informal economy contexts.

This research argues that the mainstream instrumental and consulting approach used in CP implementation programs is not appropriate for the specific characteristics of McSEs (Chapters 5 and 6) for the following reasons. First, environmental authorities do not see CP implementation as a flexible process that needs to be monitored and agreed upon in terms of the particular context of micro enterprises. Second, local authorities are reluctant to implement CP because they lack appropriate control instruments (CP policy often does not run in parallel with command-and-control, and is simply relegated to voluntary agreements). Third, local authorities appear to believe that CP is designed for formal enterprises that are already clean and want to become cleaner. Fourth, CP is generally supported by private initiatives, such as the CP centres supported by the Swiss government that are viewed as too expensive for McSEs and inappropriate to the specific social and environmental context in which change must be promoted. Fifth, CP implementation in the context of McSEs usually offers exclusive quantitative monitoring and evaluation instruments that cannot capture the initial process-related changes. Sixth, tanneries, and similar industries, lack appropriate monitoring systems as a result of which the initial balances of energy and matter are not easily determined. In essence, the principle of CP implementation (Van Berkel and Bouma, 1999) that awareness of costs and the cost-effectiveness of technologies lead companies to invest in CP do not necessarily apply to McSEs. In fact, in the context of McSEs, savings on CP may not be easily determined, or may simply not be possible at earlier stages of CP implementation.

Consequently, this research suggests the following measures:

(1) To focus on attitudinal and organizational (systemic) mechanisms, such as the possibility for long-term support; a high degree of participation; multidisciplinary capacity building programs based on learning through trial-and-error and the best available technical options; prioritising the understanding and recognition of context specific social and economic barriers for implementation; "identifying policy entrepreneurs" who take on a leadership role; and academic support to enhance the process of understanding the McSE context and raise institutional awareness on their specificities and needs.
(2) To build an institutional context for CP implementation, including appropriate command and control policies, and facilitate CP implementation by promoting continuous coordination between spatial planning and environmental policies. As shown above for San Benito, spatial planning measures often contradict environmental policies or for Cerrito, such policies may hamper practical solutions for local contexts if there is lack of political will to privilege McSEs' agenda.

This thesis concludes that while taking into account all available mechanisms for CP implementation, the systemic (attitudinal and organizational) issues must be addressed if successful and sustainable CP implementation is to be attained among McSEs. Such systemic attitudinal and organization challenges cannot be addressed simply through the instrumental (classical) approaches listed above.

7.2.4   The role of SASI Action Research in Change

During this research, the researcher developed - with support from the community of tanners - the Systematic Approach for Social Inclusion (SASI), a comprehensive approach to address the needs of McSEs in developing countries. SASI is a method, which includes principles, criteria, steps, and indicators drawn from AR, systemic negotiation, and conflict resolution, with the purpose to fill current gaps in policy instruments, stakeholder participation, and CP (Chapter 2, Tables 2-11 and 2-12, and Chapter 3). Building on these principles, the SASI objective is to integrate McSEs in society by preventing violence and supporting collective decisions, yet showing due respect for individual differences, and stimulating debates and problem-solving. The SASI approach proposes six sequential steps:

(1) identify the interests at stake, the health of the system, the urgency of matters demanding immediate action, and the initially perceived problem definition;

(2) share information between local, regional and national platforms, and foster trust among actors;
(3) build consensus on problem definition;
(4) empower communities by creating strategic alliances;
(5) promote a knowledge-based prevention platform; and
(6) implement technical solutions based on prevention approaches that are affordable to McSEs.

SASI includes a comprehensive and systematic way of dealing with the complexity of marginalized communities. SASI also helps discard narrow, preconceived notions of reality by featuring context-based strategies aimed at understanding the specificities of each setting, thereby allowing the researcher to become acquainted with the multiple barriers and challenges that constitute the reality of marginalization. It focuses on systemic analysis, recognizing unique realities, and is based on the researcher's long-term involvement. These features render SASI well suited to the special needs of McSEs.

SASI was developed to foster leadership and consensus in the complex policy-making and implementation process required in addressing McSEs, their communities and their institutions (as illustrated, e.g., in Villapinzón, Chapter 6). In fact, positive leadership resulted from "identified policy entrepreneurs" like the community leader (awarded as Best Colombian Leader in 2011) or simply by organizing big group methodologies where natural leaders like the magistrate or the senator were able to make contact with the community and positively inspire the local process.

In this research, SASI proved to be a powerful instrument in promoting CP awareness and implementation for McSEs involved in conflicts over environmental degradation.. In Villapinzón, over a period of six years, SASI fostered positive change. Through SASI a response became possible to urgently needed change in a complex situation, a vulnerable community became empowered, public agency awareness was raised, and effective, sustainable CP implementation was achieved.

The observed improvements in the conditions of the tanning community occurred by deploying conflict resolution methods such as Open Space Technologies (OST), centred on building common ground between a variety of stakeholders, understanding each other's interests, focusing on integrated solutions, and searching for technical options that fit local contexts through trial-and-error (Holman, 2004; Schein, 1996; Fisher and Ury, 1981).

SASI helped (a) empower McSEs to proactively participate in a change process, adopting approaches towards prevention and turning them into experts in their own field by offering theoretical and practical workshops stimulating an attitude of self-help (Schein, 1996); (b) the community to pool their resources by establishing strategic alliances, raising their negotiation power (Merlano, 2005); and (c)them to overcome exclusion from the policy process through transformative participation (Levin and Greenwood, 2011; Arieli et al., 2009; Van de Kerkhof, 2004; Cornwall, 2003). A climate favouring change was fostered, featuring mutual trust and learning, and joint ventures among stakeholders involved, and generating agreement to set rules for CP implementation.

In sharp contrast, during the previous two decades the tanneries faced internal disagreements, and judicial actions because of their violation of environmental and water regulation. SASI allowed technical innovation to emerge on the dehairing process by stimulating local knowledge through open and respectful debates between specialists and local communities, enabling the latter to choose their preferred technical option among a wide variety (Dieleman, 2007).

SASI proved effective as it offered a marginalized population a sense of accomplishment and self-reliance in the long-term with respect to solutions community members themselves implemented

(Figure 7.1). For example, breaking with their usual passive attitudes, tanners were actively involved in meetings with the authority where they could assert their right to implement CP (SWITCH video, 2010). The experience in Villapinzón showed that once tanners became self-reliant, the change process gained momentum and sustainable results could be obtained despite local interests opposing them (Chapter 6). Figure 7-1 expresses the key outcome that assured continuity of the process of change of a marginalized community living through conflicts.

Through conflict resolution methodologies, people develop a sense of belonging and commitment to honour their own agreements (Holman, 2004). For example, despite the standpoint of local and national actors, community members invested in the elaboration of Environmental Management Plans - requested by the environmental agency - as a sign of willingness to change (15% by tanners themselves), and in CP implementation that included adopting best practices, modifying chemical products, implementing recycling and re-using, and installing adequate primary effluent treatment facilities despite the fact that the tanneries were being closed by the environmental agency and formal financing was not an option. (The SWITCH project included training, workshops and technical advice but did not invest in technical improvements). In time, the tanners were invited to participate in trade programs targeted at small industries and received support from technical agencies aimed at capacity building in order to improve product quality.

Figure 7-1 Key effect of SASI on marginalized communities

Additionally, the SASI method holds the promise of assisting marginalized communities develop the potential to overcome their hardships as they address environmental issues, by

1. showing public authorities the need to shift from exclusive end-of-pipe policies to prevention-oriented policies and approaches, such as occurred with the 2004 and 2013 Bogotá River Court Order;
2. increasing national and regional institutional awareness of the needs of McSEs, as illustrated by the 2005 meetings held with Colombia's President; and
3. involving a leading university to pay attention to a major pollution issue hampered by social exclusion, triggering long-term involvement and holding the promise of significant implications for social science, CP dissemination, and development of the management of McSEs.

SASI allows for systemic change mechanisms to take effect and lead to successful CP implementation, even before regulatory and financial mechanisms are in place. Accordingly, SASI may contribute to assist transitions towards formalization.

Nonetheless, SASI is a very demanding process for both researcher and community. Researchers must assume a long-term commitment and learn to face and deal with power effects. E.g. in Villapinzón the researcher, because of her privileged social background and networking abilities, was able to reach the higher political levels and, consequently, counterbalance the traditional high local interests who at the beginning felt threatened and tried to block the SASI process.

A reason that makes the approach used in this research project valuable for studying complexity may also be used against it *i.e.* the high costs associated with long-term and continuous involvement. High costs are justified by the fact that the followed approach allows to keep track of the chain of events leading to change in general (combined case study method) and even behavioural change (SASI). In the specific case of Villapinzón, the costs of decades of conflict, non-action and misinformation, exceeded all other costs by an ample margin (Chapter 6).

SASI proved successful in Villapinzón despite the community's deep internal social conflict. At the start of the SASI process, the direction of change eventually achieved was by no means ensured. It encouraged the empowerment of communities, seeking long-term results in uncertain institutional settings *e.g.* policies may abruptly shift with a change in government or public agency official, and both technical and social decisions and policies are in practice contingent upon political forces or privately led interests. The success of SASI lies squarely on the capacity of empowered communities for self-reliance, resilience and endurance (Figure 7.1), despite the limitations they face *e.g.* with respect to access to technology, markets, credits, education (section 1.2.2) or due to their fragile socio-economic context that make them more exposed and likely to suffer from exploitation and abuse of power. In the specific case of Villapinzón, local tanners had been made many promises by politicians and business representatives that failed to materialise, resulting in what Lewin (1946) describes as cycles of disconfirmation, which can be understood as cycles of disappointment. Accordingly, SASI places marginalized communities at the core of the events, supporting them as the heart of the environmental, social and technical process of change. In turn, the communities become a powerful source of support for the researcher (Chapter 6).

7.3   Contributions to science

At the global level, this thesis contributed to what Delli Priscoli and Wolf (2009) identified as bringing the environmental debate beyond critique into practice, by designing and implementing solutions, prioritizing behavioural change at the level of institutions, practitioners, community members, academics, and decision-makers. At the local and national levels, it focused concern towards values such as equity, transformative participation, and conflict resolution with respect to McSEs. A proactive attitude was taken in addressing complex local problems by deploying in-depth dialogue and targeting legitimate McSEs community concerns. It was successful at changing attitudes by means of 'socializing' technology, adapting it to local needs through action research on CP implementation.

In accordance with the three sub-research questions previously presented, this research

a)   contributes to the knowledge base of the research field of McSEs in developing countries,
b)   adds value to CP implementation in the context of micro-enterprises, and
c)   has methodological implications for the social sciences.

This section develops these three contributions in detail. In order to support the social inclusion of McSEs in developing countries like Colombia, in this thesis a (1) framework for analysis of McSEs was developed based on recognizing complexity, focusing on a conflict resolution dimension, targeting a development model based on inclusion and relating theory to practice.

Recognizing complexity (Mason and Mitroff, 1981, in Van de Kerkhof, 2004; Gummesson, 2007a) was essential to generate insights on policy design and implementation based on understanding of the multidimensional phenomenon of McSEs. E.g. in Cerrito, the CP policies were defined in a participative manner, on the one hand tied to command-and-control but, on the other hand, also setting time lines for the phased implementation of the CP approach. Inspired by Cerrito, the community in Villapinzón strongly recommended their environmental authority to define a similar strategy and the authority publicly admitted their mistakes in the past.

Introducing conflict resolution methods offered alternatives to exclusive judicial appeals, which may entail expensive processes with lawyers not always accessible to McSEs or willing to truly represent them. The Villapinzón strategy undertook joint, closely consulted efforts. Since conflicts were at the heart of the problem, by setting limits to expectations and clear rules through appreciative methods, actors with negative attitudes were discouraged from interfering and hampering fieldwork. This implied a break with mainstream approaches, which do not voluntarily choose to deal with conflict. The latter was evident in Villapinzón where academics and officials were discouraging the researcher from intervening in the area: at the time that the presidential meetings took place, the director of the regional environmental agency warned the researcher of the high risk of wasting effort and money by dealing with this twenty-year-old conflict.

Addressing a development model based on inclusion demands an analysis of the values and purposes of society. The results of this thesis suggest the value of support approaches, which may generate successful outcomes, provided they are based on respecting contexts, social differences and local interests, and can target community empowerment and self-reliance. The experience shows that excluded groups integrate in the society at their own pace e.g. by offering the possibilities of establishing networks and formalizing their activities, 25% of the owners of the McSEs in Villapinzón (20 out of 80) acted as entrepreneurs willing to grow, i.e. taking risks, innovating, taking advantage of a credit line offered by the Governor while 75% (60 out of 80) expressed their interest in living in peace without conflicts with the environmental authority through being law abiding without a further growth ambition. These results support what Mel et al. (2008) have suggested which is that an identified lack of growth of the McSEs may respond also to a lack of ability or desire to grow rather than just lack of financing opportunities. Regardless of the reasons for formalization, the tanners improved their quality of life and were able to have self-determination for the following five years. In all cases, initial outcomes of the project suggested that an integrated approach proved effective. Public agencies concerned with environmental pollution were invited to coordinate actions and implement multidimensional instead of linear solutions. The formalization process became a collective effort, which brought benefits for all. This work supports Stiglitz's (2012) thesis stating that a better world will be possible once the privileged people understand that their destiny is linked to the way the great majority of people live. The legitimacy of the state grows through inclusive policy (Perry et al. 2007) especially also aimed at the smallest industries.

Relating theory (see Chapter 2) to practice was a continuous effort. The theoretical framework entailed integration and interaction of *policy instruments, action research negotiation/conflict resolution, stakeholder participation,* and *cleaner production.* Dialogue with practice constituted a paradigm shift, because it put the McSEs and underprivileged community members as agents able to become empowered, hence breaking their own cycles of exclusion.

Challenges that are difficult to overcome remain, for example, the abuse of power effects, the need for long term commitment, and for academics; in particular the distance to analyse and write while the process is continuously changing and action constantly needed.

By integrating the fields presented above (2), CP programmes which emphasize waste prevention and reduction approaches (eg. waste valuation, recycling, re-using) and which stimulate self-reliance, were able

- to be participative (participation being a means or an end) and context and actor oriented,
- to improve scientific practice, policy-making and cross-sectoral planning because of the field of stakeholder participation,
- to stimulate networks, mutual learning, collaborative working and change processes,
- to assess the risk of intervening in a process,
- to focus initially on semi-quantitative approaches,
- to put scientists, policy-makers and practitioners on equal footing because of the field of managed learning or action research,
- to focus first on socially-robust solutions before technical implementations,
- to work either bottom-up or top-down (in parallel with command-and-control),
- to deal with conflicts, power imbalances and empowerment,
- to lead negotiations and decision-making because of the field of negotiation and conflict resolution,
- to stimulate the transition to formalization,
- to strengthen institutions, and
- to stimulate innovation and systemic solutions because of the field of policy instruments.

Last but not least, this thesis enriches (3) the research field of the social sciences on detailed qualitative casework over time developing a valuable learning experience for the researcher as well as for community and institutions that asked for more than a snapshot to be able to deal with environmental issues related to McSEs in developing countries such as Colombia. The methodology responded to the complex demands of both Colombia as a layered case study, and those of underprivileged micro-tannery communities. Understanding such multi-scalar contextual complexity demanded a focus on developing description and analysis through context-orientation, actor-orientation and, hence, qualitative detailed casework over time. Because the topic of this research concerns a process of change, an action research approach became self-evident in the third case where SASI was applied. The combined case study methodology and SASI were inspired by Yin's case study approach (2009) and Burawoy (1998), and Lewin's managed learning (1946) or action research (Quental et al., 2011; Jacobs, 2010; Dobers and Halme, 2009; Fals Borda and Mora-Osejo, 2003), and Schein (1996) respectively. Given the complexity of the problems addressed, both methods crystallize the need for long-term research that well exceed whatever benefits may be obtained from short-term efforts producing quick results. Instead of limiting data collection to specific activities at certain isolated moments, ongoing intense relationships with actors may generate some micro-cases that regularly provide information and commentary, thereby enriching the research activities. In such cases, those identified actors evolve as "policy entrepreneurs" and may enrich the sources of information of the research, as for example, the technician responsible for CP implementation programmes in Cerrito, or the new leader in Villapinzón.

The literature review (Chapter 2) on scientific approaches towards communities shows that scholars on AR (action research) have long addressed issues concerned with marginalized groups, developing and applying knowledge drawn from transformative processes in the organizational field of business management, and from sociology (Brydon-Miller *et al.*, 2011; Levin and Greenwood, 2011; Faure *et al.*, 2010; Bodorkós and Pataki, 2009; Fals Borda and Mora-Osejo, 2003). Following that trend, the SASI approach created and applied in this research goes further in the sense that it integrates and improves (Chapter 2) the theories of stakeholder participation, policy instruments and cleaner production acting in the context of McSEs and constructed a more comprehensive approach on action research to complex socio-ecological dynamics in the environmental field in developing countries like Colombia. By acting together, theories are brought into action and constantly revised as shown on the contributions for the McSEs' field. SASI was presented in the previous section and constitutes a new systematic look, which builds on insights into existing transformative processes and theories such as AR, systemic negotiation, and conflict resolution, applied and analysed for six years. SASI led to effective change among a group of McSEs in Colombia and was consistent with Lewin's (1946) observation that a good theory for the social sciences should always be practical.

This thesis contributes towards enriching the role of the researcher, to be agent of change or agent of inclusion (AoI). The AoI features additional characteristics from those described for the agent of change i.e. being able to identify when to be a helper, a facilitator or a mediator, as well as being able to critically assess the risk of intervening (Schein, 1996). Moreover, the AoI is willing to commit to long-term involvement, to be open to learn from the processes, and to develop skills to act in an independent, holistic and self-reflexive manner (Chapter 6).

7.4   Recommendations, shortcomings and future perspectives

Recommendation 1: Notwithstanding different epistemological approaches, certain fields of knowledge tend to organize knowledge into separate boxes, thus leading to artificial dilemmas between scientists and practitioners, and between the social and technical sciences. This thesis recommends merging the boxes and building a more robust knowledge base, by means of systemic approaches that may be best suited for today's complex world. It invites stakeholders to get involved in social change and to support institutions that contribute to social justice, environmental protection and poverty eradication.

Recommendation 2: SASI can contribute towards integrating marginalized communities, provided it is accompanied by context-relevant capacity building approaches. The lessons learnt from this thesis may serve other excluded communities. Based on the virtues of how to choose methods for complex environmental inquiry based on the kind of research question suggesting (or not) close interaction with actors, intervention for change processes, or for leading negotiations.

Recommendation 3: With respect to the Colombian environmental framework, the country should (a) redesign policies prioritizing prevention and make them operate in parallel to command-and-control instruments; (b) include marginalised communities through participative approaches, alternative venues for conflict resolution and specific financing options; (c) strengthen the institutional capacity of the control authorities; (d) establish differentiated policies for the different informal groups; and (e) improve systemic monitoring, which can identify either structural or non-structural challenges.

Limitation 1: Integrating marginalized communities into policies and approaches entails expensive, multilevel and long-term commitment from actors and institutions. This requires awareness of the risks and responsibilities involved and demands the involvement of state agencies in the process.

Limitation 2: SASI relies on obtaining data on systemic indicators, which may be difficult to determine. To apply SASI, training is necessary so as to be able to obtain the necessary data but also to encourage a process that recognises marginalized groups as experts in their own fields and allow them to become innovators.

Future perspective 1: The present work targets Colombia. Future research initiatives should focus on other Developing Countries and build more robust conclusions relevant to both theory and practice. A future contribution to this effort may well be to develop more indicators from the common elements in the six steps of SASI.

Future perspective 2: In Colombia, an organizational framework for CP, based on effective cooperation between ministries could benefit from the fact that there is growing global interest in prevention technologies that are more accessible to McSEs. Provided there is a comprehensive approach to CP implementation, this field can also ease the transition from informality to formality and lead to effective law enforcement.

8   References

Access Initiative, 2009. The Access Initiative. http://www.accessinitiative.org/resources/168. Retrieved November 22, 2012

ACOPI, 2007.www.acopibogota.org.co Retrieved on November 19 2012

ACUEDUCTO-UNAL, 2010. Campañas rio Bogotá given on a CD.

Adelle C., Fajardo T., Pallemaerts M., Withana S., Van Den Bossche K., 2010. The external dimension of the sixth environment action programme: An evaluation of implementing policy instruments. Final report. The Institute for European Environmental Policy. 128p.

Agosin M., Larrain C., and Grau N., 2009. Industrial Policy in Chile. Working paper SDT 294. Facultad de Economía y Negocios. Universidad de Chile. 61p.

Alfaro A.J., Franco A., and Torres A., 2000. Desastres naturales y Desplazados en Colombia. De desastre natural a Catástrofe Social 7p. Universidad Javeriana Instituto Geofísico. www.ainsuca.javeriana.edu.co/geofisico/Documentos/Acofi/Desastres2000.PDF Retreived May 31st 2004.

Altham W., 2007. Benchmarking to trigger Cleaner Production in small businesses: Dry cleaning case study. Journal of Cleaner Production (15), p798-813.

Alvarez G., 2003. La planificación hidrológica y el manejo de las cuencas hidrográficas. In: Derecho de Aguas. Tomo 1. Universidad Externado de Colombia (eds), p13-59.

Alvarez M., and Durán J., 2009. Manual para Micro, Pequeña y Mediana Empresa. GTZ, CEPAL, 95p.

Amaya O., 2003. La protección del agua en Colombia dentro del marco de la Constitución política y ecológica de 1991. In: Derecho de Aguas. Tomo 1 Universidad Externado de Colombia (eds). p103-123.

Amonoo E., Acquah P., and Asmah E., 2003. The impact of interests rates on demand for credit and loan repayment by the poor and SMEs in Ghana. ILO ISBN 92-2-11489-X p50.

Andrews R., 2006. Managing the environment, managing ourselves: A history of American Environmental Policy. Yale University Press.

AnteaGroup, 2012. Colombia water sector. Market survey 2012. Final report for the Embassy of the Netherlands. p79.

Argyris, C., 2004. Reasons and rationalizations: The limits to organizational knowledge. Oxford, UK. Oxford University Press. p250.

Argyris, C. Putnam. R and McLain Smith. 1985. Action Science. Concepts methods and skills for research and intervention. Jossey-Bass. San Francisco. USA. ISBN 0-87589-665-0 p251.

Argyris, C. and Schön, D., 1996. Organizational Learning II: Theory, method and practice. Addison-Wesley. ISBN 0-201-62983-6 p180.

Arieli D., Friedman V., and Agbaria K., 2009. The paradox of participation in action research. Action research Vol 7(3): p263-290.

Arnstein S.R., 1969. A ladder for Citizen Participation. JAIP, Vol 35, No4, July 1969, p216-224.

Ashton W., Luque A., and Ehrenfeld J., 2002. Best practices in cleaner production promotion and implementation for smaller enterprises. School of forestry and environmental studies. Yale University. New Haven CT. Interamerican Development Bank. Multilateral Investment Fund. 56p.

Ataov A., 2007. Democracy to become reality: participatory planning through action research. Habitat international Vol 31. p333-344.

Audretsch D., Grilo I., and Thurik A., 2009 Explaining entrepreneurship and the role of policy: a framework. In: Audretsch D., Grilo I., and Thurik A.(eds) Handbook of research on entrepreneurship policy. ISBN 9781 84542 409. 17p.

Aziz A., Shams M. and Khan K., 2011. Participatory action research as the approach for women's empowerment. Action research 9(3).p303-323.

Azuma Y., and Grossman H., 2008. A theory of the informal sector. Economics and Politics Vol 20, March 2008. 18p.

Baas, L., 2007. To make zero emissions technologies and strategies become a reality, the lessons learned of cleaner production dissemination have to be known. Journal of Cleaner Production (14). p1205-1216.

Bacow, L., Wheeler, M., 1987. Environmental dispute resolution. Plenum Press. p371.

Bakker J., Denters B., Vrielink M., and Klok P., 2012. Citizen's initiatives: How local goverments fill their facilitative role. Local Government Studies Vol 38 # 4. p395-414.

Banrepublica, 2012. Producto Interno Bruto Colombiano http://www.banrep.gov.co/series-estadisticas/see_prod_salar_2005.html. Retrieved November 17, 2012.

Barcellos P., Galelli A., Mueller A., Dos Reis Z., and Peretti J., 2009. Collaborative Networks: An innovative approach to enhance competitiveness of small firms in Brazil. Working paper. University of Caixas do Sul. Graduate Program in Business. p6.

Barde J.P., 2000. Environmental Policy and Policy Instruments. In:Principles of Environmental and Resource Economics: A Guide for Students and Decision-makers. H. Flomer, H. L. Gable and J. B. Opschoor. Aldershot, Edward Elgar. p157-201.

Barreto L., 2001. Technological learning in Energy Optimization Models and Deployment of Emerging Technologies. Swiss Federal Institute of Technology. Zurich. Ph.D. Thesis. 295p.

Barrios R. and Siebel M., 2003. Minutes Videoconference for the lawsuit of the Bogotá River. Dec 15.

Barry B., 2008. Negotiator affect: The state of the art (and the science). Group Decis Negot Vol 17. p97-105.

Baskerville R., 1997. Distinguishing Action Research from participative case studies. Journal of Systems and information technology. Vol 1 #1.p24-43.

Beck T., and Demirgue-Kunt A., 2006. Small and medium-size enterprises: Access to finance as a growth constraint. Journal of Banking and Finance (30). p2931-2943.

Beckerman W., 1995. Small is Stupid: Blowing the Whistle on the Green. London, Duckworth Press. p190.

Beder S., 2006. Environmental principles and policies. An interdisciplinary introduction. Earthscan. ISBN 9781844074044. p304.

Berk, M., Hordijk,L., Hisschemöller, et al. 1999. Climate OptiOns for the long term (COOL). Interim phase report. NRP. Report nr. 410200028. Bilthoven. The Netherlands. p23-64.

Bernardini M., 2008. Environmental impact evaluation using a cooperative model for implementing EMS (ISO 14001) in small and medium-sized enterprises. Journal of Cleaner Production Vol 16. p1447-1461.

Best R., Ferris S. and Schiavone A., 2005. Building linkages and enhancing trust between small scale rural producers, buyers in growing markets and suppliers of critical inputs. Beyond agriculture, making markets work for the poor. Conference London. p30.

Biller, D., and Quintero, J.D., 1995. Policy Options to address informal sector contamination in urban Latin America: The case of leather Tanneries in Bogotá Colombia. Laten Dissemination note #14. World Bank. p33.

Blackman A., 2000. Informal Sector Pollution Control: What Policy Options Do We Have? World Development 28. p2067-2082.

Blackman A., 2005. Colombia's discharge fee program: Incentives for polluters or regulators? Resources for the future Discussion paper 05-31. p29.

Blackman A., 2006. Small firms and the environment in developing countries: Collective impacts, collective action. Journal of International Development Vol 20 issue 2. p247-248.

Blackman A., 2008. Can voluntary environmental regulation work in developing countries? Lessons from case studies. The Policy Studies Journal Vol 36 #1. p24.

Blackman A., 2010. Alternative pollution control policies in developing countries. Environmental Economics and Policy Vol 4# 2. p234-253.

Blackman A., 2011. Does eco-certification boost regulatory compliance in developing countries? ISO 14001 in Mexico. Resources for the future. Discussion paper 11-39. p24.

Blackman A. and Kildegaard A., 2003. Clean Technological change in developing-country industrial clusters: Mexican leather tanning. Resources for the Future. Discussion paper 03-12. p26.

Blackman A. and Kildegaard A., 2010. Clean technological change in developing country industrial clusters: Mexican leather tanning. Environmental Economics and Policy Studies 12. p115-132.

Blackman A., Morgenstern R., Montealegre L., and García JC., 2006. Review of the efficiency and effectiveness of Colombia's environmental policies. Resources for the future. p315.

Blackman A., Morgenstern R., Topping E., Hoffman S., and Sanchez-Triana E., 2007a. Actors and Institutions. In: Sánchez-Triana E., Ahmed K. and Awe Y.,(eds) 2007. Environmental Priorities and Poverty Reduction. A country environmental analysis for Colombia. Directions in Development. The World Bank. p33-65.

Blackman A., Lahiri B., Pizer W., Rivera M. and Muñoz C. 2007b. Voluntary Environmental Regulation in Developing Countries: Mexico´s clean industry program. Resources for the future. p34.

Blackman A., Uribe E., Van Hoof B. and Lyon T., 2009. Voluntary Environmental Agreements in Developing countries: The Colombian experience. Resources for the future. IADB and RFF Report June. p67.

Blackman A., Uribe E., Van Hoof B. and Lyon T., 2012. Voluntary Environmental Agreements in Developing countries: The Colombian experience. Resources for the future. Discussion paper 12-06 p51.

Blake R. and Mouton J.S., 1964. The managerial grid: key orientations for achieving production through people. Gulf Pub. Co. p340.

Bodorkós B., and Pataki G., 2009. Linking academic and local knowledge: community-based research and service learning for sustainable rural development in Hungary. Journal of Cleaner Production Vol 17. p1123-1131.

Boehmer-Christiansen, S., 1994. Politics and environmental management. Journal of environmental Planning and Management, 37(1), p69-86.

Bogdan R. and Biklen S., 1992. Qualitative research for education: an introduction to theory and methods. Allyn and Bacon, Needham Heights, MA. p 66.

Bohm P. and Russel C.S., 1985. Comparative Analysis of Alternative Policy Instruments. Handbook of Natural Resource and Energy Economics. A.V. Kneese and J. L. Sweeney. Amsterdam, Elsevier Science Publishers BV. 1.

Bourdieu P., 1998. Acts of resistance. Against the new myths of our time. Wiley publications. ISBN 978-0-7456-2217-0. 120p.

Bradford U, 2012. Unit2: History of conflict resolution. www.brad.ac.uk/acad/confres/dislearn/unit2.html. Retrieved May 9 2012.

Bravo D., Deeb A., Gonzalez H., Vanderhammen T., Herran A., Jaimes F., Jaramillo W., Mammen D., Mendoza A., Mockus A., Moncayo E., Noriega M. and Ortiz J. C., 1999. Hacia la metropolización de la Sabana de Bogotá. Por una planificación del desarrollo sostenible. CAR. p159.

Brett J., 2007. Negotiating globally. How to negotiate deals, resolve disputes, and make decisions across cultural boundaries. 2nd edition. John Wiley & Sons, Inc. ISBN : 978 07879-8836-4.

Brydon-Miller M., Kral M., Maguire P., Noffke S., and Sabhlok A., 2011. Jazz and the Banyan tree. Roots and riffs on participatory action research. In: Denzin N., and Lincoln Y. The SAGE handbook on qualitative research. p387-400.

Burawoy M., 1998. The Extended Case Method. Sociological Theory 16:1.p33.

Burawoy M., 2007. Open the social sciences: To whom and for what? Portuguese Journal of Social Sciences Vol 6 # 3. p137-146.

Burnes B., 2004. Kurt Lewin and the planned approach to change: A re-appraisal. Journal of Management Studies 41:6 977-1002.

Burton J., 1990. Conflict : Resolution and Prevention. St Martin's Press. 295p.

Camargo M., Dal Bo G., Panizzon M., Andreola J., and Ventura M.E., 2010. Collaborative networks, social capital and relationship marketing: Competitive divergences, convergences and unfoldings. Global Journal of Management and Business research. Vol 10 Issue 2. p23-35.

Canal F., 2004. Orientación del presupuesto de las corporaciones autónomas regionales CARS- en los planes de acción trianual; periodos 1995-2003. Consultant report prepared for the World Bank, Washington, DC. 30p.

Canal F., and Rodríguez M., 2008. Las corporaciones autónomas regionales, quince años después de la creación del SINA. In: Rodríguez M. (ed), Gobernabilidad, instituciones y medio ambiente en Colombia. Foro Nacional Ambiental. p303-389.

CAR, 1994. Tannery courses, p1984-1994.

CAR, 1998. Revisión documentos curtiembres. p1982-1996.

CAR, 2004 Términos de referencia para la presentación del Plan de Manejo Ambiental dirigido a la industria del cuero, October. p 10.

CAR, 2006. Acuerdo 043 Parámetros de calidad del río Bogotá

CAR, 2008. Public document on the tanneries' new approach

CAR, 2009. Plan de acción trianual.

Cárdenas M., and Mejía C., 2007. Informalidad en Colombia: Nueva evidencia. Fedesarrollo Working papers March 2007. #35. p43.

Cardona A., 2003. El régimen jurídico de las Aguas en Colombia. In: Derecho de Aguas.Tomo 1 Universidad Externado de Colombia (eds). p151-167.

Caro S., and Pinto J.A., 2007. Tránsito Informalidad- Formalidad: La hora de la inclusión. Debate político #26. Konrad Adenauer foundation. La Imprenta editores. 70p.

Carrizosa J., 2008. Prólogo. Instituciones y ambiente. In: Rodríguez M. (ed), Gobernabilidad, instituciones y medio ambiente en Colombia. Foro Nacional Ambiental. p2-64.

Chambers R., 1981. Rapid rural appraisal: rational and repertoire. Public Administration and Development Vol 1 p95-106.

Chambers R., 1994. Participatory Rural Appraisal (PAR): Analysis of experience. World Development Vol 22 issue 9 p1253-1268.

CIA, 2012. The World Fact book. Retrieved, November 19, 2012. https://www.cia.gov/library/publications/the-world-factbook/geos/co.html.

Civico A., 2012. El fin del capitalismo. El Espectador. 31 de Enero. www.elespectador.com/search/apachesolr_search/Aldo%20Civico%20desigualdad%20en%20Colombia.

Cleaver F., 1999. Paradoxes of participation: questioning participatory approaches to development. Journal of International Development Vol 11. p597-612.

Cloquell-Ballester, VA., Monteverde-Díaz, R., Cloquell Ballester, VAn,, Torres-Sibille, A ., 2008. Environmental education for small and medium-sized enterprises. Methodology and e-learning experience in the valencian region. Journal of Environmental Management 87. p507-520.

CNPML, 2004. Proyecto gestión ambiental en la industria de curtiembre en Colombia. Diagnóstico y Estategias. Centro Nacional de Producción Más Limpia. 23p.

Coenen F., 2009. Local Agenda 21: Meaningful and Effective participation? In: Coenen F., (ed). Public participation and better environmental decisions. Chapter 10. p165-179.

CONPES, 3320. Política Nacional para la recuperación del río Bogotá. p 14.

Contraloría General de la República, 2003. Estado de los recursos naturales y del Ambiente 2002-2003. Políticas, saneamiento básico y sistemas de información ambiental. Contraloría General de la República. p183.

Cooke, B. and Kothari, U.(eds), 2001. Participation. The new tyranny? Zed books. London. UK. p 36-55.

Cooperrider D.L., 2000. Positive Image, Positive Action: The Affirmative Basis of Organizing, Appreciative Inquiry: Rethinking Human Organization Toward a Positive Theory of Change. Stipes Publishing, Champagne, IL. p29–53. http://www.stipes.com/aichap2.htm.

Cornwall A., 2003. Whose Voices? Whose choices? Reflections on Gender and Participatory Development. World Development,Vol 31, 8. p1325-1342.

Corredor A., 2007. Los planteamientos sobre el comercio y sus políticas en el seminario. "Hacia una economía sostenible. Conflicto y posconflicto en Colombia. In: Ordoñez F., (eds) Stiglitz en Colombia. Reflexiones sobre sus planteamientos. Fundación Agenda Colombia. Intermedio Editores. p203-243.

Corredor C., 2007. Relaciones entre economía y violencia. In: Ordoñez F., (eds) Stiglitz en Colombia. Reflexiones sobre sus planteamientos. Fundación Agenda Colombia. Intermedio Editores. p245-272.

Court Order. Bogotá River 2004. Tribunal de Bogotá y Cundinamarca. Magistrada Nelly Yolanda de Villamizar #01-479.p496.

CRPML, 2003. Documentos oficiales CVC. Corporación Autónoma Valle del Cauca. p 6.

CRPML, 2005. Documentos oficiales CVC. Corporación Autónoma Valle del Cauca. p 5

CRPML, 2006. Sustancias contaminantes de curtiembres en El Cerrito. Valle del Cauca. Instituto Cinara. ISBN 978 958-97017-1-X. p11.

CRPML, 2007. Informe final de resultados Curtiembres. Proyecto 0201. Convenio 162-2005. p14.

CRPML, 2008. Caracterizaciones finales. Proyecto 0201. Convenio 162-2005. p 9.

Cruces G., and Ham A., 2010. Flexibilidad laboral en América Latina: las reformas pasadas y las perspectivas futuras ECLAC. p132.

CVC, 2007. Informe. Caracterización biofísica y socioeconomica. Convenio de Asociación –Proagua #82. p102.

CVNE, 2012. Colombian ministry of Education. March 2012. http://www.mineducacion.gov.co/cvn/1665/w3-article-300919.html. Retrieved November 19, 2012.

DAMA, 2005. Ventanilla ambiental curtiembres. Departamento Administrativo Medio Ambiente distrito de Bogotá. Given on a CD

DANE, 2005. Censo General 2005. www.dane.gov.co . Retrieved July 19, 2008.

DANE, 2007. Informe poblacional. www,dane.gov.co Retrieved July 19, 2008

DANE, 2009. Mission for the Conveyance of the Series Employment, Poverty and Inequality (MESEP) carried out in 2009 by the National Statistics Administrative Department (DANE, by its Spanish initials) and the National Planning Department DNP. 70p.

DANE, 2011. Pobreza monetaria y multidimensional en Colombia www.dane.gov.co/index.php?option=com_content&view=article&id=430&Itemid=66. Retrieved September 21, 2012

DANE, 2012a. Boletin de prensa September 2012 http://www.dane.gov.co/files/investigaciones/boletines/exportaciones/bol_exp_sep12.pdf. Retrieved September 12, 2012

DANE, 2012b. Boletin de prensa. August 2012 17p. http://www.dane.gov.co/files/investigaciones/boletines/ech/ech_informalidad/re_ech_informalidad_jun_ago2012.pdf Retrieved September 12, 2012

DANE, 2012c. Boletin de prensa. October 2012. Retrieved November 28 2012. http://www.dane.gov.co/files/investigaciones/boletines/mmm/bol_mmm_ago12.pdf.

Dasgupta N., 2000. Environmental Enforcement and Small Industries in India: Reworking the Problem in the Poverty Context. World Development 28(5). p945-967.

Deléage J.P., 1991.Une histoire de l'écologie. coll. Points Sciences. Ed La Découverte. p321.

Deléage J.P., 2002a. Eau et développement durable, Le Monde Diplomatique. 5p. Manières de voir, April.

Deléage J.P., 2002b. L'écologie scientifique de la nature a l'industrie ? In : Deléage JP.(eds). Ecologie de la nature á l'industrie. Ecologie et Politique # 25-2002. p.57-66.

Deléage J.P., 2007 (eds). Des Inégalités Ecologiques parmi les hommes Ecologie et Politique #35. p186.

Delli Priscoli J., 2003. Participation, Consensus Building and Conflict Management Training course. Institute for Water Resources, USACE. PCCP Publications UNESCO. p189.

Delli Priscoli J., and Wolf A.T., 2009. Managing and transforming water conflicts. International Hydrology Series. Cambridge University Press. ISBN 978-0-521-12997-8 354p.

Del-Rio-González P., 2009. The empirical analysis of the determinants for environmental technological change: A research agenda. Ecological economics (68) p. 861-878.

Denzin N. and Lincoln Y., 1994. The fifth moment. Chapter 36 In: Denzin N. and Lincoln Y. (eds). Handbook of Qualitative Research. SAGE Publications p 575-586..

Denzin N., and Lincoln Y., 2011. The SAGE handbook of qualitative research. 766p.

Development Act, 2006. www.zunia.org/uploads/media/knowledge/Small_and_Medium_Enterprises_in_India1259675103.pdf. Retrieved September 20, 2012.

Dick B., 1999.Sources of Rigour in action research: Addressing the issues of trustworthiness and credibility. Association for qualitative research conference. "Issues of rigour in qualitative Research". Duxton Hotel. Melbourne Victoria. 6-10 July 1999.

Dick B., Stringer E., and Huxham C., 2009. Final reflections, unanswered questions. Action Research Vol 7 (1) p117-120.

Dieleman H., 2007. Cleaner production and innovation theory. Social experiments as a new model to engage in cleaner production. Rev. Int. Contam. Ambiental. Vol 23. p79-94.

Dieleman H., and Huisingh D., 2006. Games by which to learn and teach about sustainable development: exploring the relevance of games and experiential learning for sustainability. Journal of Cleaner Production Vol 14 issues 9-11. p837-847.

DNP, 1996. Economía Ambiental. Revista de planeación y desarrollo. Volumen XXVII/32/april-june. # 2 p 10.

DNP, 2003. Plan Nacional de desarrollo. Departamento Nacional de Desarrollo. p284.

DNP, 2007a. Conpes 3484. Política Nacional para la transformación productiva y la promoción de las Micro, Pequeñas y Medianas Empresas: Un esfuerzo público-privado. p32.

DNP, 2007b. Cadena cuero, calzado, y manufacturas. Agenda interna para la productividad y la competitividad. Report DNP, August 2007. p41.

DNP, 2011 Conpes social # 140. Modificación a Conpes social 91 del 14 de Junio de 2005: "Metas y estrategias de Colombia para el logro de los objetivos de desarrollo del milenio-2015", 2011. www.dnp.gov.co/LinkClick.aspx?fileticket=rSQAQZqBj0Y%3d&tabid=1235. Retrieved September 21 2012.

DNP, 2012. Conpes social # 150. Metodologías oficiales y arreglos institucionales para la meddicion de la pobreza en Colombia. Mayo 2012. www.dnp.gov.co/LinkClick.aspx?fileticket=RYo14Bnq2Dg%3d&tabid=1473. Visited September 21, 2012.

Dobers P., and Halme M., 2009. Corporate social responsibility and environmental management. Corporate social responsibility and developing countries Vol 16 issue 5. p237-249.

Douglas, M., and Wildavsky, A., 1983. Risk and culture. An essay on the selection of technological and environmental dangers.University of California Press. Berkeley. USA.

Duque I., 2012. Innovación social y filantropía. Portafolio. November 15, 2012.

Duraiappah A.K., Roddy P., and Parry J., 2005. Have participatory approaches increased capa- bilities? Manitoba, Canada: International Institute for Sustainable Development. p34.

EAAB, 2000a. Síntesis del estado actual de los humedales bogotanos. EAAB and Conservación Internacional. p78.

EAAB, 2000b. Definición de los lineamientos para continuar con el saneamiento del río Bogotá. Informe Final. Unión temporal saneamiento río Bogotá. September. p132.

Echeverry C., Galindo G., Hincapié L., and Trujillo L., 2011. Política económica Desigualdad: Un freno para las metas de desarrollo y pobreza en Colombia. Revista económica Supuestos. Desigualdad: un freno para las metas de desarrollo y pobreza en Colombia. Universidad de los Andes. Facultad de Economía. ISSN: 2248-6836. p10-16.

Edelenbos J., Klok P., and van Tatenhove J., 2009. The Institutional Embedding of Interactive Policy Making. Insights from a comparative research based on eight interactive projects in The Netherlands. The American Review of Public Administration, Vol 39 issue 2, p125-148.

El Espectador, 2010. 2'121894 damnificados. December 17.

El Espectador, 2011. Devolveremos la dignidad a la gente. November 11.

El Espectador, 2012a. Producción de calzado y cuero colombiano creció 15.4%. July 31 2012.

El Espectador, 2012b. Las mayorías del Procurador. November 27.

El Espectador, 2012c. Soluciones para el río Bogotá. November 16.

El Tiempo, 1993. En San Benito, selladas cinco curtiembres por contaminar. August 19.

El Tiempo, 1994. Curtiembres, la mayor contaminación. February 26.

El Tiempo, 2004a. Curtiembres, a cumplir la norma. March 12.

El Tiempo, 2004b. El río sigue siendo una alcantarilla. March 24.

El Tiempo, 2004c. Hora cero para curtiembres y mataderos. May 15.

El Tiempo, 2004d. Los grandes costos de la pobreza. August 8.

El Tiempo, 2005a. Cierran 59 " fábricas" de cuero. January 28.

El Tiempo, 2005b. Mipymes. Fedesarrollo. August 17.

El Tiempo, 2005c. Carta Ambiental. Edition 10, August-November.

El Tiempo 2008. Blindan recursos de agua potable para evitar despilfarro. March 12.

El Tiempo, 2009. Así se ve la crisis desde la casa de un Nobel. El Tiempo March 8.

El Tiempo, 2010. CAR serán reestructuradas. December 17, 2010.

El Tiempo, 2011a. Ser trabajador informal significa ganar la mitad. August 29.

El Tiempo, 2011b. Potentados en Colombia pagan menos imporrenta. October 25.

El Tiempo, 2012 41% de la tierra en Colombia tiene menos de 100 000 dueños. Retrieved on November 16. http://m.eltiempo.com/justicia/terratenientes-en-colombia/10528066.

ENA, 2010. IDEAM. Estudio Nacional del Agua. p 39.

EnDeuda, 2012. Interview to Jose Antonio Ocampo. Newspaper from the faculty of Economics from the University of los Andes. ISSN:2322-6625 Number 3 October 2012. p9.

Engels, A., 2002. Evaluation of four science-policy interface workshops EFIEA. Amsterdam. The Netherlands. p 6-20

EPAM, 1993. Estrategia de Saneamiento del rio Bogotá. Informe especial. FONADE, EAAB. p278.

EPI, 2012. Environmental performance index and pilot trend performance index. 2012 EPI .Yale University. http://epi.yale.edu/dataexplorer/countryprofiles.

Escobar A., 1999. The invention of development. Current History November p382-386.

Estacio E., and Marks D., 2010. Critical reflections on social injustice and participatory action research: the case of the indigenous Ayta community in the Philippines. Procedia Social and Behavioral Sciences Vol 5, p548-552.

EU, 2005. SME user guide. September 20. http://ec.europa.eu/enterprise/index_en.htm.

Eversole R., 2003. Managing the pitfalls of participatory development: some insight from Australia. World Development Vol 31, 5. p781-795.

Fafchamps, M. 1994. Industrial structure and micro enterprises in Africa.The Journal of Development Areas. 29 (1), p1-30.

Fals Borda O., and Mora-Osejo L., 2003. Context and diffusion of knowledge: A critique of Eurocentrism. Action research Vol 1, 1. p20-37.

Fals Borda O., and Rahman M., (eds). 1991. Action and knowledge: breaking the monopoly with participatory action research. New York : Intermediate Technology/ Apex. p 188.

Fandl K., 2010. Beyond the invisible. The impact of trade liberalization and formalization on small businesses in Colombia. PhD Dissertation on Public Policy. George Mason University. p256.

FAO, 1995. Water sector policy review and strategy formulation. A general framework. http://www.fao.org//docrep/V7890E09.htm. Retrieved December 5 2005.

FAO, 2000. AQUASTAT Colombia Sistema de Información sobre el uso del Agua en la Agricultura y el Medio Rural de la FAO. http://www.fao.org/nr/water/aquastat/irrigationmap/col/index.stm Retrieved December 18 2010.

Faure G., Hocdé H. and Chia E., 2010. Action research methodology to reconcile product standardization and diversity of agricultural practices: A case of farmers' organizations on Costa Rica. Action research 0, 0. p1-19.

Fedesarrollo, 2006. Reporte oficial primer semestre año 2006. 79p.

Fiorino D.J., 1990. Citizen Participation and Environmental Risk: A survey of Institutional Mechanisms. US. Environmental Protection Agency. Science, Technology & Human Values. Vol 15 No2. p226-243.

Fisher R., and Ury W., 1981. Getting to Yes: Negotiation Agreement. Without giving in. Boston, Houghton Mifflin. p 3-90

Fisher R., Ury W. and Patton B., 1991.Getting to Yes: Negotiation Agreement. Without giving in. 2nd Ed .Penguin Books. New York. p 6-84

Fisher, F., 2000. Citizens, experts and the environment. The politics of local knowledge. Duke University Press. London. UK. p 89-210.

Flyvbjerg, B., 2004. Five misunderstandings about case-study research. In: Seale, C., Gobo, G., Gubrium, J F., and Silverman, eds., Qualitative Research Practice. London and Thousand Oaks, CA: Sage. p420-434.

Forero, A., 2010. Reformas, Por fin! El Espectador, September 6.

Freire, P., 1993 /1970. Pedagogy of the oppressed. New York: Herder and Herder.

Freire, P. 1982. Creating alternative research methods: Learning to do it by doing it. In: Hall B., Gillette A., and Tandom R., (eds). Creating knowledge: A monopoly? Participatory research in development. New Delhi: Society for participatory research in Asia.

Frenkel D., and Stark J., 2008. The practice of mediation. Aspen publishers. ISBN 978-0-7355-4439-0. p508.

Frijns J. and Van Vliet B.V., 1999. Small-scale industry and cleaner production. World Development 27, 1. p967-983.

Frondel M., Horbach J. and Rennings K., 2005. End-of pipe or Cleaner Production? An Empirical comparison of Environmental Innovation Decisions across OECD countries. Discussion paper #04-82 . ZEW. Center for European Economic research. p31.

Fu B., Zhuang X., Jiang G., Shi J., and Lu Y., 2007. Environmental problems and challenges in China. Environmental Science and Technology. Chinese Academy of Sciences Vol 41, 22. p7597-7602.

Fundación Agenda Colombia, 2007. La política macroeconómica y sus instrumentos. In: Ordoñez, F., (eds) Stiglitz en Colombia. Reflexiones sobre sus planteamientos, p166-201.

Funtowicz, S. and Ravetz, J. 1993. Science for the postnormal age. Futures. The journal of Policy, Planning and future Studies, 50, 7. p739-755.

Gagnon B., Lewis N., and Ferrari S., 2007. Environnement et pauvreté: regards croisés entre l'éthique et la justice environnementales. In: Deléage *et al.*, Des inégalités écologiques parmi les hommes. Ecologie et Politique Vol 35. p 79-90.

Garay L., 2002. Repensando a Colombia. Hacia un nuevo contrato social. PNUD-ACCI. p171.

García C.E.R, and Brown S., 2009. Assessing water use and quality through youth participatory research in a rural Andean watershed. Journal of Environmental Management. Vol 90, p3040-3047.

García H., Gutiérrez R., and Molano A., 2010. No es de descartar reacciones violentas. El espectador, Septiembre 5. p 1.

Gaviria A., 2011. Cambio social en Colombia durante la segunda mitad del siglo XX. CEDE documents October 30. ISSN:657 7191. p44.

Geist M., 2010. Using the Delphi method to engage stakeholders: a comparison of two studies. Evaluation and Program Planning Vol 33. p147-154.

Giacomantonio M., De Dreu C., Shalvi S., Sligte D., and Leder S., 2010a. Psychological distance boosts value-behavior correspondence in ultimatum bargaining and integrative negotiation. Journal of Experimental Social Psychology. Vol 46. p824-829.

Giacomantonio M., De Dreu C., and Mannetti L., 2010b. Now you see it, now you don't: interests, issues, and psychological distance in integrative negotiation. Journal of Personality and Social Psychology. Vol 98, 5. p761-774.

Gijzen H.J., 1998. Sustainable wastewater management via re-use – Turning Waste into Wealth. In: Garcia M., Gijzen H.J., Galvis G. (eds.), Proc. AGUA_ water and Sustainability, June 1-3. Cali, Colombia. p211-225.

Gijzen H.J., 2001. Aerobes, anaerobes and phototrophs – a winning team for wastewater management. Wat Sci Tech. Vol 44, no 8. p123-132.

Godard O. and Laurans Y., 2004. Evaluating environmental issues. Valuation as coordination in a pluralistic world. Ecole Polytechnique. International Journal of Environment and Pollution (IJEP). p37.

Greenwood, D.J., and Morten, L., 1998. Introduction to action research: social research for social change. Thousand Oaks, Ca.Sage. 274p.

Grimble R., Chan M., Aglionby J. and Quan J., 1995. Trees and Trade-offs: A Stakeholder Approach to Natural Resource Management. Gatekeeper series No SA52 Part. p17.

Guhl E., Macías L., and Giraldo C.A., 2007. Gestión Integrada del recurso hídrico en Colombia. Quinaxi. Corcas editores. p63.

Guhl E., 2008. La ciencia y la tecnología en el SINA. Dificultades, logros y recomendaciones. In: Rodríguez M. (ed), Gobernabilidad, instituciones y medio ambiente en Colombia. Foro Nacional Ambiental. p390-476.

Guio D., 2004. Water resources Management in Colombia: An institutional Analysis MSc. Thesis UNESCO-IHE, September. p158.

Gummesson, E., 2007a. Case study research and network theory: birds of a feather. Qualitative Research in Organizations and Management: An International Journal. Vol 2,3. p226-248.

Gummesson, E., 2007b Access to reality: observations on observational methods. Qualitative Market research: An International Journal. Vol 10 # 2, 130-134p.

Gunningham N., and Sinclair D., 2002. Leaders and laggards. Next generation environmental regulation. Green leaf publishing. ISBN: 1874719489 225p.

Gupta, J. 2004. Non-State Actors: Undermining or Increasing the Legitimacy and Transparency of International Environmental Law, in I.F. Dekker & W. Werner (eds.). Governance and International Legal Theory, Nova Et Vetera Iuris Gentium. p297-320.

Gupta, J. and L. Lebel (2010). Access and Allocation in Global Earth System Governance: Water and Climate Change Compared, INEA, 10(4). p377-395.

Gutteres M., Aquim P., Passos J., and Trierweiler J., 2010. Water reuse in tannery beamhouse process. Journal of Cleaner Production. 18. p1545-1552.

Gutierrez NP., 2005. Proceso de formulación de la ley del agua en Colombia. Paper presented at Externado University. December 2005. p33.

Gutierrez NP., 2006. Si a la ley del agua, no a la privatización. El Tiempo January 21 2006 2p. http://www.eltiempo.com/archivo/documento/MAM-1891111. Retrieved November 27, 2012.

Hamed M. and El Mahgary Y. 2004. Outline of a national strategy for cleaner production: The case of Egypt. Journal of Cleaner Production. 12. p327-336.

Hansen J., and Lehmann M., 2006. Agents of change: universities as development hubs. Journal of Cleaner Production. 14. p820-829.

Hayward C., Simpson L., and Woods L., 2004. Still left out in the cold: problematizing participatory research and development. Sociologia Ruralis. 44, 1. p95-108.

Herrero A., and Henderson K., 2003. El costo de la resolución de conflictos en la pequeña empresa: el caso de Perú. Sustainable development best practices series. IADB. p58.

Hillary, R., 1997. Eu environmental policy, voluntary mechanisms and the Eco management and audit scheme (EMAS). In: Hillary (ed) Environmental Management Systems and Cleaner Production. Wiley, Chichester p129-142.

Hirsch D., Abrami G., Giordano R., Liersch S., Matin N., and Schluter M., 2010. Participatory research for adaptive water management in a transition country- a case study from Uzbekistan. Ecology and Society Vol. 15, 3, article 23. p33.

Hisschemöller, M., 1993.De democratie van problemen. De relatie tussen de inhoud van beleidsprobemen en methoden van politieke besluitvorming.VU uitgeverij. Amsterdam.The Netherlands p 180.

Hisschemöller, M.,2005. Partcicipation as Knowledge production and the limits of democracy In: Maasen S,Weingart P(eds). Democratization of expertise? Exploring novel forms of scientific advice in political decision-making. Springer.Dordrecht pp 189-208.

Hisschemöller, M., and Hoppe R.,1995. Coping with unstractable controversies: The case for problem structuring in policy-design and analysis. Knowledge and policy. The international Journal of Knowledge transfer and utilization. 8(4):40-60

Holman P., 2004. Emerging in Appreciative Space. AI Appreciative Inquiry Practitioner. The International Journal of Appreciative Inquiry AI best practice. November. p48.

Holman P., 2010. Engaging emergence. Turning upheaval into opportunity. Berrett-Koehler publishers, Inc. p238.

Holman P. and Devane, T. (eds), 1999. The Change Handbook. Groups methods for shaping the future. Berrett-Koehler Publishers, Inc. p390.

Holman, P., Devane, T. and Cady, S. (eds). 2007. The Change Handbook: The definitive Resource on today´s Best Methods for Engaging Whole Systems. Berrett-Koehler publishers, Inc. p732.

Hommes R., 2011. Exportación y desarrollo industrial. El Tiempo December 16.

Howgrave-Graham A. and Van Berkel R., 2007. Assessment of Cleaner Production uptake: method development and trial with small businesses in western Australia. Journal of Cleaner Production. 15.8/9. p787-797.

Hoyos A., 2012. La cuestión agraria. El Espectador February 22.

Huisingh D., 2002. Have we achieved inherently safer and humane societies? Journal of Cleaner Production. 10, 4. p297-298.

ICEX, 2004. El mercado de cuero y marroquinería en Colombia. Oficina económica y comercial de la embajada de España. 46p http://www.plancomo.org/pdf/34/2004-Colombia.pdf. Retrieved November 28, 2012.

IDEAM, 2002. Estudio Nacional del Agua. Instituto de Hidrología, Meteorología y Estudios Ambientales. p253.

IIFT, 2012. SMEs overseas. Centre for SMEs. Indian Institute of Foreign Trade. http://www.smeiift.com/sme/SME_Overseas.asp. Retrieved September 20, 2012.

ILO, 2002. Decent work and the informal economy. Report of the Director-General presented to the 90th International Labour Conference, International Labour Office, Geneva. p 100-104.

INDERENA, 1977. Decreto 1449, Sobre la zona de protección de la ribera de los ríos Código de recursos naturales. p 5

Indexmundi, 2012. A country profile. http://www.indexmundi.com/colombia/economy_profile.html. Retrieved May 13, 2013.

Jacobs G., 2010. Conflicting demands and the power of defensive routines in participatory action research. Action Research. 8,4. p367-386.

Jaramillo AD., and Vásquez P., 2003. Proceso de concertación para la implementación de producción más limpia en curtiembres de El Cerrito. Conferencia Internacional usos múltiples del agua: para la vida y el desarrollo sostenible. Agua. p4.

Jaramillo AD., Vásquez P., and Restrepo I., 2003. Potencial de mejoramiento ambiental a través de la implementación de PML en las curtiembres de El Cerrito. Conferencia Internacional usos múltiples del agua: para la vida y el desarrollo sostenible. Agua. p4.

Jaramillo AD., 2005. Mejoramiento de la gestión ambiental sectoral en el Valle del Cauca. Una propuesta para la sostenibilidad del Centro Regional de Producción Más Limpia. Universidad Del Valle. Master thesis on Environmental Engineering. p151.

Jaramillo AD., Vásquez P., and Restrepo I., 2005. Aplicación de PML para disminuir el impacto al recurso hídrico provocado por una comunidad industrial de curtidores en una zona urbana. Seminario gestión integrada de servicios relacionados con el agua en asentamientos nucleados. 5p.

Jaramillo AF., Restrepo I., and Jaramillo AD., 2006. Solar energy installation to heat process water applied to a tannery industry employing high efficiency materials on topology developed in Colombia. Revista Energía y Computación. 14,2. p41-48.

Jansenson D. 2004. Mediation Workshop. Curso de Alta Dirección en Negociación. Desarrollo Gerencial. October. p25

JenkinsR., 2003. Has trade liberalization created pollution havens in Latin America? Cepal review 80. p 81-93.

JICA, 2002. El estudio del desarrollo sostenible del agua subterranea en la sabana de Bogota en la republica de Colombia. Yachiyo Engineering co ltd. Asia Air survey co ltd. 203p.

Jønsson J., Appel P., and Chibunda R., 2009. A matter of approach: the retort's potential to reduce mercury consumption within small-scale gold mining settelments in Tanzania. Journal of Cleaner Production. 17, 1. p77-86.

Jordan A., Wurzel R. And Zito A., 2005. The rise of the "new" policy instruments in comparative perspective: Has governance eclipsed Goverment? Political studies. 53, 3. p477-496.

Jung C.H., Krutilla K. and Boyd R., 1996. Incentives for advanced pollution abatement technology at the industry level: an evaluation of policy alternatives. Journal of Environmental Economics and Management 30, 1. 95-111.

Jutz M., 2003. Workshop 11: Cleaner Production. Haute Ecole Spécialisée Deux Bâle Suisse. 2p.

Kalmanovitz, 2012. Exportaciones, inversión y políticas. El Espectador January 30.

Kanagaraj J., Chandra N.K., and Mandal A.B., 2008. Recovery and reuse of Chromium from Chrome tanning waste wáter aiming towards zero discharge of pollution. Journal of Cleaner Production. 16, 16. p1807-1813.

Kangas A., Saarinen N., Saarikoski H., Leskinen L.A., Hujala T., and Tikkanen J., 2010. Forest policy and Economics. 12. p213-222.

Kanji N., and Greenwood L., 2001. Participatory approaches to research and development in IIED: Learning from experience. London: IIED p 51.

Karl M., 2000. Monitoring and evaluating stakeholder participation in agriculture and rural development projects: a literature review. Sustainable Development Department (SO). Food and Agriculture Organization of the United Nations (FAO). SD Dimensions. p27.

Kemmis S. and McTaggart R., 1988. The Action Research Planner, Geelong, Victoria: Deakin University Press. p 321-337.

Kemp R., and Martens P., 2007. Sustainable development: how to manage something that is subjective and never can be achieved? Sustainablility: science, pactice, & policy. ISSN 15487733. 3,2. p5-14. http//ejournal.nbii.org.

Kemp R., and Pontoglio S., 2011. The innovation effects of environmental policy instruments- A typical case of the blind men and the elephant? Journal of Ecological Economics. 72. p28-36.

Kennedy L., 1999. Cooperating for survival: Tannery pollution and joint action in the Palar Valley (India). World development. 27, 9. p1673-1691.

Ker Rault P.A., 2008. Public participation in integrated water management. A wicked process for a complex societal problem. Which type of public participation for which type of water management challenges in the Levant? School of Applied Sciences. PhD thesis. Cranfield University. p384.

Kevany K., 2010. Water, women, waste, wisdom, and wealth -An energizing international collaboration, action research, and education project. Journal of Cleaner Production. 18, *Call for papers.* p1769-1771.

Kevany K. and Huisingh D., 2013. A review of progress in empowerment of women in rural water mangement decisionmaking processes. Journal of Cleaner Production. Vol 60 Special volume Water, women, waste, wisdom and wealth p53-64.

Kindlein W., Alves L.H., and Seadi A., 2008. Proposal of wet blue leather remainder and synthetic fabrics reuse. Journal of Cleaner Production. 16, 16 p1711-1716.

Kisito F., Mutikanga H., Ngirane-Katashaya G., and Thunvik R., 2009. Development of decision support tools for decentralized urban water supply management in Uganda: an action research approach. Computers, Environment and Urban Systems. 33. p122-137.

Kolkman MJ.,and Van der Veen A., 2005. Mental modelmapping, a new tool to analyse the use of information in decision-making in integrated water management. Phys Chem Earth 30(4-5): 317-332.

Kubr M., (ed) 1986. Management consulting: a guide to the profession. International Labour Office. p611.

La República, 2009. Curtiembres del Valle alistan su ingreso a nuevos mercados 12/4/2009.

Latorre E., 1996. Empresa y Medio Ambiente en Colombia. Fescol. CEREC p282.

Law on Land Restitution, 2011. Office of the Presidency of Colombia. June 2011. p112.

Law 1508, 2012. Public-Private Alliances. Retrieved November 14, 2012. http://wsp.presidencia.gov.co/Normativa/Leyes/Documents/Ley150810012012.pdf.

Lax D. and Sebenius J., 1991. "Power of Alternatives" In: The Power of Alternatives or the Limits to Negotiation. Cambridge: The Program on Negotiation at Harvard Law School. p97-113.

Le Van Khoa, 2006. Greening Small and Medium-sized enterprises: Evaluating Environmental policy in Vietnam. PhD thesis. Wageningen University. ISBN 90 8504-482-0. p247.

Levin M., and Greenwood D., 2011. Revitalizing universities by reinventing the social sciences. In: Denzin N., and Lincoln Y (eds). The SAGE handbook of qualitative research. p27-41.

Lewin K., 1946. Action Research and minority problems. Journal of Social Issues, 2, p34-46.

Lewis N., 2001. La gestion intégrée de l'eau en France :critique sociologique à partir d'une étude de terrain (bassin Loire-Bretagne). Université d'Orléans, France. Ph.D. Thesis. p514.

Lincoln Y., and González E., 2008. The search for emerging decolonizing methodologies in qualitative research. Further strategies for liberatory and democratic inquiry. Qualitative Inquiry. 14,5. p784-805.

Lippit G., and Lippit R., 1978. The consulting process in action. University associates. ISBN 0883901412. p130.

Londoño B., 2008. Las organizaciones no gubernamentales ambientales colombianas y su ejercicio de las herramientas de participación institucionalizada. In: Rodríguez M. (ed), Gobernabilidad, instituciones y medio ambiente en Colombia. Foro Nacional Ambiental. p523-546.

López C., 2011. Seis países: sólo uno en vía a la modernidad. Portafolio 28 March. 1p.

Lozano R., and Huisingh D., 2011. Inter-linking issues and dimensions in sustainability reporting. Journal of Cleaner Production. 19, 2/3. 99-107.

Machado A., 2012. Tierra y desarrollo humano. www.arcoiris.com.co/2012/11/la-tierra-y-el-desarrollo-humano/?utm_source=rss&utm_medium=rss&utm_campaign=la-tierra-y-el-desarrollo-humano. Retrieved on November 16, 2012.

MADS, 2012. Ministry of Environment and Sustainable Development. http://www.minambiente.gov.co//contenido/contenido.aspx?catID=463&conID=1077. Retrieved November 10 2012.

Mangun W. and Henning D., 1999. Managing the environmental crisis, Duke University Press, Durham and London. 392p.

Márquez G., 1997. Notas para una historia de la ecología y su relación con el movimiento ambiental colombiano. In: Ecofondo (ed) Se hace camino al andar. Aportes para una historia del movimiento ambiental en Colombia. ISBN: 958-95626-1-2.

Martinez L., Gerritsen P., Cuevas R., and Rosales J., 2006. Incorporating principles of sustainable development in research and education in western Mexico. Journal of Cleaner Production. 14. p1003-1009.

MAVDT, 1997a. CP national policy. August. p43.

MAVDT, 1997b. Presentation from former vice-minister of environment.

MAVDT, 2006. Report tanning industry. Diana Moreno. p 15

MAVDT, 2010. National policy for Integrated Water Resources Management. Viceministry of Environment. p 5.

Mayer I., 1997. Debating technologies. A methodological contribution to the design and evaluation of participatory policy analysis. Tilburg. The Netherlands. Chapter 4.

MCIT, 2004. Law on SMEs #905.

MCIT, 2008. Targets for year 2010. https://www.mincomercio.gov.co. Retrieved August 2008.

MCIT, 2012.. https://www.mincomercio.gov.co/publicaciones.php?id=10422. Retrieved November 19, 2012

McKenzie-Mohr D., and Smith W., 2006. Fostering Sustainable Behavior. New Society Publishers. Gabriola Island, BC, Canada.p160.

McMillan J. and Woodruff C., 2002. The central role of entrepreneurs in transition economies. Journal of economics perspectives. 16, 3. p153-170.

McTaggart R., 1996. 'Issues for participatory action researchers' in O. Zuber-Skerritt (ed.) New Directions in Action Research, London: Falmer Press. p 243-55.

Mehta L., Leach M., and Scoones I., 2001. Editorial: Environmental governance in an uncertain world. IDS Bulletin Vol 32, 4. Brighton: Institute of Development Studies, p14.

Mejia R., López A., and Molina A., 2007. Experiences in developing collaborative engineering environments: an action research approach. Computers in industry. 58. p329-346.

Mel S., McKenzie D., and Woodruff C., 2008. Who are the microenterprise owners? Evidence from Sri Lanka on Tokman V de Soto. Institute for the study of labor. IZA. DP #3511. p35.

Merlano A., 2005. Módulo Estrategias Negociación. UNIANDES.

Merlano J., and Negret C., 2006. Del conflicto a la conciliación. Periodicas edt. ISBN 9789588129723 p175.

Merrey D.J, Meizen-Dick R., Mollinga P.P. and Karar E., 2007. Policy and Institutional Reform The art of the Possible. In: Molden (ed) Water for food, water for life Comprehensive Assessment of Water Management in Agriculture. London: Earthscan Chapter 5. p193-232.

MHETA, 2003. Camino hacia un país de propietarios con desarrollo sostenible. Plan sectoral 2002-2006. Embajada de Colombia en los Países Bajos. p52.

Milliman S. R. and Prince R., 1989. Firm incentives to promote technological change in pollution control. Journal of Environmental Economics and Management 17,3. p247-265.

Mipymes, 2012. www.mipymes.gov.co. Retrieved November 27, 2012.

Mitchell C., 2006. Beyond barriers: examining root causes behind commonly cited Cleaner Production barriers in Vietnam. Journal of Cleaner Production. 14. p1576-1585.

Mitroff I., Mason R., and Barabba V., 1983. The 1980 census: policymaking amid turbulence. Lexington books. Massachusetts USA. p 48.

Molano A., 2010. Zonas de reserva campesina. Un análisis desde el Magdalena Medio. El Espectador September 5.

Mollinga P., 2008. Water, politics and development: Framing a political sociology of water resources management.Water Alternatives. 1, 1. p7-23.

Mondragón-Vélez C., Peña X., and Wills D., 2010. Labor market rigidities and informality in Colombia. CEDE institute. Document 2010-7. Universidad de los Andes. ISSN: 1657-5334. p32.

Montalvo C., 2003. Sustainable production and consumption systems—cooperation for   change: assessing and simulating the willingness of the firm to adopt/develop cleaner technologies. The case of the In-Bond industry in northern Mexico. Journal of Cleaner Production. 11,4. p411-426.

Montalvo C. and Kemp R., 2008. Cleaner technologies diffusion: Case studies, modeling and policy. Journal of Cleaner Production. 16 Supplement 1. S1-S6.

Moors EHM., Mulder K., and Vergragt PJ., 2005. Towards cleaner production: barriers and strategies in the base metals producing industry. Journal of Cleaner Production. 13, 7. p657-668.

Moran S., and Ritov I., 2007. Experience in integrative negotiations: what needs to be learned? Journal of experimental social psychology. 43. p77-90.

Morse S., 2008. Post-sustainable Development. Sustainable Development, 16. p341-352.

Nazer, D., Rashed, M.A, Siebel, M., 2006. Development of a modified method for reducing the environmental impact of the unhairing-liming process in the leather tanning industry. Journal of Cleaner Production. 14, 1. p65-74.

Nentjes A.,1988. An economic model of innovation in pollution control technology. Annual meeting of the American Association of Environmental and Resource Economists, New York. p 28-30.

Nhapi I., 2004. Options for Wastewater Management in Harare, Zimbabwe. Balkema Publishers. Ph.D. Thesis UNESCO-IHE. p167.

OAS, Unit for Sustainable Development and Environment, 2001. Inter-American Strategy for the promotion of Public Participation in Decision-Making for Sustainable Development. p47.

Ocampo J.A., 2002. Rethinking the Development Agenda. Cambridge Journal of Economics, 26, 3. May.

Ocampo J.A., 2003. Development and the global order In: Chang, H., (ed). Rethinking Develpoment Economics: An introduction. p10.

Ocampo J.A., 2010. Retos de la economía para Santos. Portafolio August 2. 1 p.

OECD Directorate, 2012 www.oecd.org/env/environmentalpolicytoolsandevaluation/extendedproducerresponsibility.htm Retrieved September 20 2012

Ogliastri E., 1999-2000. Actas de seminario en negociación. Universidad de los Andes. Bogotá. 18p.

Ogliastri E., 1999. "Una Introducción a la Negociación Internacional." Bogotá. Monografías, Facultad de Administración, Universidad de Los Andes, Segunda Edición. 58 p.

O'Hogain S., 2008. Reed bed sewage treatment and community development/participation. In: Vymazal J., (ed) Wastewater Treatment, Plant Dynamics and Management in Constructed and Natural Wetlands: Springer Science+Business Media B.V. Chapter 12. p135-148.

Ojeda D., 2004. Diagnóstico Ambiental por vertimientos de residuos de curtiembres al río Bogotá en el corredor industrial Villapinzón- Chocontá. Cuenca Alta del río Bogotá. Universidad Nacional de Colombia. Facultad de ciencias humanas Departamento de Geografía. Informe de pasantía para grado de Geógrafo. October. p131.

Oosterveer P., Kamolsiripichaiporn S., and Rasiah R. 2006. The "greening" of industry and development in southeast Asia: Perspectives on industrial transformation and environment regulation. Environment, Development and Sustainability (8). p217-227.

Opschoor J. B. and Turner R., 1994. Economic incentives and environmental policies: principles and practice. Dordrecht, Kluwer Academic Publications p 173.

Owen H., 1997. Expanding Our Now: the Story of Open Space Technology, Berrett-Koehler, San Francisco, CA. p 11-140.

Pan American Health Organization, 2010. Convenio cooperacion tecnica no. 637 de 2009 ministerio de la proteccion social – organización panamericana de la salud. Informe técnico Dinámica poblacional y caracterización de la zona habitacional aledaña a la cuenca media del Río Tunjuelito en el sector de curtiembres del barrio San Benito (Documento de la fase I), Bogotá, D.C. p33.

Pateman C., 1970. Participation and Democratic theory. Cambridge: Cambridge University Press. p 125.

Perry G., 2003. La visita de Stiglitz. www.eltiempo.com, March 14.

Perry G., Maloney W., Arias O., Fajnzylber P., Mason A., and Saavedra-Chanduvi J., 2007.Informality: Exit y exclusion. World Bank Latin American and Caribbean studies. WB. p248.

Pietroni D., Van Kleef G., Rubaltelli E., Rumiati R., 2009. When happiness pays in negotiation. The interpersonal effects of "exit option": directed emotions Mind Soc. 8. p77-92.

Pinto J.A, 2006. Por una economía social y ecológica de mercado en Colombia. Fundación Konrad Adenauer. p156.

PND, 2010. Plan Nacional de desarrollo 2010-2014 "Prosperidad para todos". www.dnp.gov.co/Default.aspx?tabid=337 Retrieved September 21, 2012.

PNUD (United Nations Program on Development), 2010. Regional report on human development for Latin America and the Caribbean: Actuar sobre el futuro: romper la transmisión intergeneracional de la desigualdad. San José Costa Rica. http://www.idhalc-actuarsobreelfuturo.org/site/informe.php. Retrieved November 18, 2012.

Portafolio, 2005. El 59 por ciento de los trabajadores del país son informales. Portafolio, año 12, 2179. Periódico El Tiempo. September 2.

Portafolio, 2010. Usuarios de agua: Solución a problemas de sequía? Portafolio January, 18. 1p.

Posada-Carbó E., 2012. Alejados del fracaso? El Tiempo. August 3.

Poveda L., and Sánchez M., 2009. Propuesta para el diseño, estructuración e implementación del departamento de gestión ambiental en las industrias de curtiembres localizadas en el barrio San Benito en Bogotá, D.C. Hacia una producción más limpia con la adecuada administración de los recursos naturales. Universidad Minuto de Dios. Business Administration Faculty. Thesis. p75.

Procuraduría, 2008. Panorama y perspectiva sobre la gestión ambiental de los ecosistemas de páramo. Procuraduría Delegada para Asuntos Ambientales y Agrarios 135p. www.fundacion-ecoan.org/Documentos/Eventos/libro5.pdf #page=135. Retrieved February 18, 2013.

PROPEL, 1995. Curtigran case study .INEM Casebook. Case Studies in Environmental Management in SMEs. German Agency for Technical Cooperation. GTZ. 3p.

Quental N., Lourenco J., and Nunes da Silva F., 2011. Sustainable development: policy, goals, targets and political cycles. Sustainable Development. 19, 1. p15-29.

Raiffa, H., Richardson, J., and Metcalfe, D. 2002. Negotiation Analysis. The Science and Art of Collaborative Decision Making. Harvard University Press. p548.

Raven R., Van den Bosch S., and Weterings R., 2010. Transitions and strategic niche mangement: towards a competence kit for practitioners. International Journal of Technology Management. 51.,1. p57-74.

Ravetz, J. 1999. What is post-normal science. Futures. The journal of policy, planning and future studies, 31, 7, p647-653.

Rawls J., 1971. A theory of justice. Harvard University Press. Cambridge. USA. p538.

Reason P., 1994. Three approaches to participative inquiry. In: Denzin N., and Lincoln Y., (eds). Handbook of qualitative research. SAGE publications. ISBN: 0-8039-4679. p324-339.

Reason P., 2006. Choice and quality in action research practice. Journal of Management inquiry. 15, 2. p187-203.

Reed M., 2008. A stakeholder participation for environmental management: A literature review. Biological conservation. 141. p2417-2431.

Reed M., Graves A., Dandy N., Psthumus H., Hubacek K., Morris J., Prell C., Quinn C., and Stringer L., 2009. Who's in and why? A typology of stakeholder analysis methods for natural resource management. Journal of environmental management. 90. p1933-1949.

Reed M., Buenemann M., Althopheng J., Akhtar-Schuster M., Bachmann F., Basr=tin G., Biga H., Chanda R., Dougill J., Essahli W., Evely A. C., Fleskens L., Geeson N., Glass J., Hessel R., Holden J., Ioris A., Fruger B., Liniger H.P., Mphinyane W., Nainggolan D., Perkins J., Raymond C., Ritsema C. J., Schwilch G., Sebego R., Seely M., Stinger L., Thomas R., Twomlow S., and Verzandvoort S., 2011. Croos-scale monitoring and assessment of land degradation and sustainable land management: a methodological framework for knowledge management. Land Degrad & Develop. 22. p261-271.

Reijnders L., 2003. Policies influencing cleaner production: the role of prices and regulation. Journal of Cleaner Production. 11. p333-338.

Revell A., Stokes D., and Chen H., 2010. Small businesses and the environment: turning over a new leaf? Business strategy and the environment. 19, 5. p273-288.

Reyes, A., 2009. Guerreros y campesinos. El despojo de la tierra en Colombia. Editorial Norma. 378p.

Rivela B., Moreira M., Bornhardt C., Méndez R., and Feijoo G., 2004. Life cycle assessment as a tool for environmental improvement of the tannery industry in developing countries. Environ Sci Technol. 38. p1901-1909.

Robinson L., and Berkes F., 2011. Multi-level participation for building adaptive capacity: Formal agency-community interactions in northern Kenya. Global Environmental Change. 21, 4. p1185-1194.

Rocha R., 2007. Un marco de referencia. In: Ordóñez, F., (eds)Stiglitz en Colombia. Reflexiones sobre sus planteamientos. Fundación Agenda Colombia. Intermedio Editores, p127-163.

Rodríguez G., 2003. La realidad de la PYME colombiana. Fundación FUNDES. ISBN: 958 335278-0. p207.

Rodríguez M., 2008. Declive de las instituciones y la política ambiental de América Latina y el Caribe. In: Rodríguez M. (ed), Gobernabilidad, instituciones y medio ambiente en Colombia. Foro Nacional Ambiental. p65-99.

Rodriguez, M., 2009. Agua Riqueza de Colombia. Villegas Ed. 221p.

Rosegrant M.W., Cai X. and Cline S.A., 2002. Global Water Outlook to 2025. Averting an impending crisis. A 2020 Vision for Food, Agriculture, and the Environment Initiative. International Food Policy Research Institute. Washington D.C, U.S.A. International Water Management Institute. Colombo, Sri Lanka. p25.

Rowe G. and Frewer L.J., 2000. Public Participation Methods: A framework for Evaluation. Science, Technology & Human Values, Vol.25 No1. Sage publications Inc. p26.

Rowe G. and Frewer L.J., 2004. Evaluating public participation exercises: A research agenda. Science, Technology & Human Values. 29,4. p512-556.

Saini R., 2006. Las PYMES, sector en crecimiento. Istmo. II. p6.

Sakr D., Baas L., El-Haggar S., and Huisingh D., 2011. Critical success and limiting factors for eco-industrial parks: global trends and Egyptian context. Journal of Cleaner Production. 19, 11. p1158-1169.

Salazar M., 1991. Young labourers in Bogotá: Breaking authoritarian ramparts. In: Fals Borda O., and Rahman A., (eds). Action and knowledge: breaking the monopoly with participatory action research. New York : Intermediate Technology/ Apex. p 54-63.

Samper D., 2008. Qué se hicieron los pobres? El Tiempo June 11.

Sánchez-Triana E., Ahmed K. and Awe Y., 2007 (eds). Environmental Priorities and Poverty Reduction. A country environmental analysis for Colombia. Directions in Development. The World Bank. p483.

Saner, R. 2003. El experto negociador. Estrategias, tácticas, motivación, comportamiento, delegación efectiva. Leadership. Ediciones Gestion Barcelona. p281.

Santamaría M., and Rozo S., 2009. Qualitative and quantitative analysis of firm informality in Colombia. Desarrollo y sociedad. ISSN 0120-3584. p269-296.

Santos T., and Camacho L.A., 2011. Modelación dinámica de calidad del agua con efluentes de curtiembres. Estudio de caso cuenca alta del rio Bogotá. Una herramienta de planeación. Agua 2011. September 30 p10.

Sayer A., 1992. Method in social science. A realist approach. 2nd Edition. Routledge. p313.

Sayer A., 2010. Method in social science. A realist approach. Revised 2nd Edition. Routledge. 313p.

Schaper, M., 2003. Areas de oportunidad en el sector ambiental de ALC. Project Presentation ECLAC-GTZ September 4. Cartagena. Colombia. 10 p.

Schein E., 1968. "Personal change through Interpersonal Relationships". In Bennis, W.G. p 406-426.

Schein E., 1988. Process consultation: Its role in organization development 1. Addison-Wesley. p204.

Schein E., 1992. Organizational Culture and Leadership .2nd Edition. San Francisco, Jossey-Bass Inc. p350.

Schein E., 1996. Kurt Lewin's Change Theory in the Field and in the Classroom: Notes Toward a Model of Managed Learning', Systems Practice, p34.

Schein E., 2010. Organizational Culture and Leadership. 4th Edition. John Wiley and Sons. p 415.

Schmoch U., Rammer C., and Legler H., 2006. National systems of innovation in comparison: structure and performance indicators for knowledge societies. Berlin: Springer. p 265-286.

SDA, 2006. Inventario de carácter ambiental en el sector comprendido entre las calles 57a y 59b sur y las carreras 19b bis y 16b (AK Tunjuelito) del barrio San Benito, localidad de Tunjuelito en Bogotá D.C. Convenio 014-2006. Bogotá Sin Indiferencia. Alcaldia Mayor de Bogotá. VIDEO.

Selfa T., and Endter-Wada J., 2008. The politics of community-based conservation in natural resources management: a focus for international comparative analysis. Environment and Planning. 40, 4, p948-965.

Semana, 2006. De cara al río. Revista Semana. December 4-11 p94-95.

Semana, 2010. Las tierras del agua. Medio Ambiente. Revista Semana. February 8-15, p48-49.

Semana, 2011. Desigualdad extrema. Colombia es el país más desigual de América Latina y el cuarto del mundo. Revista Semana. 14-21. March. p32-34

Semana, 2012. El poder de las regiones. Revista Semana. 17-24 September 2012 p:92-93

Shapiro D.L., 2006. Teaching students how to use emotions as they negotiate. Negotiation Journal Vol 22 Issue 1 p:105-109.

Shi H., Chertow M., Song Y., 2010. Developing country experience with eco-industrial parks: a case study of the Tianjin Economic –Technological Development Area in China. Journal of Cleaner Production Vol 18 Issue 3 p: 191-199.

Shmueli D., Warfield W., and Kaufman S., 2009. Enhancing Community Leadership negotiaton skills to build civic capacity. Negotiation journal 249-266p.

Siaminwe L., Chinsembu K., and Syakalima M. 2005. Policy and operational constraints for the implementation of Cleaner Production. Journal of Cleaner Production (13). p. 1037-1047

Siebel M.A. and Gijzen H.J., 2002. Application of cleaner production concepts in urban water management in: Workshops Alcue – Sustainable development and urbanization: from knowledge to action, Science & Technology, Vol.3.

SINA II, 2005. Estructuración de la zona ecoeficiente –ecozona- para el sector de curtiembres de los municipios de Villapinzón y Chocontá. Departamento de Cundinamarca. Ministerio de Ambiente, Vivienda y Desarrollo Territorial- Corporación CAR. p37.

Sivakumar V., Chandrasekaran F., and Swaminathan G., 2009. Towards cleaner degreasing methods in industries: ultrasound-assisted aqueous degreasing process in leather making. Journal of cleaner production, 17, 1 p101-104.

Smith M.K., 2001. 'Kurt Lewin, groups, experiential learning and action research', the encyclopedia of informal education, http://www.infed.org/thinkers/et-lewin.htm. Retrieved September 13 2010.

Soni P., 2006. Global solutions meeting local needs. Climate change policy instruments for diffusion of cleaner technologies in small scale industry in India. Ph.D.-Thesis final draft. Vrije Universiteit. Amsterdam. p253.

Steele F., 1975. Consulting for Organizational change. University of Massachusetts Press 202p.

Stephanou C., and Rodriguez C., 2008. Bank financing to small and medium-sized enterprises (SMEs) in Colombia. Policy research working paper #4481. The World Bank Latin America and the Caribbean region. Financial and private sector development unit. p46.

Steiner G. and Posch A., 2006. Higher education for sustainability by means of transdisciplinary case studies: an innovative approach for solving complex, real-world problems. Journal of Cleaner Production. 14. p877-890.

Sterner T., 2003. Policy instruments for environmental and natural resource management. Resources for the future. World Bank ISBN:0-8213-5381-0 p500.

Stiglitz J., 2002. El malestar de la globalización. Buenos Aires. Taurus. 447p.

Stiglitz J., 2003. Hacia una nueva agenda para America Latina. Cepal. #80. August p 7-40.

Stiglitz J., 2011. El malestar de la globalización. Buenos Aires. Taurus. 447p.

Stiglitz J., 2012. The price of inequality: How today's divided society endangers our future. Norton and Company Inc. p340.

Stringer L., Reed M., and Dougill J., 2007. Implementing the UNCCD: Participatory challenges. Natural Resources Forum, 31, p198-211.

Susskind L., and Field P., 1996. Dealing with an angry public. The Mutual Gains Approach to Resolving Disputes. The MIT-Harvard Public Disputes Program. The Free Press. p268.

Susskind L., Mnookin R., Fuller B., and Rozdeiczer L., 2003. Teaching multiparty negotiation: a workbook. Cambridge, MA: Program on Negotiation. p 408.

Sverisson A., and Van Dijk M.P., (eds) 2000. Local economies in turmoil. International political economy series. p208.

TANDEM, 2005. Curso de Alta Dirección en Negociación. Universidad de los Andes, Desarrollo Gerencial. April 2004- March 2005. Teaching material 130 p.

TANDEM, 2010. Negotiation and Conflict Resolution Manual. Mc Gill University. 35p.

Terry J., and Khatri K., 2009. People, pigs and pollution-Experiences with applying participatory learning and action (PLA) methodology to identify problems of pig-waste management at the village level in Fiji. Journal of Cleaner Production. 17. p1393-1400.

Thabrew L., Wiek A., and Ries R., 2009. Environmental decision making in multi-stakeholder contexts: applicability of life cycle thinking in development, planning and implementation. Journal of Cleaner Production. 17. p 67-76.

The Economist, 2011. Inequality: The rich and the rest. Printed edition January.

Thompson L., 2009. The Mind and heart of the Negotiator. Fourth edition. Kellog School of Management, North Western University. Prentice Hall editors. p411.

Tokman, V.E., 2007. Modernizing the informal sector. DESA. Economic & Social Affairs. Working paper. 42. 15p.

Triana H., 2000. Documentos Acueducto de Bogotá. Alternativas técnicas para el río Bogotá Given on a CD.

ULeeds, 2012. www.konsult.leeds.ac.uk/public/level1/sec09/index.htm. Retrieved September 12, 2012.

UNCTAD, 2008. Small businesses as a way out of poverty. http://unctad.org/en/Docs/diaeed2008d4_en.pdf. Retrieved September 20.

UNDP, 2003. Human Development Report. Human Development Indicators. Millenium Development goals A compact among nations to end human poverty. United Nations Development Programme. Oxford University Press 358p.

UNDP, 2007. Marco de Asistencia de las Naciones Unidas Para El Desarrollo. 2008-2012. p37.

UNDP, 2010. Acting on the future: breaking intergenerational transmission of inequality. Regional Human Development Report. Regional bureau for Latin America and the Caribbean. New York. 206p.

UNEP, 1991. Tanneries and the environment. A technical guide to reducing the environmental impact of tannery operations. Technical report series 4. Paris. France. Industry and environment. United Nations Publications. p15-39.

UNEP, 1994. Government Strategies and Policies for Cleaner Production, UNEP Industry & Environment. Paris. http://www.unido.org/cp.html Retrieved December 9 2007

UNEP, 1992. Agenda 21 . Retrieved October, 4, 2012. http://www.unep.org/Documents.Multilingual/Default.asp?DocumentID=52.

UNEP, 1999. International Cleaner Production Information Clearinghouse. CD Version 1.0, Paris, France.

UNEP, 2006. Environmental agreements and Cleaner Production. Questions and answers. United Nations Environment Programme. p28.

UNEP, 2010. Annual report. United Nations Environment Programme. p124.

UNIANDES, 2002a. Aplicación de un modelo numérico para la priorización de la gestión de aguas residuales domésticas en Colombia. Informe final. Ministry of Environment. May. p208.

UNIANDES, 2002b. Campañas calidad del río Bogotá. Facultad de Ingeniería. Universidad de los Andes. Given on a CD.

UNIDO. OECD, 2004. Effective policies for small business. A guide for the policy review process and strategic plans for micro, small and medium enterprise development. UNIDO. OECD. Centre for private sector development Istanbul. p110.

UNIDO, 2005. Private Sector Development. The Support Programmes of SMEs Branch. Working paper 15. Wilfried Luetkenhorst. United Nations Industrial Development Organizations. December. p43.

UNIDO, 2009. Industrial Development Report. United Nations Industrial Development Organizations. p146.

Ury W., 2007. The power of a positive No. Save the deal, save the relationship and still say No. A Bantam book ISBN: 978-0-553-38426-0. p257.

Ury, W., Brett, J., and Goldberg, S., 1993. Getting Disputes Resolved. PON Program on negotiation at Harvard Law School. p199.

Usdept, 2010. Colombia: A country study. Federal Research Division. Library of Congress Ed. Hudson. p421.

USEPA, 1992. Facility pollution prevention planning guide. United States Environmental Protection Agency, Cincinnati, EPA 600/R92/088. 143p.

USITC, 2010. Small and Medium-sized enterprises. Overview of participation in US exports. USITC publication 4125. Investigation# 332-508. January 91p.

Van Berkel R. 2007. Cleaner production and ecoefficiency initiatives in Western Australia 1996-2004.The Journal of Cleaner Production. 15, 8-9. p741-755.

Van Berkel R., 2010. Evolution and diversification of National Cleaner Production Centres (NCPCs). Journal of Environmental Management. 91. p1556-1565.

Van Berkel R., 2011. Evaluation of the global implementation of the UNIDO-UNEP National Cleaner Production Centres (NCPC) programme. Clean Techn Environ Policy. 13. p161-175.

Van Berkel R. and Bouma J., 1999. Promoting cleaner investments in developing countries: A status report on key issues and potential strategies. Paris, United Nations Environment Programme.

Van de Kerkhof, M., 2004. Debating Climate Change. A study of Stakeholder Participation in an Integrated Assessment of Long-Term Climate Policy in the Netherlands. Lemma Publishers. Utrecht. **316 p**

Van der Zaag P., 2005. Integrated Water Resources Management : relevant concept or irrelevant buzzword? A capacity building and research agenda for Southern Africa. Physics and Chemistry of the Earth 30. Elsevier publications. p867-871.

Van der Zaag P. and Bolding A., 2005. Water governance in the Pungwe river basin: institutional limits to the upscaling of hydraulic infrastructure. Paper prepared for the session "Transboundary water governance: lessons learned in Southern Africa" of the 6th Open Meeting of the Human Dimensions of Global Environmental Change Research Community, 12 October. University of Bonn, Bonn. p12.

Van der Zaag P. and Savenije H.H.G, 2004. Principles of Integrated Water Resources Management. Delft, October. p118.

Van der Zaag P., Bos A., Odendaal A. and Savenije H.H.G., 2003. Educating water for peace: The new water managers as first-line conflict preventors. Paper prepared for the UNESCO-Green Cross" From potential conflict to Cooperation Potential: Water for Peace" sessions; 3rd World Water Forum, Shiga, Japan, 20-21 March. p16.

Van Herk S., Zevenbergen C., Ashley R. and Rijke J., 2011. Learning and action alliances for the integration of flood risk management into urban palnning: a new framework from empirical evidence from The Netherlands. Environmental science & policy. 14. p543-554.

Van Hoof, B., 2005. Políticas e instrumentos para mejorar la gestión ambiental de las PYMES en Colombia y promover su oferta en materia de bienes y servicios ambientales. CEPAL. p77

Van Hoof B., and Herrera C.M., 2007. La evolución y el futuro de la producción más limpia en Colombia. Revista de ingeniería de la Universidad de los Andes. 26. ISSN 0121-4993. p101-119.

Vásquez L., 2012. Las curtiembres en el barrio San Benito de Bogotá: Un análisis bioético en la perspectiva de Hans Jonas. Thesis for Magister in Bioethics. Javeriana University. p75.

Velásquez F., and González E., 2003. Qué ha pasado con la participación ciudadana en Colombia? Fundación Corona. p455.

Villegas N., 2012. Santiago y Lima dejaron atrás a Bogotá. El Centro Unido. Bogotá y Cundinamarca deben trabajar de la mano para enderezar el rumbo. Revista Semana. p2.

Vinke-de Kruijf J., Hommes S., and Bouma G.,2010. Stakeholder participation in the distribution of freshwater in the Netherlands. Irrigation Drainage Systems 24 DOI 10.1007/s10795-010-9097-3 P 248-263.

Vishnudas S., Savenije H., Van der Zaag P., Kumar C.E., and Anil K.R., 2008. Sustainability analysis of two participatory watershed projects in Kerala. Physics and Chemistry of the Earth. 33. p1-12.

Visvanathan C., and Kumar S., 1999. Issues for better implementation of Cleaner Production in Asian small and medium industries. Journal of Cleaner Production. 17, 2. p127-134.

Walker H., and Preuss L., 2008. Fostering sustainability through sourcing from small businesses: public sector perspectives. Journal of Cleaner production. 16, 15 October 2008. p1600-1609.

Wang M., Webber M., Finlayson B., and Barnett J., 2008. Rural industries and water pollution in China. Journal of environmental Management. 86. p648-659.

WB,2006. Participation web site. http://www.worldbank.org/participation/ participation/participation.htm. Retrieved October 5, 2012.

WB, 2010. Colombia. Informality in Colombia. Implications for workers welfare and firm productivity. Colombia and Mexico country management unit. Human development department. Report 42698-CO. March 2010. p157.

WB, 2011a. http://data.worldbank.org/country/colombia. Retrieved September 21, 2012.

WB, 2011b. A break with history: Fifteen years of inequality reduction in Latin America. http://siteresources.worldbank.org/INTLACREGTOPPOVANA/Resources/840442-1291127079993/Inequality_Reduction.pdf. Retrieved September 21, 2012.

WB, 2012. Movilidad económica y crecimiento de la clase media en América Latina. http://www.slideshare.net/lnsincensura/informe-movilidad-econmica-y-crecimiento-clase-media-en-amrica-latina-banco-mundial. Retrieved November 19, 2012.

Webler T. and Renn O., 1995. A brief primer on participation: Philosophy and practice. In: Renn, O., Webler, T., and Wiedmann, P. (eds). Fairness and competence in citizen participation. Evaluating models for environmental discourse. Kluwer Academic Publishers. Dordrecht. The Netherlands. p17-33.

Welbourne T. and Pardo M., 2009. Relational capital: Strategic advantage for small and medium-size enterprises through negotiation and collaboration. Group decision negotiation 18: p483-497.

Welp M., Vega-Leinert A., Stoll-Kleemann S., and Jaeger C., 2006. Science-based stakeholder dialogues: theories and tools. Gobal Environmental Change. 16. p170-181.

White S., 2008. A conceptual framework to guide research on pivate sector development in developing countries. IDRC working papers on globalization, growth and poverty #6. p84.

WHO, 2000. Global Water Supply and Sanitation Assessment 2000 Report, World Health Organization, Geneva, Switzerland. 75p.

WHO/UNICEF, 2010. Progress on sanitation and drinking water 2010 update. World Health Organization p60.

Whyte, W.F. (ed). 1991. Participatory action research. Sage publications. London. UK. p 247.

Whyte, W.F., 1998. Rethinking sociology. Applied and basic research. The American Sociologist. Vol 29, 1, p16-19.

Wiber M., Charles A., Kearney J., and Berkes F., 2009. Enhancing community empowerment through participatory fisheries research. Marine Policy. 33. p172-179.

Wieczorek A., Hekkert M., and Smiths R., 2010. Systemic policy instruments and their role in addressing sustainability cahllenges. Making innovation work for society: linking, leveraging, and learning. Globelics 2010. 8th International Conference 1-3 November. Malaysia. p34.

Wieczorek A. and Hekkert M., 2012. Systemic instruments for systemic innovation problems: A framework for policy makers and innovation scholars. Science and Public Policy 39, 1. p74:87.

Wilkinson R.G., Wilkinson R., and Pickett K., 2010. The Spirit Level: Why equality is better for every one. London, England: Penguin. p 400.

Winter R., 1987. Action-Research and the Nature of Social Inquiry. Professional innovation and educational work, Ashgate Publishing Company p 167.

WIPO, 2012. International Property Arbitration. Retrieved May 14 2012 www.wipo.int/amc/en/center/advantages.html

Withney D. and Trosten-Bloom A., 2003. The Power of Appreciative Inquiry. Berret-Koehler Publishers. p252.

Wolf A., 2000. Indigenous approaches to water conflict negotiations and implications for international waters. International Negotiation: A journal of Theory and Practice. 5,2. p357-373.

Wolf A., 2008. Healing the enlightenment rift: rationality, spirituality and shared waters. Journal of International Affairs Columbia University. 61, 2. p51-73.

World Bank, 1996a. Reflections: What is participation. The World Bank Participation Sourcebook. Chapter I. p7.

World Bank, 1996b. Identifying Stakeholders. The World Bank Participation Sourcebook. Chapter III. p5.

Worrell E. and Van Berkel R, 2001. Technology transfer of energy efficient technologies in industry: A review of trends and policy issues. Energy Policy 29, 1. p29-43.

Wynne, B., 1994. Scientific knowledge and the global environment. In: Redclift, M. and Benton T. (eds). Social theory and the global environment. Routledge. London. UK p169-190.

WWAP, 2003. Water for people, Water for Life. The United Nations World water Development Report. United Nations Educational, Scientific and Cultural Organization (UNESCO) & Berghahn Books, Barcelona. p 576.

WWAP, 2006.Water: A shared responsibility. The United Nations World Water Development Report 2. Unesco publishing p 584

WWAP, 2009. Water in a changing world. The United Nations World Water Development Report 3. UNESCO publishing p 318.

Yin R., 1994. Case Study Research. Design and Methods. Second Edition. SAGE publications. p170.

Yin R., 2009. Case Study Research. Design and Methods. Fourth Edition. SAGE publications. p240.

Annexes

Annex to Chapter 4

4.1[63] Micro-cases at national level

Date	Institution	Position	#
2004-2012	Senate	Head of Senate (Former Representative)	*mc*1
2004-2012	Ministry	Head of tanneries	*mc*2
2004-2019	Regional Court	Magistrate on the Bogotá River Court Order	*mc*3

Year	Head of Senate	mc1
2004 2012	I am very concerned about the last events in Villapinzón. A month ago, I was asked to attend a meeting at the Office of the Presidency where CAR asked for political support to undertake forced closures of the tanneries in the upper river basin. CAR explained that the tanners had lost all their legal chances and the proof was that all their legal terms had expired. Unfortunately, we realized CAR was on the right track. I say unfortunately, because I am a Representative for this province and I would not like these kinds of measures to happen because they cause severe harm to the already difficult social and economic situation in the area. Besides, I am very sensitive to these environmental issues. I am working on a water law for Colombia (the law faced opposition because it was thought to privilege private interests). I was born at the lower basin of the Bogotá River that has been suffering from the unsustainable uses of the river water for decades from communities such as Villapinzón and Bogotá-with eight million inhabitants- that has not implemented IWRM and whose only sewer plant is ineffective. Our Bogotá River is a dead river that should not be used for human consumption. Unfortunately, farmers have no other options when it comes to irrigation of their crops. As a politician, I am invited to attend multiple meetings every week. It is difficult to decide which to attend. Usually, I must concentrate on the most important ones in terms of the kind of people (high decision makers or powerful supporters of my campaign) that are taking part. Last week, I found a fax inviting me to a local meeting in Villapinzón. It appealed to me that it was written on a positive attitude to support the recovery of the Bogotá River. The latter was in contrast to the meeting at the Office of the Presidency where it was decided that the closures were the "only way out". I decided to cancel all other appointments and drove an hour and a half from Bogotá. At the meeting, I was surprised to listen to the tanners' arguments. They seemed reasonable to me. It seemed to me that they were willing to commit to the recovery of the river. An academic institution facilitated the meeting and the possibility to implement CP could work out.	

---

[63] 4.1 refers to the corresponding section in Chapter 4.

The Colombian Constitution has taken into account diverse participative mechanisms but it has not specified how to implement them. Real participation from CARs and NGOs is no implemented in practice. Since Colombia has been ruled by authoritarianism and clientelism, participation has been looked with suspicion. I felt bad that in twenty years of the tanner's conflict, I had never listened to the tanners' arguments. They are humble people needing support and capacity building. I told them I had the power to help them on the co-financing issue of their CP project, provided CAR approves.

At the end of the meeting, my assistant went to the responsible person asking her if we could meet in private later on. We started a relationship that has lasted all these years. We went together to the Chamber of Commerce in order to look for financial support for the CP project. I helped them to attend the President's public councils with communities on the weekends and draw attention from the president to help the tanners to avoid or handle the forced closures with the best possible way. When Monica brought the idea to have a lawyer dealing with the penal fines the tanners could not afford to pay, we went to el Consejo de la Judicatura that is in charge of evaluating judges and magistrates in order to ask their opinion on presenting an innovative proposal switching peal fines for environmental work. On December 2010, as President of the Senate, I was invited by the SWITCH project to tell how three women from different backgrounds managed to change the everlasting conflict in the upper basin of the Bogotá River at a meeting in Zaragoza for UN. I always tell people that what draw my attention in this case was the academic focus and commitment. Academic support on relevant issues in the long term lacks in Colombia. 7 years have passed by and it was very rewarding to see the tanner's leader becoming empowered. As marginalized, they are always vulnerable but this has been an important step towards inclusion.

As politicians, it is very rewarding to offer support to serious endeavours like this one. When we help, we never know the real motives behind the scenes. Colombia has serious corruption problems. CARs face a paradoxical situation because of their autonomy, which seems to have caused lack of governance on relevant national issues.

Year	Head of tanneries	mc2
2004-2007	I was starting to work at the ministry when I was told to go to a meeting on tanneries, that was taking place in Villapinzón. The conflict was at its highest because the new CAR director had decided to close the tanneries and other actors such as the ministry, the presidency, the governorship were supporting that decision. I was organizing monthly national tannery meetings and the only ones absent were the tanners from Villapinzón. Other tanners say that they are very difficult to deal with, that they do not want to change. My perception of the problem in that region started to change after the first meeting in June 2004. I realized that the tanners were vulnerable people needing encouragement and empowerment. An interesting process was aiming at implementing CP and afterwards, at building the secondary treatment, which would be less expensive. I was named by the Minister to represent the Ministry in all the meetings on this regard. There were meetings in presidency,	

	at CAR, at the Governorship, with the mayor and with all the actors in Villapinzón.
	Even though the closures could not be avoided, CAR accepted that the tanners could implement CP provided they presented their environment plans and asked for all the legal permits.
2008-2011	An European project has supported the tanners, their leader has become a more structured person and the group of tanners, once with their tanneries opened, were successfully implementing CP. Unfortunately, since 2004, we have had 5 different ministers, and CAR has also been lead by 4. They are politicians that know little about environmental issues and take long to grab the essential goals to be pursued.
	The coordination between ministries is lacking. The common agenda between the ministry of industry and environment for example has not met for the last three years.

Year	Magistrate on Court Order mc3
2004	When I was named responsible to integrate all popular pleas on the Bogotá river into one, I never thought that this topic would become so relevant in my life.
	I have always believed that better results can be reached through collective agreements that acknowledge people's legitimate interests. Based on that, the process on the Court Order aimed at recognizing all the actors involved on the pollution of the river and at building initial agreements with them.
	After a year of work, I found that there were marginalized communities such as miners, tanners or brick workers that needed to be heard, but because of limited education and financial resources were either not represented by any lawyer or either unable to participate in court appealings,. In search for a better understanding of their complex situations, I decided to visit myself all the localities where vulnerable communities had caused important environmental impacts. In Villapinzón, for instance, a leader told me that they just needed money to build their treatment plant. Once in Court, I met a researcher that talked to me about cleaner production, about adopting best practices.
	I finished up understanding the tanners' problems participating at big meetings organized by the researcher in Villapinzón. I had the chance to realize that there has been actors that had manipulated their situation and which had even made a living out of their conflict with the authority.
	I have to rule also on the other conflicts found along the river. The environmental agency CAR for instance, should not be operating the waste water treatment plants at the municipalities. They should be controlling and encouraging communities to abide the law.

4.2 In-depth interviews Chapter 4

The format for the open questions was:

Open questions
1. Please tell me how is Colombia performing with respect to its environment
2. How important has been water management?
3. From your point of view, please present what you consider have been the positive outcomes and the major draw-backs if any in environmental policy?
4. What solutions do you propose?
5. Have marginalized communities like the tanners in Villapinzón been taken into account in the design of such solutions?

Annex to Chapter 5

5.1 In-depth interviews Chapter 5

CERRITO

Year	Head National Tannery Committee	Nt
2004	Which and what are the mechanisms supporting CP? (a) a *regulatory framework* that recognizes their needs, (b) access to *technology*, (c) *economic incentives* and *financial support*, and (d) access to sound *information* on all related issues like legal and technical options	
2009	How did the successful mechanisms operate? This community has had the same persons in charge of the CP program for five years and the approach has remained. The head of CP body is a supportive person. They have dealt first with social and economic impending issues. They prioritized the water recovery over land planning and ordering (river bank properties were supposed to be relocated). Trust was built and consensus possible. The academia has inspired the process transforming it highly participative and based on trial-and-error. CP was put on command-and-control through the *acuerdo de obligaciones.*. Standard limits were calculated on loads and not just concentrations. Tannery experts have been brought even from Brazil.	

Year	Head CP Project	HdCP
2004	Which and what are the mechanisms supporting CP? (a) a *regulatory framework* that recognizes their needs, (b) access to *technology*, (c) *economic incentives* and *financial support*, and (d) access to sound *information* on all related issues like legal and technical options She insisted on leading a high degree participative and trial-and-error oriented process, on contrasting with the main stream consulting on smaller industries.	
2009	How did the successful mechanisms operate? There has been a permanent multidisciplinary approach supported by Universidad del Valle. The social component is considered highly relevant. I considered myself a supporter more than anything but I do not do things for them. I always tell them to decide what ever is better for them. People are happy because we have brought tannery experts for them to choose. The tanners are applying and learning from their own experiences. They have built consensus on the causes of their problems. The biggest tanner has been a positive leader. The authority has stayed on backstage all the time supporting me. CP was tied to command-and-control and the standards	

	fixed on loads. The policies on zone planning have not interfered with environmental priorities.

Year	Technician in charge	**Tech**
2004	Which and what are the mechanisms supporting CP? (a) a *regulatory framework* that recognizes their needs, (b) access to *technology*, (c) *economic incentives* and *financial support*, and (d) access to sound *information* on all related issues like legal and technical options	
2009	How did the successful mechanisms operate? Success was possible because of long time support from authorities, of making CP implementation a requirement by law, of working through trial-and-error and participatively,, of prioritizing the tanners'impending social and economic challenges and of supporting hence, trust building, of inviting tannery experts to discuss and solve technical questions, of the role of the head of CP body as supporter or catalyser, of avoiding overlapping sensible land conflicts with the environmental solutions, of creating consensus from the side of the tanners on their problems, of having support from academia..	

Year	Officer regional environmental authority	**Rg**
2004	Which and what are the mechanisms supporting CP? (a) a *regulatory framework* that recognizes their needs, (b) access to *technology*, (c) *economic incentives* and *financial support*, and (d) access to sound *information* on all related issues like legal and technical options	
2009	How did the successful mechanisms operate? In the past, the relationship with the tanners was very linear. Today with the CP body, we have had the opportunity to support processes not only through punitive approaches but participative. Their role has been essential. There has been long term commitment from the head of the CP body and she turned out to ask for multidisciplinary support from Academia. I found out after many years as an engineer, that the social and economic issues must be prioritized in order to work with local communities. Less than a year ago, I took a course at the university to become a facilitator. Despite the above, as authorities we must keep on being strict and enhance law abiding. In this case, CP implementation was binded to a regulatory instrument that has allowed us to monitor the tanners' performance. The tanners had had visits from tannery experts and they had had the opportunity to learn by experiencing technical processes. We put aside for a while solving the riverbank issue.	

Year	Officer local environmental authority	**Lc**

2004	Which and what are the mechanisms supporting CP?
	(a) a *regulatory framework* that recognizes their needs, (b) access to *technology*, (c) *economic incentives* and *financial support*, and (d) access to sound *information* on all related issues like legal and technical options
2009	How did the successful mechanisms operate?
	My relationship with them is not always the best because they try to hide their real environmental impacts. They are finally showing good results once we told them this was their last chance to integrate into formal and clean activities. We told them they could go through harsh processes like Villapinzón if they do not obey. On the one hand, CP was tied to regulatory instruments. On the other hand, the CP body offered participative processes based on trial-and-error (this is good because they cannot blame anybody for teaching them wrong), and consensus on their problems. Academia has inspired the CP Body; they are socially oriented. They warned us not to overlap land uses with the environmental recovery plans in order to avoid more conflicts. The head of CP body has committed at long term basis, we all had.

Year	Tanner 1	T1
2004	Which and what are the mechanisms supporting CP?	
	(a) a *regulatory framework* that recognizes their needs, (b) access to *technology*, (c) *economic incentives* and *financial support*, and (d) access to sound *information* on all related issues like legal and technical options	
2009	How did the successful mechanisms operate?	
	How did the successful mechanisms operate?	
	By taking into account our realities before any technical implementation; by letting us participate effectively and make our own decisions; by organizing workshops that are 70% trial-and-error and 30% theory; by helping us build a common understanding of our problems; by writing clear rules that force us to be serious and be able to measure our impacts on loads; by allowing us being on the river bank, provided we clean the river; by calling tannery experts; by creating the CP body that through the head, has been respectful to us and that has invited academia to help us with workshops that had never been organized before in this village	

Year	Tanner 2	T2
2004	Which and what are the mechanisms supporting CP?	
	(a) a *regulatory framework* that recognizes their needs, (b) access to *technology*, (c) *economic incentives* and *financial support*, and (d) access to sound *information* on all related issues like legal and technical options	

2009	How did the successful mechanisms operate?
	By taking into account our realities before any technical implementation; by letting us participate effectively and make our own decisions; by organizing workshops that are 70% trial-and-error and 30% theory; by helping us build a common understanding of our problems; by writing clear rules that force us to be serious and be able to measure our impacts on loads; by allowing us being on the river bank, provided we clean the river; by calling tannery experts; by creating the CP body that through the head, has been respectful to us and that has invited academia to help us with workshops that had never been organized before in this village.

Year	Tanner 3	T3
2004	Which and what are the mechanisms supporting CP?	
	(a) a *regulatory framework* that recognizes their needs, (b) access to *technology*, (c) *economic incentives* and *financial support*, and (d) access to sound *information* on all related issues like legal and technical options	
2009	How did the successful mechanisms operate?	
	By taking into account our realities before any technical implementation; by letting us participate effectively and make our own decisions; by organizing workshops that are 70% trial-and-error and 30% theory; by helping us build a common understanding of our problems; by writing clear rules that force us to be serious and be able to measure our impacts on loads; by allowing us being on the river bank, provided we clean the river; by calling tannery experts; by creating the CP body that through the head, has been respectful to us	

Year	Tanner 4	T4
2004	Which and what are the mechanisms supporting CP?	
	(a) a *regulatory framework* that recognizes their needs, (b) access to *technology*, (c) *economic incentives* and *financial support*, and (d) access to sound *information* on all related issues like legal and technical options	
2009	How did the successful mechanisms operate?	
	By taking into account our realities before any technical implementation; by letting us participate effectively and make our own decisions; by organizing workshops that are 70% trial-and-error and 30% theory; by helping us build a common understanding of our problems; by writing clear rules that force us to be serious and be able to measure our impacts on loads; by allowing us being on the river bank, provided we clean the river; by calling tannery experts; by creating the CP body that through the head, has been respectful to us	

SAN BENITO

Year	Head national tannery committee	Nt
2004	Which and what are the mechanisms supporting CP? a) a *regulatory framework* that recognizes their needs, (b) access to *technology*, (c) *economic incentives* and *financial support*, and (d) access to sound *information* on all related issues like legal and technical options	
2009	How did the successful mechanisms operate? Associative work, focusing on social and economic issues before handling the technical ones by trial-and-error, and establishing leadership from the authority were important Enhancing participation and building consensus are also relevant drivers Promoting green champions is essential	

Year	Head CP Project	HdCP
2004	Which and what are the mechanisms supporting CP? a) a *regulatory framework* that recognizes their needs, (b) access to *technology*, (c) *economic incentives* and *financial support*, and (d) access to sound *information* on all related issues like legal and technical options	
2009	How did the successful mechanisms operate? Associative work, focusing on social and economic issues before handling the technical ones by trial-and-error, and establishing leadership from the authority were important Enhancing participation and building consensus are also relevant drivers Promoting green champions is essential The role of the CP body should be of facilitator	

Year	Technician in charge	Tech
2004	Which and what are the mechanisms supporting CP? a) a *regulatory framework* that recognizes their needs, (b) access to *technology*, (c) *economic incentives* and *financial support*, and (d) access to sound *information* on all related issues like legal and technical options	
2009	How did the successful mechanisms operate? Associative work, focusing on social and economic issues before handling the technical ones by trial-and-error, and establishing leadership from the authority were important	

Year	Officer regional environmental authority	Rg
2004	Which and what are the mechanisms supporting CP? a) a *regulatory framework* that recognizes their needs, (b) access to *technology*, (c) *economic incentives* and *financial support*, and (d) access to sound *information* on all related issues like legal and technical options	
2009	How did the successful mechanisms operate? Associative work, focusing on social and economic issues before handling the technical ones by trial-and-error, and establishing leadership from the authority were important Enhancing participation and building consensus are also relevant drivers	

Year	Officer local environmental authority	Lc
2004	Which and what are the mechanisms supporting CP? a) a *regulatory framework* that recognizes their needs, (b) access to *technology*, (c) *economic incentives* and *financial support*, and (d) access to sound *information* on all related issues like legal and technical options	
2009	How did the successful mechanisms operate? Associative work, focusing on social and economic issues before handling the technical ones, by trial-and-error, and establishing leadership from the authority were important Enhancing participation and building consensus are also relevant drivers	

Year	Tanner 1	T1 (80T)
2004	Which and what are the mechanisms supporting CP? a) a *regulatory framework* that recognizes their needs, (b) access to *technology*, (c) *economic incentives* and *financial support*, and (d) access to sound *information* on all related issues like legal and technical options Clear rules and coordination	
2009	How did the successful mechanisms operate? We need associative work at long term basis Mayor issues are (1) adapting the programs to the social needs, (2) working in a clearly participative manner stimulating trial-and-error and the choice of technology, group efforts and collaborative work, (3) stimulating a unified vision, before any technical issues are considered, and (4) building a strong leadership on the part of the urban environmental authority.	

Year	Tanner 2	T2 (80T)
2004	Which and what are the mechanisms supporting CP? a) a *regulatory framework* that recognizes their needs, (b) access to *technology*, (c) *economic incentives* and *financial support*, and (d) access to sound *information* on all related issues like legal and technical options Clear rules and coordination	
2009	How did the successful mechanisms operate? We need associative work at long term basis Mayor issues are (1) adapting the programs to the social needs, (2) working in a clearly participative manner stimulating trial-and-error and the choice of technology, group efforts and collaborative work, (3) stimulating a unified vision, before any technical issues are considered, and (4) building a strong leadership on the part of the urban environmental authority.	

Year	Tanner 3	T3 (80T)
2004	Which and what are the mechanisms supporting CP? a) a *regulatory framework* that recognizes their needs, (b) access to *technology*, (c) *economic incentives* and *financial support*, and (d) access to sound *information* on all related issues like legal and technical options Clear rules and coordination	
2009	How did the successful mechanisms operate? We need associative work at long term basis Mayor issues are (1) adapting the programs to the social needs, (2) working in a clearly participative manner stimulating trial-and-error and the choice of technology, group efforts and collaborative work, (3) stimulating a unified vision, before any technical issues are considered, and (4) building a strong leadership on the part of the urban environmental authority.	

Year	Tanner 4	T4
2004	Which and what are the mechanisms supporting CP? a) a *regulatory framework* that recognizes their needs, (b) access to *technology*, (c) *economic incentives* and *financial support*, and (d) access to sound *information* on all related issues like legal and technical options Clear rules and coordination	
2009	How did the successful mechanisms operate?	

	We are not optimistic because we do not have support in the long term

Year	Tanner 5	T5
2004	Which and what are the mechanisms supporting CP? a) a *regulatory framework* that recognizes their needs, (b) access to *technology*, (c) *economic incentives* and *financial support*, and (d) access to sound *information* on all related issues like legal and technical options Clear rules and coordination	
2009	How did the successful mechanisms operate? We are not optimistic because we do not have support in the long term	

Year	Tanner 6	T6
2004	Which and what are the mechanisms supporting CP? a) a *regulatory framework* that recognizes their needs, (b) access to *technology*, (c) *economic incentives* and *financial support*, and (d) access to sound *information* on all related issues like legal and technical options Clear rules and coordination	
2009	How did the successful mechanisms operate? There were none. We face serious financial and land problems	

Year	Tanner 7	T7
2004	Which and what are the mechanisms supporting CP? a) a *regulatory framework* that recognizes their needs, (b) access to *technology*, (c) *economic incentives* and *financial support*, and (d) access to sound *information* on all related issues like legal and technical options Clear rules and coordination	
2009	How did the successful mechanisms operate? There were none. We face serious financial and land problems	

Year	Tanner 8	T8

2004	Which and what are the mechanisms supporting CP?
	a) a *regulatory framework* that recognizes their needs, (b) access to *technology*, (c) *economic incentives* and *financial support*, and (d) access to sound *information* on all related issues like legal and technical options
	Clear rules and coordination
2009	How did the successful mechanisms operate?
	There were none. We face serious financial and land problems

Year	Tanner 9	T9
2004-	Which and what are the mechanisms supporting CP?	
	a) a *regulatory framework* that recognizes their needs, (b) access to *technology*, (c) *economic incentives* and *financial support*, and (d) access to sound *information* on all related issues like legal and technical options	
	Clear rules and coordination	
2009	How did the successful mechanisms operate?	
	There were none. We face serious financial and land problems	

5.2 Micro-cases at regional & local level

Date	Institution	Position	#
2004-2010	CRPML	Head CP project	*mc4*
2004-2006	SDA	Head CP project	*mc5*

Year	Head of CP project CRPML
2004	When I arrived at CVC, I was told that I would face a very difficult challenge: To work out a solution based on CP for the tanneries in Cerrito. I was told that they were going to be connected to a wastewater treatment plant and these industries needed to adjust their discharges to that.
	Since I had just finished a postgraduate course on environmental management, I learned about giving importance to social issues before considering solving the technical ones.
	At the time I was hired, the CVC director's belief was that the environmental agency should offer support to the smallest firms and not to the big ones through a process that should go beyond the traditional consulting one, that is known for not getting involved with the clients.

	We designed then a learning process based on trial-and-error: our workshops turned out being 70% practical, 30% theoretical. The biggest tannery in the area opened its doors for organizing the workshops there and students from the university were encouraged to put their knowledge in practice there.
	Even though the CRPML is an independent body, it works hand-in-hand with CVC. I make very clear to the tanners that the process of CP implementation belongs to them and that I am just a facilitator and a supporter of their own decisions.
	I think I have been lucky to work with people in the agency that are highly motivated, academically sound, and that have been for a long time in their posts. The latter has made that the design of a regulatory tool to CP implementation was creative and effective and that spatial planning issues were left aside provided the environmental impacts were worked out.
	It was very rewarding to get support from the Municipality plan because the whole community was asked to participate in the planning.
	Even though the process has not been without difficulties, the results have been quite promising. It was very rewarding to be able to support the process itself in Villapinzón thanks to the relationship that developed between Monica and me.
2005-2009	I found very disappointing that the smallest industries hardly get the credits they need. Access to financing is a critical issue.

Year	Head of CP project (Three different ones) SDA
2004	Because of the limited space in Bogotá, the tanneries in San Benito have been trying to relocate to rural areas close to Bogotá.
	Some tanneries have bought nearby areas to build their own end-of-pipe system. Most of tanneries cannot afford to do that because of high costs and lack of space.
	I have been teaching and making practical workshops on CP. Some tanneries have been adopting it. The majority has not.
	All officers had resigned. Forced closures take place. The HdCP feels he cannot talk to the tanners.
2006-2008	New officers are designed, the urban agency is modified.
	SDA signs an agreement with a Commercial body to stimulate the production of green products. I was giving the freedom to stimulate CP through an industrial park for making the most polluting phases.
2009	The spatial planning problems cause us serious problems to any of these strategies. The social issues get very difficult to face also unless there is political will to face them.

Annex to Chapter 6

6.1 Micro-cases at regional & local level

Date	Institution	Position	#
2004-2013	ACURTIR	Positive Tanners' leader	*mc6*
2004-2008	CAR	Head of technical department (CAR officer in Bogotá)	*mc7*

Year	Tanner's leader	mc6
2004-2005	We are very concerned about the last events in Villapinzón. We have requested to get reliable information from CAR, but we have also asked to be heard. We certainly want to clean our river but we want to keep our identity as small and micro industries. We do not want to become simple employees of the only medium-size tannery in the area.	

We feel we have done everything the agency asks us to do. Each time a new director comes along, new directives arise contradicting the former ones. In the late 1980's, we were taught from CAR how to use chemicals. In the late 1990's we were blamed for the use of those chemicals and some closures were carried out. In the year 2000, we were asked to present PMAs –environmental planning. We did present them but never got an answer regarding their approval until 2004 when they said they had been of a poor quality. We had signed contracts with technicians that came knocking at our doors. It is difficult for us to determine who is a good or a bad technician. The use of chemicals is being taught by a great variety of salesmen that visit Villapinzón. Each time a crisis arises, we even need to pay lawyers that are very expensive and that do not show us effective results. Our representative told us he was a lawyer and we found out recently that he was not. Whenever there are meetings, we are told what should be done, that we are ignorant and unable to solve our problems. All gatherings are claim oriented.

At the same time as the announcement of forced closures occurred a month ago, a researcher from a university overseas came along asking us about our perception of the conflict and our interests. This is the first time in twenty years that someone bothers to ask us, the owners of the smallest industries. All the people that come here go first to see the largest tanner in the area and they do not get an accurate understanding of our situation. After some time we decided to ask this researcher whether she could help us in solving our conflict. The first thing we discussed was regarding the impending problems of the tanners on the riverbank. I went with her to visit the *Procuradora* to inform her of the property rights. To my surprise, she was not aware of. The researcher talked about a process called Cleaner Production and we decided to visit the community of tanners called Cerrito where a participative CP project was taking place. We found tanners, as small as ourselves, preventing discharges into the river and collaborating with the largest industry. If we manage to convince CAR regarding CP, the investments could be affordable and we could maintain our independency. Thanks to the representative of the Chamber of Representatives, many paths opened to us. CCB supported our CP project. We had many meetings at the Presidency with all the actors. The Court Order on the Bogotá River made CP implementation our way to go. The researcher and the Magistrate had established a relationship and it was possible to invite her to our local

	meetings and big groups methodologies. Mónica helped us to write invitations on positive and motivating terms. Our meetings were not based on claims and fights any more.

An EU project could be developed in the area but we still need to motivate all kind of important decision makers in order to receive their support.

We asked CAR to open the industries in order to be able to implement technological changes. We contacted, with support from the researcher, the representative for co-financing. Unfortunately we distrust greatly the agency CAR and we need to convince it to approve our project.

Of the five issues to be discussed at meetings, all are important to us. Even though the issue on the river bank does not affect me directly we will support our friends on the river bank to find a sustainable solution. The most pressing ones are the technical solution, the fines, closures and co-financing. |
| 2006-2010 | Despite closures we were able to implement CP because we received support from overseas for technical matters and from UNAL Finally, in 2008, the authority recognized our effort and supported SWITCH. We voted for a candidate for mayor that won and were finally able to formalize our spatial planning and to determine the size of the river bank.

Nevertheless, the properties on the river bank have not been bought and we still need to deal with this issue. Financing has been unaffordable for us. |
| 2011-2013 | Even though there has been international and national recognition to our efforts, as SWITCH is coming to an end, negative actors have been trying to abort our process. A new candidate for mayor is manipulating information and wants to invest on a an end-of-pipe system that is not convenient for us. The tanners that went to real change are silently resisting the negative influences. |

Year	Head of technical department	mc7

| Early 2004 | We are tired of being considered inefficient. The new director wants to improve our image. She announced she was going to clean-up the river banks from invaders. She announced forced closures to all the dirty industries in the province. The new director found herself with an *Acción de Cumplimiento* (Compliance action) from the *Procurador* on the issue of the tanneries of Villapinzón. We need to show effective results with regard to the recovery of the Bogotá River on a short term basis, besides we are tired of those tanners that have no willingness to change. |

The tanners of Villapinzón are aggressive and not to be trusted. Whenever we need to go to their place, they even harm us. They hate the CAR.

We had spent millions of pesos trying to convince them to implement new and clean technologies. We even worked on a project on an industrial park and they fought against it.

The time to cooperate has now ended. Either they invest on new technologies or they will close permanently.

It is impossible to deal with the smaller industries. They do not know how to work cooperatively. Our director believes they are bound to disappear.

Now that we started to close them for good and not any more as we did in the late 1990's, they asked for a last meeting. They said they have a project that will work out for them based on CP. Our director decided to fix that last meeting to show the public opinion how democratic she is. Our director thinks CP should only privilege clean industries to become *cleaner*. We are not going to support any effort.

We will surely listen to them but will not cooperate with them because we cannot trust them. We know they do not have the ability to implement any good solution, nor will they have support from anyone. All the authorities are supporting us.

Of the five issues to be discussed at this meeting, three are important to us. Even though the tanners from the river bank had just being found to be the legal owners of the premises, it is not our duty to solve their problems. Regarding the technical solutions, end-of-pipe approaches are more convenient to us for control purposes and we do not need to bother with behavioural change. If we decide that the size of the river bank should be fixed in terms of the floods occurring over a span of the last 100 years, more tanners would be definitely closed.

Even though we are not stepping back with respect to the forced closures, we are now being asked by the Governorship, *Procuraduría,* and the Presidency to take into account that UNESCO is now supporting the tanners on CP solutions. If they do not present first our legal requirements we will not bother to look at their CP project. We have been very clear at the meetings on the Presidency (our director did not allowed us to go to the first one that took place).

The meeting at the beginning of the year formalized the closures. Instead of having positive comments on the media, the reports are negative towards our intervention. The Magistrate on the Court order is even suggesting mediating. Our director is suspicious on the real intentions of Monica Sanz and the tanner's leader. Local social unrest may result from this process. I

	personally think that Monica is being naïve supporting the tanners' community. I met Mónica more than a year ago when I was in charge of the conflict on the wastewater utility of Bogotá and she furnished us with valuable documents on the French case of the city of Grenoble that helped us interrupt our own contract with the French in Bogotá.
Sep 2004	For the first time in this conflict, we received good quality PMAs in Spanish or EMPs. They are going to be technically approved. Legal approval will take longer because of the inter-related zonal planning issues that entail local powerful political interests. Reopening the industries is not a option yet.
Jan 2005	I have told Monica many times that I have no tools to monitor CP.
Aug 2005 Early 2006	It was time for me to look for new professional opportunities. This job entails too much stress and not good salary. I decided to be an independent consultant. I met Monica at the ministry by coincidence and we discussed our past difficulties working for different organisations and purposes in the same context. I was pleased to hear about the good results of the project in the area.

6.2 In-depth interviews Chapter 6

Stakeholders interviewed in Villapinzón in 2004

Type of Stakeholder	#
CAR officer in Bogotá	1
Regional CAR officer	1
Public Prosecutor	1
Big tanner	1
Mayor	1
Tanners	3
Leader at time of conflict	1
Officer at Governorship	1
Ministry Head of tanneries	1
Judge (Magistrate) Court order	1
Total number of interviewees	**12**

The format for the open questions was:

Open questions
1. Why is it that the conflict on the tanneries in this region has been so difficult? What are the reasons?
2. What are the most important issues to be faced?
3. Are there opportunities to participate and have a real say on the problems?
4. How is the relationship between members of community, officers, politicians, technicians, etc.
5. What are the solutions?

6.3 Evidence of the action research in relation to the SASI stages.

6.3.1 Official support from the director of CVC to a project on CP implementation chosen by the tanners

Tanners asking for support

Villapinzón, 04 de Mayo de 2004

Señora:
ANA DORLY JARAMILLO
Centro Regional de producción más Limpia
C.V.C
Cali Col.

Después de nuestra visita como representantes de las Fábricas de Curtiembres de Villapinzón, donde presenciamos el acompañamiento que ustedes realizan a los curtidores de esa zona en El Cerrito – Valle, queremos manifestar nuestra grata admiración por el trabajo que ustedes realizan en pro de esa comunidad.

Por tal motivo, respetuosamente nos permitimos solicitar se tramite una asesoria oficial por parte de ustedes a nuestras industrias que atraviesan desde hace ya 20 años una crisis profunda.

La seriedad, profesionalismo y experiencia de ustedes en este campo estamos seguros le daría a nuestra situación un giro definitivo.

Agrademos de antemano su gentileza y colaboración.

Atentamente,

REPRESENTANTES DE LOS CURTIEMBRES
VILLAPINZÓN – CUNDI.

CRPML asking permission to CVC

Santiago de Cali, Mayo 05 de 2004
crpml, 054-2004

EFPML
Centro Regional de
Producción Mas Limpia

Doctor:
JULIAN CAMILO ARIAS RENGIFO
Presidente Junta Directiva
CRPML

Respetado doctor:

Como es de su conocimiento, el sector curtidor de Villapinzon ha sido anunciado por la CAR del cierre inminente de sus empresas debido a la contaminación generada por sus procesos productivos. Los empresarios al enterarse del proyecto de PML que se adelanta en El Cerrito, solicitaron una reunión con los curtidores y el CRPML para conocer el proceso de concertacion con la CVC, el papel del CRPML y los resultados alcanzados hasta la fecha.

La reunión se realizó el lunes 3 de mayo. Asistieron, Eduardo Gonzalez empresario. Mónica Sanz ingeniera facilitadora y Evidalia Fernández como representante de 130 curtidores de Villapinzón, acompañaron tres empresarios de El Cerrito y la directora del CRPML. Una vez terminada la visita los empresarios enviaron al Centro regional la carta adjunta.

Solicito muy comedidamente su opinión respecto a esta solicitud.

Agradezco su atención

Ana Dorly Jaramillo
Directora CRPML

Official support by the director of CVC

DIRECCION GENERAL

Santiago de Cali, 28 de junio de 2004

100-05-633-2004

Doctora
GLORIA LUCÍA ALVAREZ PINZON
Directora
CAR
Bogotá, D.C.

Referencia: Proyecto *"Producción mas Limpia en las Curtiembres de Cundinamarca".*

Asunto: Aval CVC

La CVC ha apoyado el proyecto "Producción mas Limpia en las Curtiembres de El Cerrito", el cual ha sido ejecutado por el Centro Regional de Producción mas Limpia (CRPML) a través del convenio 089 CVC-Centro Nacional de Producción mas limpia y Tecnologías Ambientales (CNPMLTA).

En la actualidad, los 21 curtidores de El cerrito están implementando acciones de prevención de contaminación ambiental. Es importante destacar que la metodología participativa utilizada por el CRPML que ha generado espacios de concertación entre la CVC y los empresarios, involucrando a las diferentes instituciones relacionadas con el sector curtidor, ha contribuido a la obtención de estos resultados.

Esta experiencia está siendo replicada por la CVC, en las curtiembres de Cartago y Guacari. De esta manera, todas las curtiembres del Valle del Cauca estaran trabajando por mejorar su desempeño ambiental.

Debido a la importancia del sector curtidor para el país, al interés de la CVC por fortalecer las relaciones con las diferentes instituciones del SINA y a la experiencia adquirida con el CRPML, la CVC respalda el proyecto denominado *"Producción mas Limpia en las Curtiembres de Cundinamarca".*

Este proyecto ha sido preparado por el Centro Regional de Producción mas Limpia, la Universidad del Valle - el Instituto Cinara, UNESCO-IHE y la Corporación de Desarrollo Productivo del Cuero Calzado y Marroquinería con el apoyo de la corporación Autónoma Regional del valle del Cauca CVC.

Agradezco su atención.

Cordialmente,

JULIAN CAMILO ARIAS RENGIFO
Director General

Proyecto: Ana Dorly Jaramillo S
Revisó: Jorge Enrique Garcia
Beatriz Eugenia Orozco

6.3.2 Letter to Court Order on solution to Villapinzón

Mayo 7/04

Bogotá, Mayo 5 de 2004

En calidad de estudiante de doctorado de la universidad de Orleáns, Francia en proceso de traslado a UNESCO_IHE de Holanda alerto sobre el manejo que se viene dando al conflicto de las curtiembres de Villapinzón. Me encuentro realizando el trabajo práctico con la comunidad desde el mes de diciembre del año pasado.

La CAR se dispone a cerrar las curtiembres argumentando incumplimiento, cuando muchas de ellas si presentaron en su momento los planes de manejo que se les exigían y nunca obtuvieron respuesta alguna.
Critico la ausencia de acompañamiento a comunidades como estas, donde el nivel de educación y de formación tecnológica son deficientes y propongo la creación de una figura en Cundinamarca enfocada a la educación ambiental.
En este trabajo se denuncia que el argumento del no cumplimiento de las normas planteado por la autoridad ambiental CAR para proceder por la fuerza a solucionar el problema de contaminación, es en muchos casos un reflejo de la falta de acompañamiento, y no de falta de voluntad.
Se plantea también que hay una absoluta ausencia de facilitadores en los conflictos ambientales y que estas posiciones bien podrían adoptarse por las instituciones académicas.

Plantea el estudio que el caso de las curtiembres del Cerrito Valle constituye un ejemplo a nivel nacional. En esta zona del país, después de 20 años de infructuosos ensayos de sólo comando y control, el Centro Regional de Producción Más Limpia, apoyado por la autoridad ambiental del Valle del Cauca, CVC, emprendió un plan a corto, mediano y largo plazo donde luego de una adecuada capacitación, los industriales se comprometieron a implementar procesos limpios que redundan en ahorros a nivel individual y a nivel de grupo. Este último ahorro es definitivo al momento de estimar el tamaño de la planta de tratamiento o PTAR a ser construida en el mediano plazo. Los industriales en estos casos no sólo se comprometen en sus procesos sino que apoyan financieramente según sus posibilidades, determinadas de manera conjunta con el Centro. En el Cerrito, comenzado el proceso de implementación, se vislumbran ya inclusive hasta posibilidades de asociación entre los mismos industriales para enfrentar el reto del TLC o Tratado de Libre Comercio.

Dice este estudio que a pesar de que la constitución del 91 creó mecanismos de participación ciudadana, estos no se han llevado efectivamente a la práctica y las comunidades no son verdaderas forjadoras de su propia realidad. Ejemplo palpable de esto es Villapinzón, donde la fijación de la ronda del río nunca fue concertada con la comunidad.

Este trabajo hace un llamado de alerta sobre las difíciles consecuencias que acarrearán las medidas de fuerza que planea aplicar la CAR a partir del 15 de mayo del presente año. Con estas medidas el conflicto toma mayores dimensiones, empeora el clima de hostilidad y afecta principalmente a los más vulnerables.

El estudio propone que en una primera instancia se utilicen los recursos financieros existentes en lograr procesos limpios que mitiguen efectivamente la contaminación. Se plantea que la sóla relocalización aumenta los costos y los escasos recursos no alcanzan para la realización de procesos limpios. Lo único que se logra es desplazar el problema.

Concretamente se propone que dado el ambiente actual, adverso a la formulación de proyectos que sean realmente sostenibles en el largo plazo, se interrumpan los actos administrativos por un período de 3 meses mientras:

Se hace un manejo profesional del conflicto aplicando metodologías de concertación a grandes grupos. Este proceso inicial toma tres días.
Una vez trabajados los puntos cruciales, se elaborarán los proyectos técnicos que tenga a bien determinar el ejercicio de resolución de conflictos.

Todo este proceso tiene el aval de UNESCO, entidad que garantizará seriedad y compromiso. UNESCO-IHE son expertos mundiales en recuperación de las aguas y en manejo de sus conflictos.

El centro regional de producción limpia del Valle propone hacer un acompañamiento, basado en su exitoso plan piloto en el Cerrito.

La multinacional Tandem insourcing, liderada por un profesor de la universidad de Mc Gill orientaría las tecnologías de resolución de conflictos.

Una carta de la comunidad de Villapinzón acompaña este planteamiento.

6.4 Last three steps of SASI

6.4.1 Invitation to first big group methodologies

TANDEM

Bogotá, mayo 31 de 2004.

Dra. Gloria Lucia Álvarez
Directora de La Car
Dr. Mauricio Bayona
Subdirector Ambiental

Estimados señores,

Con el fin de poder implementar el proyecto que nos hemos propuesto trabajando de manera general y unificada en la recuperación del Río Bogotá en la comunidad de las curtiembres de Chocontá y Villapinzón, los queremos invitar a un ejercicio de "Espacio Abierto" el día 07 de junio de 8:30 AM a 4:30 PM en Villapinzón. Ese día Tandem facilitará el ejercicio trabajando con todos los actores implicados en el proceso de Recuperación del Río Bogotá para abordar la siguiente pregunta:

¿CÓMO LOGRAR VALORIZAR EL MUNICIPIO MEDIANTE UNA CONVIVENCIA ARMONICA CUMPLIENDO CON LOS REQUERIMIENTOS DE LA AUTORIDAD AMBIENTAL?

Este trabajo se va a desarrollar bajo una dinámica de participación libre fundamentada en el entusiasmo y la responsabilidad de los individuos y del grupo. Los participantes tendrán total libertad de expresión para hablar de procesos respetando a las personas. La única limitante para los planes de acción propuestos es la de cumplir con los requerimientos de la ley ambiental.

¡Contamos con su activa participación para hacer de este un proceso exitoso!

Cordialmente,

Carolina Naranjo E.
Consultora

TANDEM insourcing: Negociación, Resolución de Conflictos y Transformación de Grupos y Sistemas.

COLOMBIA:
Bogotá D.C., Colombia

Tel: (571) 296 708- 635 6330
Fax: (571) 296 7001
E-mail: TANDEMADR@TANDEMADR.COM
www.TANDEMADR.com

NEW YORK:
Address: 1 Liberty Plaza; Suite 3500
New York, NY 10006
Tel: (212) 233-9087;
Fax:(212) 964-9200

E-mail: Diansenson@tandemadr.com

6.4.2 Court Order including this research's effort

Court Order ruling on tanners implementing CP p 410:

fuente hídrica de la que utilizan un gran volumen de agua, como también por filtración afectan los suelos y las aguas subterráneas.

*6.9.7.3. Por todo ello, lo prioritario en este momento es la inversión por la CAR en los programas que incentiven los procesos de producción más limpia de unos y otros y de ser necesario, ser el ente articulador para gestionar un convenio entre el municipio (quien dentro de su propuesta de pacto señala el proceso de curtición como una de sus metas ambientales, como quedó consignado en el aparte correspondiente) y el departamento la construcción de la planta de tratamiento de cromo y demás metales que pueden ser reutilizados en el proceso de producción, logrando el aporte de tales entes como la inversión por los curtidores cooperados mediante la financiación de los costos de todas las obras que demande la producción más limpia de las curtiembres, con el fin de se haga realidad la implementación de acciones de PML por la autoridad ambiental encargada de jalonar dicho objetivo, respetando y promoviendo las empresas familiares para hacer más competitivo el sector curtidor no solo a nivel mundial sino nacional, como ha sido el objetivo que se trazó en las mesas de trabajo de todos los involucrados en el tema, presididas y auspiciadas por la Magistrada Sustanciadora, en orden al respeto del derecho al trabajo de las clases menos favorecidas, a las que no obstante haber causado un grave daño al medio ambiente, por las necesidades básicas insatisfechas a que los ha llevado un Estado negligente que en forma permisiva dio lugar a dichos asentamientos humanos (*tras los que existe un sector fuerte de la producción de los Municipios de Villapinzón, Chocontá y el D.C. se ha prevalido para acumular su riqueza*), de lo que no se puede decir si unos u otros tienen un mayor grado de culpa y de responsabilidad cuando son las propias autoridades que no ponen remedio a esa invasión y degradación de los recursos naturales.*

6.9.7.4. Con el mismo criterio y bajo las órdenes acabadas de impartir los curtidores de Chocontá deberán acoger las políticas de la PML con las

Naming the researcher on the Verification Committee, p 459 and 460:

6.24. DE LA CONSTITUCIÓN DEL COMITÉ PARA LA VERIFICACIÓN DEL CUMPLIMIENTO DE LAS OBLIGACIONES CONTRAÍDAS EN LOS DIFERENTES PACTOS DE CUMPLIMIENTO COMO DE LAS DEMÁS QUE SE IMPONEN EN ESTA SENTENCIA.

Para los efectos, se conformará un comité integrado por la señora MÓNICA SAENZ Bióloga, con Maestría en Administración de Empresas y Doctorado en Medio Ambiente de la universidad de Orleáns, Francia, la que para los efectos de su título ha enfocado su tesis hacia el problema de la solución de la descontaminación del río Bogotá y sobre los procesos de producción más limpia y quien por su propia iniciativa ha asumido la problemática de las curtiembres de Villapinzón, como también ha procurado la colaboración de la UNESCO para la solución de la descontaminación del río, instancia con la que la Magistrado Sustanciadora en compañía de los técnicos del DAMA, de la CAR y de la EAAB tuvo una conferencia televisiva poniendo de presente los problemas de la contaminación en orden a que con sus conocimientos en la materia proporcionaran mejores elementos de juicio para esta sentencia sobre los temas discutidos en el proceso, la que mostró su voluntad y envió uno de sus técnicos que puso de presente su interés en prestar la ayuda de tales expertos (ver fl. 50053 a 5054, c.9 y escrito que sobre el tema la doctora Saenz radicó ante la CAR en el mes de junio anterior y que hizo llegar al proceso estando el expediente al despacho).

6.4.3 Adding to the Court Order at Consejo de Estado, August 2013

Bogotá, Agosto 12 de 2013

Dr:

Marco Antonio Velilla

Honorable Magistrado

Consejo de Estado

Como miembro del comité de verificación de la sentencia del río Bogotá de Agosto de 2004 de la magistrada Nelly Yolanda Villamizar de Peñaranda (Acción Popular # 01-479), quisiera dejar constancia que se llevaron a cabo acciones muy concretas desde el 2004 hasta el 2011 con 84 curtidores pequeños de Villapinzón con el fin de resolver definitivamente su problemática en la Cuenca Alta del río Bogotá.

Los curtidores pequeños me manifestaron en el año 2004, como estudiante de doctorado del Instituto UNESCO-IHE de Educación para el Agua, que ellos tenían voluntad de cambio para solucionar el conflicto que los agobiaba por más de dos décadas. En aras de buscar soluciones técnicas al alcance de sus posibilidades, viajaron conmigo con recursos propios, a conocer el proceso que se adelantaba en Cerrito Valle por parte de la Corporación CVC y del Centro Regional de Producción Más Limpia. Allá las autoridades habían resuelto alinear a los curtidores en la implementación de la Producción Más limpia (PML), en desarrollar acciones de acompañamiento y en hacerle a la PML seguimiento de comando y control gracias a unas *resoluciones de obligaciones* para cada curtiembre, según su situación, y que se establecieron definiendo los límites en función de carga y no sólo de concentración (de esta manera se evita que las industrias hagan diluciones para aparentar cumplir con la norma)(Adjuntos la de la CVC y la propuesta a la CAR en 2005). Las autoridades del Valle buscaban que el tratamiento secundario representado por una planta biológica colectiva fuese menos costoso gracias a la implementación de PML que representa múltiples estrategias en los procesos industriales en prevención de contaminantes y por lo tanto, grandes ahorros en el tratamiento del final de tubo. Esta hipótesis se comprobó a finales del año 2009 cuando la CVC pudo ahorrar más de la mitad de la inversión de la planta que se habría instalado sin la implementación de PML por los curtidores.

Inspirados por el modelo del Valle del Cauca, buscamos ayuda para elaborar Planes de Manejo Ambiental (PMA) que cumplieran con los requisitos de la CAR de Cundinamarca pero basados en PML y tratamiento primario en cada curtiembre como primer paso para cumplir con la normatividad. La Cámara de Comercio de Bogotá fue la primera institución en apoyar. Se acordó que la Cámara suministraba el 85% de los recursos para la elaboración de estos planes y los curtidores aportaron el restante 15%. (Esto consta en el proyecto de la Cámara de 2004-2005 llamado Implementación de los Planes de Manejo Ambiental en las curtiembres de Villapinzón y Chocontá). Hubo 120 solicitudes de elaboración pero como el aporte de dinero de la Cámara era limitado, se lograron 64 planes. 24 curtidores más elaboraron sus PMA inspirados en el mismo modelo, logrando 88 PMAs que por primera vez fueron aprobados técnicamente por la autoridad en la historia de este conflicto. La aprobación legal de estos PMAs tomó más tiempo por problemas de aplicación y de interferencia con la definición de la ronda hidráulica del río y de la zona industrial. (Lo anterior consta en un documento adjunto de la CAR de Marzo 31 de 2010 que relaciona el estado de las curtiembres).

Se llevaron a cabo para esa época, cinco reuniones en la Presidencia de la República ya que se habían definido los cierres de las industrias y era muy difícil implementar los cambios con las industrias cerradas. A pesar de los cierres, la Unión Europea resolvió que el esfuerzo era serio y valía la pena de ser apoyado e incorporó a Villapinzón en un gran proyecto de manejo integral de aguas donde Villapinzón daba ejemplo en cómo hacer sostenible la micro-industria en condiciones precarias y de marginalidad (proyecto SWITCH 2006-2011). Se logró el apoyo de la Universidad Nacional, posteriormente de Colciencias y de la misma CAR.

Desde el año 2006, y a pesar de los cierres, los curtidores buscaron créditos formales e informales para lograr sus implementaciones. Se realizaron 33 talleres de capacitación en PML y se lograron grandes mejoras en los vertimientos, separación de contaminantes, recirculación y agotamiento del Cromo, ahorro de agua (como consta en los documentos finales del proyecto SWITCH y en el video del conflicto que se adjuntan). Para el día Mundial del Agua en Suráfrica del año 2011, esta iniciativa fue premiada como una de las mejores 6 por la ONU (artículo del Espectador adjunto). La líder comunitaria fue premiada como una de las 35 líderes de Colombia por La Fundación Liderazgo y Democracia y la revista Semana (como consta en la revista adjunta).

Desafortunadamente, intereses contrarios a la sostenibilidad de los pequeños curtidores han querido tomar la delantera en plantear propuestas que no convienen a la capacidad de asimilación del río en esa zona ni a apoyar una microindustria con buenas prácticas.

En la zona, los curtidores que estuvieron con SWITCH buscaron soluciones colectivas a los residuos sólidos que valen la pena ser apoyadas e invirtieron como mínimo de 60 a 70 millones de pesos en los cambios tecnológicos. El parque industrial en Villapinzón ya existe de hecho y se puede asimilar a los *clusters* de cuero de tipo artesanal que son tan reconocidos en Italia y España. La propuesta de un nuevo parque industrial sólo atraería nuevos curtidores y excluiría a los locales.

Se necesita es conducir las aguas que están pretratadas y separadas según sus características, a una(s) planta(s) de tratamiento biológico que será(n)[64] más económica según cumplan con las metas de PML, la definición de acuerdos de obligaciones con la corporación y la ocupación de cauce colectivo.

Quienes se encuentran en la ronda del río están interesados en que les compren su predio para decidir si siguen en la actividad y se asocian con otros o no. Según sus planteamientos ni ellos necesitan de un parque al que sería muy costoso pertenecer.

Los curtidores están ya capacitados en PML pero el acompañamiento se hace necesario para las etapas siguientes y para los curtidores que no han implementado cambio alguno. Con estas medidas se puede premiar a quienes tienen voluntad de cambio e identificar a los que no la tienen y sancionarlos de manera efectiva y contundente.

Debe limitarse la llegada de nuevos curtidores que pueden venir de San Benito por problemas de espacio y que ocasionarían daños irreparables al río en esa zona de reserva forestal.

---

[64] Según estudios se pueden planear una o tres plantas de tratamiento biológico

La Universidad Nacional ha manifestado la voluntad, junto con el instituto CEINNOVA, de acompañar dicho proceso y ayudar también a desarrollar mejores productos derivados del cuero ya que su sostenibilidad económica después de tantos obstáculos constituye un gran desafío.

Los curtidores han dejado de ser las ovejas negras de la contaminación del río Bogotá. No podemos dejar que el proceso retroceda….

Mónica Sanz

Cc: 39 683 812

Bióloga Universidad De Los Andes

MBA   Universidad De Los Andes

Especialización Gestión de Recursos Naturales Universidad De Orléans de Francia

Especialización en Resolución de Conflictos y Negociación Universidad De Los Andes

Doctorado en Sostenibilidad de la micro-industria en Colombia UNESCO-IHE de Holanda (próxima defensa de tesis)

6.4.4 Claiming rights with support of prestigious lawyers

Bogotá, D.C., junio 9 de 2005

Doctora
CLARA INÉS VARGAS HERNÁNDEZ
Magistrado
Honorable Corte Constitucional
Ciudad.-

Ref: **Rad. No.** T-1132230. **ACCIÓN DE TUTELA** de **MANUEL DE JESÚS TORRES SEGURA** y **EVIDALIA FERNÁNDEZ DE TORRES** contra la **DIRECTORA** de la **CORPORACIÓN AUTÓNOMA REGIONAL DE CUNDINAMARCA**

Respetado Dra. Vargas:

MANUEL DE JESÚS TORRES SEGURA, mayor de edad, domiciliado en el municipio de Villapinzón (Cundinamarca), identificado con la Cédula de Ciudadanía No. 3.002.033 de Chocontá, y **EVIDALIA FERNÁNDEZ DE TORRES**, mayor de edad, domiciliada en el municipio de Villapinzón (Cundinamarca), identificada con la Cédula de Ciudadanía No. 21.101.705 de Villapinzón, actuando en nombre propio y en representación de nuestros hijos Diego Aldemar (22 años e incapaz por ser persona con discapacidades psíquicas y físicas), Wilmer Manuel (11 años) y Maritza Jazmín (7 años), en nuestra condición de legítimos propietarios de la **EMPRESA DE CURTIDO DEL CUERO "LA PRADERA"**, ubicada en la Finca Carboneles, Vereda Chigualá Bajo, del municipio de Villapinzón (Cundinamarca), en ejercicio del **DERECHO FUNDAMENTAL DE PETICIÓN**, garantizado por el artículo 23 de la Constitución Política de Colombia y reglamentado por los artículos 9° y siguientes del Código Contencioso Administrativo, de manera respetuosa nos dirigimos a usted a fin de solicitar la **SELECCIÓN PARA REVISIÓN** de la acción de tutela de la referencia, de conformidad con los fundamentos que se exponen a continuación.

Nuestra solicitud de que la acción de tutela de la referencia sea seleccionada para revisión por la H. Corte Constitucional se funda (*i*) en razones de naturaleza constitucional, derivadas del desconocimiento de clara jurisprudencia de la Corte Constitucional por parte de los jueces de instancia; y (*ii*) en razones de justicia material. Los fundamentos que sustentan nuestra petición de selección se explican a continuación.

Tanto la Juez Civil del Circuito de Chocontá como la Sala Civil del Tribunal Superior de Cundinamarca, en las sentencias de primera y segunda instancia proferidas en el presente proceso de tutela, desconocieron claramente la jurisprudencia de la Corte Constitucional relativa a varias cuestiones.

En primer lugar, los jueces de instancia ignoraron por completo la doctrina constitucional en materia de la especial protección que el Estado colombiano está en obligación de deparar a los niños, a las personas con discapacidad y a los ciudadanos en situación de vulnerabilidad social por su condición económica. En este sentido, la juez *a-quo* y el tribunal de segunda instancia no tuvieron en cuenta (*i*) que somos trabajadores que derivamos nuestro único sustento del curtido del cuero y carecemos de recursos económicos alternativos a los que genera nuestro negocio familiar de curtición, hoy clausurado por orden de la Directora de la CAR; (*ii*) que varios de nuestros hijos son niños cuya alimentación, salud y educación —todos derechos fundamentales de los niños— dependen, por entero, de los recursos que genera el negocio familiar hoy cerrado por la CAR; y (*iii*) que uno de nuestros hijos es una persona con discapacidad (autismo y retraso psico-motor) cuyo sustento y atención en salud dependen, por completo, de los recursos que genera nuestra curtiembre.

Segundo, las sentencias de primera y segunda instancia desconocieron la doctrina constitucional conforme a la cual los jueces de tutela tienen la obligación de desplegar una actividad probatoria que permita determinar si la autoridad demandada vulneró con sus actuaciones el derecho fundamental al mínimo vital. En contravención de esta obligación, los jueces de instancia —y, particularmente, la Juez Civil de Circuito de Chocontá— negaron las pruebas que solicitamos a fin de demostrar que el cierre temporal de nuestra curtiembre decretado por la CAR vulneraba nuestro derecho fundamental al mínimo vital (que fuéramos oídos en declaración y que se decretara una inspección judicial a nuestro negocio de curtido del cuero), con el argumento de que la acción de tutela implica un procedimiento judicial que, por su premura, no admite este tipo de actividad probatoria.

Finalmente, los falladores de tutela ignoraron por completo nuestros argumentos en torno a la vulneración del principio constitucional de la confianza legítima por parte de la CAR. En efecto, no les mereció ninguna consideración el hecho de que la compleja situación en la que operan las curtiembres de Villapinzón y Chocontá en la actualidad obedece, en gran medida, a graves omisiones de la CAR en el cumplimiento de sus competencias a lo largo de los años.

Por una parte, la CAR, a partir de finales de los años 70, consideró que el negocio de la curtición en el norte de Cundinamarca debía volverse más eficiente y debían abandonarse las técnicas tradicionales de curtido del cuero —más respetuosas del ambiente— que aprendimos de nuestros antepasados. Por ello, nos capacitó en técnicas altamente contaminantes del curtido del cuero, tal como lo ponen de presente las cartillas de capacitación incluidas en el expediente contentivo de la acción de tutela de la referencia. *Hoy en día, la CAR nos sanciona por utilizar técnicas de curtición que ella misma nos enseñó a utilizar.*

De otro lado, la CAR sólo comenzó a ejercer sus competencias de protección del ambiente y los recursos naturales a partir de 1999, cuando, por lo menos desde 1974, el ordenamiento jurídico colombiano establecía la prohibición de contaminar y otorgaba a las corporaciones autónomas regionales un sinnúmero de competencias para garantizar las normas ambientales.

Los suscritos, junto con un gran número de curtidores de Villapinzón vinculados en la Asociación de Curtidores de Villapinzón (ACURTIR), nos hemos asociado para poner en marcha un proyecto de producción limpia de curtido del cuero denominado *Implementación de planes de manejo ambiental y empresarial en las curtiembres de Villapinzón y Chocontá*. Una vez finalizado, este proyecto nos permitirá llevar a cabo nuestro oficio tradicional de la curtición de conformidad con todas las normas ambientales y de manera respetuosa con el ambiente y los recursos naturales. La seriedad y efectividad de este proyecto resulta puesta en evidencia por el hecho de que la Cámara de Comercio de Bogotá haya decidido invertir en el mismo mediante su financiación integral. Para poder llevar a cabo este proyecto a buen término, es necesario que nuestra curtiembre esté en operación. Por lo tanto, al mantener cerrado nuestro negocio, la CAR frustra la posibilidad de finalizar el proyecto antes señalado, y, por tanto, nuestra única posibilidad de seguir desempeñando el único oficio que conocemos y que nos brinda los recursos que permiten nuestro sustento básico.

Ninguna de las familias curtidoras de Villapinzón ha negado el carácter contaminante del curtido del cuero ni ha solicitado que se hagan concesiones violatorias del ordenamiento jurídico. Precisamente por ésto nos organizamos para poder desarrollar el proyecto de producción limpia antes señalado y continuar desempeñando nuestro oficio de conformidad con la ley. De este modo, la única pretensión que elevamos a través de la acción de tutela de la referencia consistió en que el amparo constitucional fuera concedido de manera transitoria mientras finalizábamos el proyecto *Implementación de planes de manejo ambiental y empresarial en las curtiembres de Villapinzón y Chocontá*.

Como se mencionó anteriormente, nuestra solicitud de que la H. Corte Constitucional seleccione la acción de tutela de la referencia también se funda en razones de justicia material. En primer lugar, somos personas que carecemos de recursos económicos distintos a los generados por nuestra curtiembre, nuestro nivel de alfabetización es muy bajo y nuestros conocimientos jurídicos son en extremo escasos. Nuestra vida cotidiana trascurre trabajando en nuestra curtiembre, en la vereda Chigualá Bajo de Villapinzón, sin tener grandes contactos con el derecho o con las entidades del Estado. En esta medida, nuestra capacidad para enfrentar jurídicamente procesos administrativos tan complejos como el ambiental es prácticamente nula.

Sólo hasta el año 2004, gracias a un programa de servicios jurídicos no remunerados de una universidad de la ciudad de Bogotá, pudimos acceder a una asesoría jurídica de calidad, prestada por abogados de esa misma ciudad. Sólo desde ese momento supimos que teníamos derechos fundamentales que no podían ser desconocidos por la CAR y que debían ser tenidos en cuenta en el procedimiento administrativo sancionatorio que culminó con la orden de cierre temporal de nuestra curtiembre. Antes de esto, tuvimos que confrontar ese procedimiento sin saber realmente cuáles eran sus implicaciones y sin conocer verdaderamente las defensas que el ordenamiento jurídico ponía a nuestra disposición.

Los suscritos demandantes somos curtidores por tradición familiar: aprendimos a curtir el cuero de nuestros padres, quienes, a su turno, aprendieron de nuestros abuelos. No conocemos ningún otro oficio u ocupación. La tradición curtidora del norte de Cundinamarca, proveniente de los propios Muiscas, según dicen algunos, y el sustento mismo de muchísimas familias que derivan sus recursos esenciales de ese oficio, están a punto de llegar a su fin, no sólo debido a las actuaciones de la CAR sino a los embates de proyectos de curtido del cuero de envergadura industrial que buscan hacer de lado a las pequeñas curtiembres y conseguir el monopolio del curtido del cuero en los municipios de Villapinzón y Chocontá.

De la señora magistrada,

MANUEL DE JESÚS TORRES SEGURA
C.C. No. 3.002.033 de Chocontá

EVIDALIA FERNÁNDEZ DE TORRES
C.C. No. 21.101.705 de Villapinzón

6.4.5 Struggling to avoid closures

Bogotá D.C.

Doctora
GLORIA LUCIA ALVAREZ
Directora Ejecutiva
CORPORACION AUTONOMA REGIONAL DE CUNDINAMARCA –CAR-
La Ciudad

[sello ilegible]

SUBDIRECCION de Gestion Ambiental.

Ref.   Solicitud de suspensión a la orden de cierre de
         industrias de Curtiembre de Villapinzón

Cordial saludo, Dr. Gloria Lucía

Sea esta la oportunidad para solicitarle a nombre de la ASOCIACION DE CURTIDORES DE VILLAPINZON Y CHOCONTA -"ACURTIR"-, de la cual soy su representante legal, se sirva ordenar a quién corresponda suspender el cierre de 13 industriales de Villapinzón, diligencias que se ejecutaran el lunes próximo tal como fue anunciado en Audiencia de Información celebrada el 14 de enero del año en curso en el municipio de Villapinzón, presidida por el Dr. Agustín Cortes y el señor Alcalde de dicha localidad, por las siguientes razones:

1. Si bien es cierto que las resoluciones sancionatorias se encuentran debidamente ejecutoriadas, no es menos cierto, que los industriales conscientes de la problemática ambiental y social que los rodea se han visto motivados con el plan de apoyo empresarial que inició la Cámara de Comercio de Bogotá a finales del año anterior en su I Fase de diagnóstico técnico – jurídico y este año seguramente comenzará una II Fase.

2. Se busca a través del diagnóstico técnico - jurídico sustentar y acompañar en la II Fase al industrial en la implementación de los PMA y legalizarce definitivamente. Los esfuerzos hechos anteriormente han fracasado por falta precisamente de un acompañamiento institucional serio como lo es la CCB.

3. Dr. Gloria Lucía, más del 56% de los Industriales han presentado ante la CAR el PMA, unos están para complementar y corregir y otros para implementar. Un 16% al 20% han solicitado los correspondientes permisos de vertimientos y últimamente un 80% concesión de aguas.

4. Necesitamos de su apoyo concendiendonos de manera *urgente* un plazo mínimo de cuatro (4) meses para implementar los PMA. Doctora, permítanos demostrarle la otra realidad del nuevo gremio.

5. La implementación implica la coordinación entre el industrial y la autoridad ambiental para ejecutar las decisiones tomadas con ocasión a las sanciones y medidas preventivas contenidas en los diferentes actos administrativos que hoy son objeto de aplicación, allí se ordenan caracterizaciones, medición de caudales etc.

Atentamente,

Evidalia Fernandez de Torres
EVIDALIA FERNANDEZ DE TORRES
C.C No. 21.101.705 de Villapinzón
Representante legal - NIT. 832003853-8

Notificaciones: Vereda Chigualá, finca Carbóneles, municipio de Villapinzó

Anexos: certificado de existencia y representación de la asociación y u reposa en su Despacho

c.c. Alcaldía y Personería de Villapinzón; Procuraduría y Fiscalía.

[sello] 0 8 MAR. 2005

gómez-pinzón Abogados
NIT. 800.175.087-3
RECIBIDO PARA ESTUDIO
[firma] Juan Lu Ortiz

[anotaciones manuscritas]
Acen...
cumplimiento
AC- 325

CIERRE TEMPORAL YA REALIZADO A CURTIEMBRES QUE PASAN A LA SEGUNDA FASE DEL PROYECTO Y ESTÁN UBICADOS LEJOS DE LA RONDA DEL RÍO

1. Hernando Lizarazo Rubiano
2. Manuel de Jesús Torres Segura
3. Victor Manuel Contreras Casallas
4. Alicia Bernal Contreras
5. Argemiro Casallas Silva
6. Adolfo Barrero
7. José del Carmen Fernández Dimate
8. Fabio Fernández López
9. Nelson Fernández López
10. José Omar Farfán Prieto
11. Guillermo Barrero
12. Pedro Parra Guauque   Curtiembre Moncerrate
13. Ana Isabel Bernal Mondragón
14. Octaviano Lizarazo Fernández
15. Raúl Arévalo Rubiano

CIERRE TEMPORAL A CURTIEMBRES QUE ENTRAN A PRIMERA FASE DEL PROYECTO Y ESTÁN UBICADOS LEJOS DE LA RONDA DEL RÍO

1. Gonzalo Fernández Rodríguez
2. Floriberto Farfan Melo
3. Eduardo González Medina
4. Paulino Monroy Barrero
5. Luz Yanire Gil de Barrero
6. Alcides y Parmenio López Castro
7. Alcibiades Barrero
8. Gonzalo López Castro
9. Eli Romero

CIERRE DEFINITIVO A CURTIEMBRES QUE SE ENCUENTRAN DENTRO DE LA RONDA DEL RÍO

1. Hector Hugo Casallas Lizarazo - No tiene terreno
2. Blanca Leonor Lizarazo Fernández - *Si*, tiene terreno para la reubicación
3. Emeramo Ruiz Castiblanco - *Si*, tiene terreno para la reubicación
4. José del Carmen Barrero Buitrago - No tiene terreno
5. José Arcangel Rincón - No tiene terreno
6. Merardo Barrero Lizarazo   - No tiene terreno
7. Marco Antonio Lizarazo Rubiano - No tiene terreno
8. Luis Barrero Fernández - No tiene bombos solamente una rebajadora -no tiene terreno
9. Leonilde Casallas de Lizarazo - *Si*, tiene terreno para la reubicación

PARA UN TOTAL DE 27 CURTIEMBRES QUE NO PRESENTAN DIFICULTAD PARA PARTICIPAR EN EL PROYECTO YA QUE ELLAS CUENTAN CON TERRENOS SUFICIENTE.

NOTA. De las 58 curtiembres que fueron selladas solamente reporto 33 que pertenecen a la Asociación ACURTIR

Atentamente,

EVIDALIA FERNÁNDEZ

Técnica ...

Bogotá D.C

Doctora
GLORIA LUCIA ALVAREZ
Directora Ejecutiva
CORPORACION AUTONOMA REGIONAL DE CUNDINAMARCA - CAR
La Ciudad

Ref. Solicitud ...515-115... de ... Enero 20 de 2005

En consideración a la ... de la referencia sobre la solicitud de levantamiento de manera transitoria de las medidas ... de CIERRE TEMPORAL en contra de ... curtidores de Villapinzón ejecutadas el ... 24 de Enero del año en curso, y ... considerando ... tratando en el Consejo Comunitario de Chía el día 19 de febrero de los corrientes, teniendo en cuenta para ello un antecedente similar como lo es la CONCERTACION entre la Corporación CVC y las curtiembres del Cerro ... Cauca en el año 2003, la cual y con todos respecto, consideramos puede servir ... base para la siguiente propuesta, que sometemos a su ...

1. Establecer PACTOS DE CUMPLIMIENTO entre la CAR y ... cada curtidor respondiendo a los cuatro (4) puntos generales exigidos en los diferentes actos administrativos que ... obran, para desarrollarlos en su mayor parte dentro del cronograma de doce (12) semanas definido en el proyecto de la Camara de Comercio de Bogotá -CCB-Segunda Fase denominado igual que la anterior Implementación de los Planes de Manejo Ambiental y Empresarial en las curtiembres de Villapinzón y Chocontá. Los pactos responderán a cada ... particular de cada curtidor, así:

   a.) En cada ... los curtidores que tienen los permisos exigidos en los actos Administrativos y a complementar ... adicional entregar los Planes de Manejo Ambiental conforme al acto administrativo ... para los que ya lo han presentado y los que no ... deben ... presentarlo conforme a los términos de referencia diseñados por esa ... el cual comprende los doce (12) puntos a evaluar, dentro del cronograma del proyecto ... la cual es de doce (12) semanas (propuesta de la CCB).

   b.) Una vez ... los PMA se procederá a su implementación conforme al cronograma que se aprueba igualmente, para luego continuar con la adecuación del tratamiento sanitario (Punto Cuarto) según la situación particular de cada curtido, es decir desde el punto de vista del estado técnico (obras de infraestructura para cumplir con el artículo primero del Acuerdo 008 de 2004) y el tamaño ... es funcionando se puede decir, (aprobarán la ... cual), el tiempo para este punto oscilará entre 8 y 12 semanas más.

   • ANEXAMOS DOCUMENTO DE LA CCB

Atentamente

EVIDALIA FERNANDEZ DE TORRES
C.C.No. 21,181.255 de Villapinzón
Representante Legal de ACURTIR
NIT: 832040524-8

Notificaciones: Vereda Chasmaque Junta Comunidad municipio de Villapinzón. Teléfonos 8-...-... Celular 316-8872692

6.4.6 Presenting the EMPs in the middle of closures: support from CCB

RESOLUCIÓN No.   2460   DE

0 1 DIC 2005

(Por la cual se establece un Plan de Manejo Ambiental)

La Directora General de la Corporación Autónoma Regional de Cundinamarca – CAR, en ejercicio de las facultades legales que le confiere la Ley 99 del 22 de diciembre de 1993 y,

CONSIDERANDO

Que mediante Resolución No. 554 del 25 de octubre de 1999, la Corporación inició un tramite administrativo de carácter sancionatorio contra el señor MANUEL DE JESÚS TORRES SEGURA, en calidad de propietario de la industria de cuero ubicada en el predio Carboneles, localizada en la vereda Chigualá, jurisdicción del Municipio de Villapinzón, Cundinamarca, imponiéndole como medidas preventivas, la suspensión de vertimientos y la presentación de un Plan de Manejo Ambiental.

Que mediante Resolución No. 903 del 15 de septiembre de 2004, la Corporación impuso una sanción a la citada industria de CIERRE TEMPORAL, hasta tanto, cumpla con la aprobación por parte de la CAR del Plan de Manejo Ambiental respectivo, tramite y obtenga la concesión de aguas, diligencie el formulario de solicitud de permiso de vertimientos y construya el sistema de tratamiento primario de aguas residuales industriales y éste haya sido aprobado por la CAR.

Que mediante radicación No. 17237 de junio 10 de 2005, el señor MANUEL DE JESÚS TORRES SEGURA, identificado con cédula de ciudadanía No. 3'002.033 de Chocontá, presentó el documento denominado Plan de Manejo Ambiental para la industria de curtido de pieles de su propiedad denominada La Pradera, localizado en el predio Carboneles, de la vereda Chiguala, jurisdicción del municipio de Villapinzón, Cundinamarca, el cual contiene, entre otros, la solicitud de permiso de vertimientos para la citada industria de curtido de pieles.

Que la Corporación evaluó el Plan de Manejo Ambiental presentado cuyo resultado se consignó en el Conceptos Técnicos SGAC No 1165 de 31 de agosto de 2005, el cual recomendó complementar el citado Plan de Manejo Ambiental.

Que mediante Oficio No. 8794 del 1 de septiembre de 2005, la Corporación informa al señor MANUEL DE JESÚS TORRES SEGURA, que debe complementar el Plan Manejo presentado, atendiendo las respectivas observaciones.

Que mediante radicación No. 19099 del 23 de septiembre de 2005, el señor MANUEL DE JESÚS TORRES SEGURA, presenta la complementación al citado Plan de Manejo ambiental.

Que la Corporación evaluó la complementación presentada de la cual se rindió el Informe Técnico OTSNYA 1482 de 13 de octubre de 2005, del que se conceptúa técnicamente lo siguiente:

- *Presenta cuadro de eficiencias del sistema de tratamiento primario. Asiendo comparación al sistema de tratamiento con buenas y malas practicas donde se plantean remociones mayores al 80%. También las cantidades y calidades de agua que se esperan en cada proceso.*

6.4.7 Support from CCB

Corporación Autónoma Regional de Cundinamarca - CAR
Subdirección de Gestión Ambiental Compartida
República de Colombia

Bogotá, D.C.,

Doctor
MAURICIO MOLINA
Director de Desarrollo Empresarial
Cámara de Comercio de Bogotá
Avenida Eldorado N° 68D – 35
Bogotá, D.C.

Referencia: Respuesta y aclaraciones al documento *"Implementación de Planes de Manejo Ambiental y Empresarial en las Curtiembres de Villapinzón y Chocontá. Segunda Fase".*

Cordial Saludo Doctor Molina:

Acusamos recibo de su comunicación radicada con el número 0910-1 del 31 de Enero pasado, mediante la cual presenta el documento de la referencia, según compromiso adquirido por esa institución con la Presidencia de la República.

De acuerdo con el documento presentado, puede concluirse que éste corresponde a la formulación de un Plan de Acción para la *"Implementación de Planes de Manejo Ambiental y Empresarial en las Curtiembres de Villapinzón y Chocontá. Segunda Fase"*, cuyo objetivo general permite contribuir al aumento de la competitividad empresarial, mejorando el desempeño ambiental y empresarial de las curtiembres de Villapinzón y Chocontá, a través de políticas de producción más limpia, fortalecimiento empresarial, asociatividad y resolución de conflictos; de tal forma que al ser desarrollado, de acuerdo con el cronograma presentado, se obtendrían los siguientes Productos:

- 118 empresas con solicitud de concesión de agua y permisos de vertimiento ante la CAR.
- 63 empresas con Planes de Manejo Ambiental (PMA) actualizados y presentados ante la Corporación, cumpliendo con los elementos y

Corporación Autónoma Regional de Cundinamarca - CAR
Subdirección de Gestión Ambiental Compartida
República de Colombia

requerimientos mencionados en los Actos Administrativos y Términos de Referencia (TR) proferidos por la CAR.

➢ 118 empresas capacitadas en buenas prácticas de manufactura y con seguimiento al procesos de Implementación de las mismas.

➢ 118 empresas vivenciando cambios a través de talleres de resolución de conflictos y planteando soluciones de cambio a futuro para lograra armonía en la región.

➢ 27 empresas que no participaron en la Fase I, con diagnósticos y Planes de Acción, 9 de ellas con PMA.

➢ 60 empresas fortalecidas a nivel empresarial.

➢ 72 empresas con apoyo para identificación de fuentes de financiación para la implementación de los PMA.

Todo lo anterior en un periodo de **12 semanas** calendario, a partir de la segunda de febrero de 2005.

Al respecto, queda claro que el total de curtidores comprometidos con el proyecto es de ciento dieciocho (118) según lo manifestado y verificado anteriormente en las cartas compromiso suscritas por cada uno de los curtidores adhirentes al proyecto (según reposa en los archivos de la Corporación, con radicado N° 10437-1 del 5 de octubre de 2004), que solicitarán y tramitarán los permisos ambientales respectivos y de los cuales únicamente 72 van a presentar nuevos PMA y/o ajustes a los mismos, los que estaremos esperando para su evaluación y concepto.

Adicionalmente, quiero reiterarle, de acuerdo con lo previsto en los actos administrativos proferidos por la Oficina Territorial Sabana Norte y Almeidas de la Corporación, para los curtidores de Villapinzón y Chocontá, que cada acto es único e individual y que amerita un análisis exhaustivo y específico para cada una de las industrias de curtiembre en cuestión y que, en términos generales prevé: recomendar que se ordenen que las medidas sancionatorias pertinentes se mantengan, hasta tanto se cumpla con lo siguiente:

☞ La aprobación por parte de la Corporación del PMA ajustado, según las exigencias del acto administrativo específico para cada curtiembre.

☞ El trámite y obtención del permiso de concesión de aguas, el trámite del permiso de vertimientos (con los formularios debidamente diligenciados – anexar documentación requerida en los mismos).

☞ Se haya construido por parte del peticionario el sistema de tratamiento primario de las aguas residuales industriales y una vez construido éste haya sido aprobado por la CAR.

858 9126
C. 1

Corporación Autónoma Regional de Cundinamarca - CAR
Subdirección de Gestión Ambiental Compartida
República de Colombia

☞ En el caso de que aplique, se hayan reubicado las instalaciones y área de producción de la curtiembre (infraestructura y equipo) fuera de la ronda de protección del Río Bogotá (30 metros).

Agradezco a usted el mantenernos informados de sus avances, a fin de poder hacer un mejor acompañamiento en la formulación e implementación del proyecto.

Cordialmente,

Original Firmado
MAURICIO BAYONA PULIDO

MAURICIO BAYONA PULIDO
Subdirector Gestión Ambiental Compartida

COPIA.   Dr. Cesar Clavijo Rios. Jefe Oficina Territorial Sabana Norte y Almeidas. Zipaquirá
Dra. Mónica Sanz, Carrera 11C N° 109 – 87 Bogotá, D.C.

N°s Radicación0910-1 de 31.01.05
Elaboró: Edgar Erazo
Revisó: Ing. Mauricio Bayona P
Anexos: SIN
Expediente: SIN
Fecha: 7 de Febrero de 2005.

6.4.8 Results project and struggles with land issues

2. Actividad de Cambio Grupal

Resultados Ejercicio de Espacio Abierto Villapinzón
Mayo 26 de 2005

Como resultado de la fase 2 del proyecto de La Cámara de Comercio de Bogotá sobre la problemática de las curtiembres de Villapinzón y Chocontá, por el que se comenzó la entrega de los puntos exigidos por la Corporación Ambiental CAR, a saber: Solicitudes colectivas de concesiones de agua, solicitud de permisos de vertimientos y planes de manejo ambiental, se realizó un ejercicio de resolución de conflictos el día 26 de Mayo en la Normal Superior de Cundinamarca. El objetivo de este ejercicio era el de generar y agilizar las soluciones, gracias a la participación de todos los actores en el conflicto, de puntos fundamentales que comprometen el corto plazo de la implementación de las soluciones colectivas para la región y el sector curtidor.

Asistió la Presidencia de la República, la Procuraduría, La Dra Nancy Patricia Gutiérrez, La Corporación CAR, El Ministerio de Vivienda, Ambiente y Desarrollo Territorial, La Alcaldía de Villapinzón, La ventanilla Ambiental de la CAR, La Cámara de Comercio de Bogotá, La asociación de curtidores de Villapinzón y Chocontá ACURTIR, La ONG Madre Tierra, Los concejales de Villapinzón, La Gobernación de Cundinamarca y UNESCO-IHE de Holanda. En total asistieron 125 personas y tuvo una duración de 5 horas.

La metodología utilizada fue la de Espacio Abierto y la de las 4D, trabajando el futuro como punto de partida y utilizando siempre las herramientas apreciativas.

Se trabajaron los temas relacionados con el levantamiento del cierre y con la compra de los lotes para los curtidores que se encuentran en la ronda del río.

Los avances y compromisos adquiridos al respecto fueron los siguientes:

LEVANTAMIENTO DE LOS CIERRES:

El Director de la corporación Sabana Norte se comprometió a realizar un plan de contingencia enfocado a dar prioridad y celeridad al estudio de los resultados entregados por los curtidores (Concesión de aguas, Vertimientos y Estudio de los Planes de Manejo Ambiental) y de manera especial a aquellos que se encuentran cerrados.
Se comprometió también a que una vez la Cámara dé el concepto técnico por el cual considera pertinente modificar la resolución de levantamiento de cierre en el punto de construcción del sistema de tratamiento primario como requisito para levantar la medida, la Corporación estudiará su modificación.

El día martes se presentarán dichos puntos ante el Consejo Directivo de la Corporación.

Posterior a esto, se realizará una reunión la semana entrante con el gremio curtidor en fecha que está por definir.

La Procuraduría a su vez se comprometió a hacer un seguimiento especial a este proceso.

La dra Nancy Patricia Gutiérrez se comprometió a su vez a apoyar este tema el día miércoles con la Directora de la Corporación.

LOTES DE LA RONDA DEL RÍO:

Se definió el que la compra es potestad de La Corporación CAR en coordinación con la Gobernación quien tiene los dineros y el municipio.

Se definió que los curtidores hiciesen llegar sus fotocopias de títulos y certificados de Tradición y Libertad a la alcaldía a partir del día 27 de Mayo para que se realice su estudio de títulos. El avalúo lo realiza la Gobernación. El levantamiento topográfico ya se llevó a cabo.

Nota: se anexa una copia de una de las invitaciones que se realizaron para el Espacio Abierto en documento adjunto.

6.4.9 Support from SWITCH

LANZAMIENTO PROYECTO SWITCH EN COLOMBIA
Lunes 12 de junio de 2006, 10:00 a.m.

Agenda

1. *Palabras del Señor Vice-Rector de la Universidad Nacional de Colombia Fernando Montenegro. 10: 00 10:10*
2. *Palabras del Señor Rector de la Universidad del Valle, Dr Iván Enrique Ramos Calderón 10:12-10:22*
3. *Palabras del representante de la Unión Europea, 10:24-10:29*
4. *Video conferencia con el Director de UNESCO-IHE Dr Richard Meganck 10:31- 10:41*
5. *Presentación del Proyecto Internacional SWITCH por parte del Dr Maarten Siebel de UNESCO - IHE de Holanda. 10:43-10:58*
6. *Preguntas 11:00- 11:20*
7. *Presentación del componente del Proyecto en la Universidad Nacional de Colombia. Profesora Laura Cecilia Osorio, Directora del Instituto de Estudios Ambientales IDEA 11:22-11:34*
8. *Palabras de la Cámara de Comercio de Bogotá. 11:36-11:41*
9. *Presentación del componente del Proyecto en la Universidad del Valle. Profesor Alberto Galvis 11:43-11:55*
10. *Palabras del Director del IDEAM Dr Carlos Costa 11:57-12:02*
11. *Palabras de la Senadora Nancy Patricia Gutiérrez 12:04-12:09*
12. *Palabras del Ministerio de Ambiente, Vivienda y Desarrollo Territorial 12:11-12:21*
13. *Intervenciones-Preguntas 12:23-12:40*
14. *Copa de Vino.*

6.4.10 Cofinancing SWITCH

Bogotá, 12 de septiembre de 2006.

Señores:

Centro Internacional de Investigaciones para el Desarrollo
Programa Pobreza Urbana y Medio Ambiente
Iniciativa de Investigación Ciudades Focales
"Alcanzando los ODM barrio a barrio"

Respetados señores:

Por medio de la presente manifestamos nuestro compromiso de apoyar el Equipo Ciudad del proyecto "MANEJO INTEGRAL DE CURTIEMBRES EN LA RONDA DEL RIO BOGOTÁ, MUNICIPIO DE VILLAPINZON Y CHOCONTÁ, CUNDINAMARCA" para trabajar juntos durante la duración total del proyecto.

Con nuestra firma, nos comprometemos a apoyar al proyecto de la siguiente forma:

Organización	Aportes
Laura Cecilia Osorio Muñoz Directora Instituto de Estudios Ambientales IDEA Universidad Nacional de Colombia	Coordinación del equipo ciudad Apoyo en especie con la participación de expertos investigadores y profesores de la Universidad Nacional. Estudiantes de la Universidad Nacional (maestrías) Instalaciones, personal administrativo, gastos administrativos (servicios públicos, comunicaciones, entre otros)
Evidalia Fernández Representante Legal Asociación de Curtidores ACURTIR	Apoyo al proyecto tanto en mano de obra como en tiempo y espacio en las curtiembres para realizar pruebas y todo lo concerniente. Sostenibilidad permanente de tipo social. 100 curtidores asociados apoyando todo el proceso en Villapinzón y Chocontá.
Unión Europea Proyecto SWITCH Operado por UNESCO-IHE	$ 158 820 Euros repartidos así: $ 89 520 Investigación y Desarrollo $ 69 300 Demostración

Actores involucrados	
Ministerio de Ambiente, Vivienda y Desarrollo Territorial	Acompañamiento en tanto que es la máxima autoridad ambiental nacional
[firma] DIRECTOR DESARROLLO EMPRESARIAL Cámara de Comercio	Ente asesor en materia de competitividad y asociatividad
[firma] DIRECTORA Corporación Autónoma Regional CAR	Acompañamiento en tanto que es la máxima autoridad regional
[firma] Alcalde Mpal. Alejandro Galte Alcaldía Municipio de Villapinzón y Chocontá	Acompañamiento en tanto que es el máximo poder político en el municipio
3 c 42.643 Vjpinzon *[firma]* ORLANDO MOLINA Presidencia Concejo de Villapinzón	Acompañamiento al poder político en la región
Senado de la República Senadora Nancy Patricia Gutiérrez	Acompañamiento como rama legislativa nacional
Oficina Jurídica de la Presidencia de la República *[firma]*	Acompañamiento como máximo ente ejecutivo a nivel nacional

[firma] DIRECTORA DEL AMBIENTE GOBERNACION DE CUNDINAMARCA.

ACOMPAÑAMIENTO COMO PODER POLITICO DE LA REGIÓN.

Los mencionados actores vienen acompañando los procesos de apoyo al sector curtiembres en el Municipio de Villapinzon en la medida de sus competencias, de forma decidida desde hace dos años. Se llevaron a cabo los procesos de legalización de las curtiembres con la generación de los Planes de Manejo Ambiental, resolución de conflictos y Plan Especial de Curtiembres En la actualidad se está desarrollando el proyecto SINA II.

Corporación Autónoma Regional de Cundinamarca - CAR
Dirección General
República de Colombia

Bogotá D.C.

Doctor
JUAN FRANCISCO MIRANDA MIRANDA
Director General
COLCIENCIAS
Transversal 9 A - BIS No.132-28
Ciudad

Ref : **Comité Evaluador Convocatoria # 403**

Respetado doctor:

La Corporación Autónoma Regional de Cundinamarca CAR viene adelantando desde hace tres años un proceso decidido de solución al conflicto ambiental con el sector de las curtiembres de Villapinzón. Por más de veinte años ellas han representado uno de los mayores conflictos ambientales en la cuenca alta del río Bogotá. De manera paralela, los curtidores se han organizado alrededor de una única asociación y actualmente se encuentran en fase de implementación de Producción Más Limpia con apoyo en primera instancia de la Cámara de Comercio de Bogotá y ahora de la Unión Europea y la Universidad Nacional.

Como autoridad ambiental, celebramos el que los curtidores busquen apoyo de una entidad como Colciencias para seguir adelantando sus cambios. Mediante este apoyo, se estimula en la región una adecuada e innovadora solución que busca estimular la competitividad en los procesos así como fortalecer el manejo conjunto y la valorización de los residuos sólidos.

La Corporación apoya esta iniciativa de suscribir esta convocatoria #403 bajo el título "Plan Estratégico en las Curtiembres de Villapinzón y Chocontá: Competitividad e innovación en la cadena productiva del cuero".

Atentamente,

EDGAR ALFONSO BEJARANO MENDEZ
Director General

Preparó: Sergio A. Piñeros Botero
Fecha: Abril 11de 2007

Bogotá, D. C. Carrera 7 No. 36–45 Conmutador: 320 9000 Ext. 1855   www.car.gov.co
Fax: 287 1772 Correo electrónico: direccion_general@car.gov.co

Bogotá, 12 de abril de 2007

Señores
Colciencias
Convocatoria 403 de 2007. Formulación de programas estratégicos, proyectos de investigación, desarrollo tecnológico e innovación tecnológica.

Referencia: Participación proyecto SWITCH, Universidad Nacional y UNESCO – IHE, en la mesa regional del curtido de Cueros – Cundinamarca.

La Corporación Autónoma Regional de Cundinamarca - CAR y la Cámara de Comercio de Bogotá - CCB en desarrollo del convenio No. 329 de 2006, firmado para continuar operando la Ventanilla Ambiental CAR, han diseñado una serie de estrategias para la implementación de la Política de Producción Más Limpia, en los diferentes sectores productivos priorizados por las dos instituciones. Como parte del desarrollo de este programa la ventanilla tiene a su cargo la secretaría técnica de la mesa regional del Sector de Curtido y Preparado de Cuero.

El objetivo principal de esta mesa es articular regionalmente los lineamientos nacionales concertados a través de la Mesa Nacional del Sector de Curtido y Preparado de Cuero en cabeza del Ministerio de Ambiente Vivienda y Desarrollo Territorial – MAVDT. Estos lineamientos contemplan el mejoramiento ambiental y productivo de este sector con el fin de minimizar sus impactos ambientales negativos y aumentar su competitividad. Esta mesa esta conformada por el sector productivo, la autoridad ambiental nacional y regional así como otras instituciones públicas y privadas que tienen algún tipo de incidencia sobre el desarrollo ambiental y productivo de este sector.

La Ventanilla Ambiental de la CAR ejerce la secretaría técnica de esta mesa, por lo que se permite certificar, con base en registros documentales, la participación de las siguientes instituciones:

1. UNESCO – IHE participa desde febrero de 2006.
2. Universidad Nacional de Colombia participa desde junio de 2006.
3. Proyecto SWITCH participa desde junio de 2006.

Atentamente

Juan Pablo Romero Rodríguez
Profesional Ventanilla Ambiental CAR
Sector Industria.

Ramón Leal Leal
Coordinador Ventanilla Ambiental CAR

Ventanilla Ambiental CAR
Corporación Ambiental Empresarial
Teléfono: 3830300 ext. 2697 y 2655
Fax. 3830690 ext. 2697
Correo electrónico: ventanillaindustriacae@ccb.org.co

Signing formal agreement between university and CAR

ESTUDIO DE OPORTUNIDAD Y CONVENIENCIA

**CONVENIO INTERADMINISTRATIVO DE ASOCIACIÓN
ENTRE LA CORPORACIÓN Y LA UNIVERSIDAD NACIONAL DE COLOMBIA, A TRAVÉS
DEL INSTITUTO DE ESTUDIOS AMBIENTALES (IDEA) PARA BRINDAR CAPACITACIÓN
Y ACOMPAÑAMIENTO TÉCNICO EN PRODUCCIÓN MÁS LIMPIA AL SUBSECTOR
CURTIEMBRES**

I. INFORMACIÓN GENERAL

1. **DEPENDENCIA SOLICITANTE:** Subdirección de Desarrollo Ambiental Sostenible
2. **PROGRAMA:** Desarrollo Productivo Sostenible
3. **PROYECTO:** 11. Apoyo y Promoción de la Producción Más Limpia y los Mercados Verdes.
4. **META:** 5. Suscribir un (1) convenio de Producción Más Limpia con el subsector curtiembres para la mejora del desempeño ambiental y realizar el respectivo seguimiento anual.
5. **VALOR:** Se estima en la suma de DOSCIENTOS SESENTA Y TRES MILLONES OCHOCIENTOS CUARENTA MIL PESOS MONEDA CORRIENTE ($263.840.000), PESOS MONEDA CORRIENTE.
6. **OBJETO:** La Asociación entre la Corporación y la Universidad Nacional de Colombia, a través del Instituto de Estudios Ambientales (IDEA) para brindar capacitación y acompañamiento técnico en Producción Más Limpia al subsector curtiembres.

II. DEFINICIÓN DE LA NECESIDAD

El subsector curtiembres en la jurisdicción de la CAR se compone básicamente de 182 unidades productivas distribuidas en dos Provincias del Departamento de Cundinamarca, como son:

- *Provincia Almeidas y Municipio de Guatavita:* En los municipios de Villapinzón (125 unidades) y Chocontá (47 unidades).
- *Provincia Sabana Centro:* En el municipio de Cogua (10 unidades productivas).

Como principal área de concentración de la actividad curtidora, se identifica un corredor de aproximadamente 8 kilómetros a lo largo de las zonas rurales de los municipios de Chocontá y Villapinzón. Esta zona, como se presenta en la siguiente figura, presenta como principales límites el río Bogotá y la doble calzada Briceño-Tunja-Sogamoso (BTS), a lo largo de los cuales se asienta la industria curtidora en unidades productivas denominadas curtiembres.

Caracterización productiva y organizacional:

Las actividades productivas desarrolladas por las curtiembres consisten en el procesamiento de pieles animales (frescas o saladas) con el fin de convertirlas en un material durable (no putrescible) denominado cuero, destinado para la producción de calzado y marroquinería, principalmente. En el caso de la actividad desarrollada en la jurisdicción de la CAR, esta se desarrolla primordialmente empleando pieles bovinas, siendo el forro y la tula los productos terminados predominantes.

El desarrollo de operaciones tiene lugar en empresas de carácter familiar, cuyas instalaciones que se componen primordialmente de plantas de producción de un nivel, construidas en ladrillo, distribuidas en un área de proceso generalmente de carácter abierto (ausencia de divisiones

1

VIII. VALOR TOTAL Y APORTES

El valor total necesario para la realización del objeto del convenio se estima en la suma de DOSCIENTOS SESENTA Y TRES MILLONES OCHOCIENTOS CUARENTA MIL PESOS MONEDA CORRIENTE ($263.840.000), representados en aportes en dinero, bienes y servicios de la siguiente forma:

1. La Corporación Autónoma Regional de Cundinamarca –CAR, aporta la suma de CIENTO VEINTE MILLONES PESOS MONEDA CORRIENTE ($120.000.000).

2. El Instituto de Estudios Ambientales (IDEA) realizará un aporte total de CIENTO CUARENTA Y TRES MILLONES OCHOCIENTOS CUARENTA MIL PESOS MONEDA CORRIENTE ($143.840.000), representado en:

   ➢ Recursos en efectivo, con cargo al Convenio SWITCH, por una suma de SESENTA Y SIETE MILLONES DOSCIENTOS MIL PESOS MONEDA CORRIENTE ($67.200.000).

   ➢ Aportes en especie, representados profesionales y estudiantes que para el efecto designe la Universidad como apoyo al desarrollo del proyecto, los cuales se estiman en la suma de SETENTA Y SEIS MILLONES SEISCIENTOS CUARENTA MIL PESOS MONEDA CORRIENTE ($76.640.000).

Atentamente,

Ing. Carlos Alberto Pérez E.
Profesional Especializado SDAS

Vo.Bo. Ing. José Miguel Rincón V.
Subdirector Desarrollo Ambiental Sostenible

11

UNIVERSIDAD
NACIONAL
DE COLOMBIA
SEDE BOGOTÁ
INSTITUTO DE ESTUDIOS AMBIENTALES
– IDEA –

ACTA DE MUTUO ACUERDO

Considerando

Que la Universidad Nacional de Colombia, como parte de las 32 instituciones a nivel mundial involucradas en el proyecto SWITCH, estudia soluciones a las problemáticas industriales que afecten la calidad del agua. Para ello se encuentra adelantando una investigación en la implementación de estrategias en producción más limpia en industrias de curtiembre en el municipio de Villapinzón, Cundinamarca.

Que la Corporación Autónoma Regional de Cundinamarca CAR en ejercicio de sus funciones, ha celebrado un Convenio de Cooperación con el Instituto de Estudios Ambientales IDEA de la Universidad Nacional de Colombia, con el objeto de "brindar capacitación y acompañamiento técnico en Producción más Limpia al subsector curtiembres" en Villapinzón.

Doce curtidores de Villapinzón fueron seleccionados para participar directamente en el proyecto, por su compromiso frente al cambio para lograr un desarrollo sostenible, su participación activa y su liderazgo en el sector.

Reunidas las partes en el municipio de Villapinzón, han acordado lo siguiente:

- Trabajar en forma integrada en el proyecto de investigación en producción más limpia PML.
- El IDEA de la Universidad Nacional realizará acompañamiento técnico y capacitación en la implementación de PML en las 12 curtiembres, en cuatro pasos: 1. Informe de estado actual de las curtiembres seleccionadas. 2. Definición de estrategias por implementar de PML en las curtiembres. 3. Pruebas de PML con 12 industriales seleccionados, 30 curtidores y operarios invitados para la capacitación. 4. Seguimiento a la implementación de PML. Adicionalmente investigará opciones para el manejo de los residuos sólidos.
- El IDEA de la Universidad Nacional se compromete a desarrollar la investigación y entregar los resultados a los industriales, los cuales serán conocidos también a nivel mundial como parte del proyecto SWITCH.

Proyecto SWITCH El manejo del agua en las ciudades del mañana.
Instituto de Estudios Ambientales IDEA - Universidad Nacional de Colombia
Calle 44 N° 45-67 Unidad Camilo Torres Bloque B2 ofc 4. Teléfonos: 3165113 -3165085 - 3165000 Exts. 10556 - 10563
Bogotá - Colombia
e-mail: insestam_bog@unal.edu.co

1

UNIVERSIDAD
NACIONAL
DE COLOMBIA
S E D E   B O G O T Á
INSTITUTO DE ESTUDIOS AMBIENTALES
- IDEA -

- Los industriales se comprometen a brindar la información necesaria y confiable para el buen desarrollo de la investigación, que redundará en su propio beneficio. Adicionalmente se comprometen a realizar las pruebas de PML y permitir el acceso de otros curtidores a los talleres prácticos en sus curtiembres.
- La Corporación Autónoma Regional CAR se compromete a realizar el acompañamiento, seguimiento y control al convenio celebrado con la Universidad, participando de forma activa en las actividades propuestas.

Para constancia de lo anterior, se firma el día 24 de abril de 2008, en el municipio de Villapinzón, Cundinamarca, Colombia.

Por el IDEA de la Universidad Nacional de Colombia

Por la Corporación Autónoma Regional de Cundinamarca CAR

Laura Cecilia Osorio Muñoz
Directora Proyecto SWITCH
C.C: 24 315 381
Tel. 3165000 ext. 10556

Claudia Patricia Torres
Corporación Autónoma Regional CAR
C.C.: 52055578
Tel. 320 90 00   1429

Curtidores participantes

Raúl Arévalo
Curtiembre
C.C.3240907
Tel: 3158777908

Rosendo Castiblanco
Curtiembre Rosendo Castiblanco
C.C.80466318 Upinzón
Tel 315 8319935

Victor Manuel Contreras
Curtiembre El Tauro
C.C.3241194
Tel3152021792

Gustavo Fernández
Curtiembre Montecarlo
C.C. 3241 993
Tel 312 582 88 68

Proyecto SWITCH El manejo del agua en las ciudades del mañana.
Instituto de Estudios Ambientales IDEA - Universidad Nacional de Colombia
Calle 44 N° 45-67 Unidad Camilo Torres Bloque B2 ofc 4. Teléfonos: 3165113 -3165085 - 3165000 Exts. 10556 - 10563
Bogotá - Colombia
e-mail: insestam_bog@unal.edu.co

UNIVERSIDAD
NACIONAL
DE COLOMBIA
SEDE BOGOTÁ
INSTITUTO DE ESTUDIOS AMBIENTALES
- IDEA -

Leovíceldo Fernández Rodríguez
Curtiembre LEOFER
C.C: 3'000998 *Choconta*
Tel 856 5576

Ananías López
Curtiembre VILLAMAR
C.C: 524157o *de V/p.*
Tel 3144132900

Gonzalo López
Curtiembre VILLALUZ
C.C: 3003383 *cdeChoatr.*
Tel# 3157819746

Claudia Patricia Marín
Curtiembre CPM.
C.C: 20.363.134 *Choconta*
Tel 3125260061.

Paulino Monroy
Curtiembre Villasol
C.C: 3240749 *Vzan*
Tel 317 329 1898

Efraín Humberto Rodríguez
Curtiembre Efraín Humberto Rodríguez
C.C: 80394951 *Lifar*
Tel 3138314848

Merardo Rodríguez
Curtiembre Merardo Rodríguez
C.C: 3003465-
Tel 3183089259

Manuel Torres Segura
Curtiembre La Pradera
C.C: 3.002035
Tel 312 13121756

Proyecto SWITCH El manejo del agua en las ciudades del mañana.
Instituto de Estudios Ambientales IDEA - Universidad Nacional de Colombia
Calle 44 N° 45-67 Unidad Camilo Torres Bloque B2 ofc 4. Teléfonos: 3165113 -3165085 - 3165000 Exts. 10555 - 10563
Bogotá - Colombia
e-mail: insestam_bog@unal.edu.co

UNIVERSIDAD
NACIONAL
DE COLOMBIA
SEDE BOGOTÁ
INSTITUTO DE ESTUDIOS AMBIENTALES
- IDEA -

ORDEN DEL DIA
24 de abril de 2008
Villapinzón, Cundinamarca.

1. Antecedentes al proyecto CAR. Cámara de Comercio, SWITCH. Mónica Sanz
2. Presentación del proyecto CAR
   2.a Alcances. Laura Osorio
   2.b. Objetivos. Laura Osorio
   2.c. Componentes
      Componente técnico y ambiental. Sofia Duarte y Carlos Escamilla
      Componente Económico. Jose Stalin Rojas
      Componente Institucional . Monica Sanz
      Componente de Asociatividad. Monica Sanz y Jose Stalin Rojas
3. CAR, Claudia Torres
4. Metodología. Tania
5. Presentación del equipo de trabajo. Tania
6. Firma del acta de mutuo acuerdo
7. Siguiente paso: programación de las primeras visitas de seguimiento a todo el proceso productivo.

Proyecto SWITCH El manejo del agua en las ciudades del mañana.   1
Instituto de Estudios Ambientales IDEA - Universidad Nacional de Colombia
Calle 44 N° 45-67 Unidad Camilo Torres Bloque B2 ofc 4. **Teléfonos: 3165113 -3165085 - 3165000 Exts. 10556 - 10563**
Bogotá - Colombia
e-mail: insestam_bog@unal.edu.co

República de Colombia

Tribunal Superior del Distrito Judicial de Cundinamarca

Sala Penal

Bogotá, Veintiséis (26) de Junio de dos mil ocho (2008)

Mag. Ponente:	Edwar Enrique Martinez Perez
Radicado:	25183-31-04-001-2003-00050-01.
Procedente:	Juzgado 01 de Ejecución de Penas y Medidas de Seguridad de Girardot– Cundinamarca -
Denunciante:	De Oficio
Procesado:	Manuel de Jesús Torres Segura
Delito:	Contaminación Ambiental
Asunto:	Apelación Auto Interlocutorio.
Decisión:	Revoca
Aprobado:	Acta No.155

I. VISTOS:

El recurso de apelación interpuesto por el procesado MANUEL DE JESÚS TORRES SEGURA, contra la providencia interlocutoria por medio de la cual el Juzgado Penal del Círcuito de Chocontá -Cundinamarca, le negó la solicitud de amortización de la pena de multa mediante trabajo.

II. HECHOS RELEVANTES

Fueron plasmados en el fallo de primera instancia en la siguiente forma:

"Se procede a través de estas diligencias, en razón de los hechos presuntamente constitutivos del delito de contaminación al medio ambiente en los que pudo haber incurrido el acá procesado MANUEL DE JESÚS TORRES SEGURA como propietario de una industria procesadora de cueros, ubicada en la vereda Chigualá, Finca Carboneles, industria de curtiembres la pradera, comprensión municipal de Villapinzón, al proceder a verter los residuos químicos sin ninguna clase de tratamiento al río Bogotá, lo que ha venido desarrollando desde hace bastante tiempo ya, pero específicamente desde cuando este asunto fue tratado por la Corporación Autónoma Regional de Cundinamarca, C.A.R., para los días 23 de Abril y 6 de Septiembre de 1999.

III. ACTUACIÓN PROCESAL:

1. Mediante sentencia de fecha 13 de febrero de 2004, proferida por el Juzgado Penal del Circuito de Chocontá, se condenó al señor MANUEL DE JESÚS TORRES SEGURA, a la pena principal de veinticuatro (24) meses de prisión y multa de cien (100) salarios mínimos legales mensuales vigentes como autor material responsable del delito de contaminación ambiental, concediéndole dentro del mismo fallo la Suspensión Condicional de la Ejecución de la Pena por un periodo de dos (2) años, y disponiendo que se le concedería al procesado un termino de veinticuatro (24) meses, a partir de la ejecutoria de dicha providencia para cancelar la multa impuesta.

2. Impugnada dicha decisión, fue confirmada por esta Colegiatura, mediante sentencia de fecha 25 de mayo de 2004, quedando ejecutoriada el día 15 de julio de 2004.

proceda conforme a lo reglado en el numeral 7° del artículo 39 de la Ley 599 de 2000.

Por lo expuesto, el Tribunal Superior del Distrito Judicial de Cundinamarca, Sala de Decisión Penal,

RESUELVE:

PRIMERO: - REVOCAR el auto impugnado, por las específicas razones aquí consignadas, a fin de que se surta la actividad probatoria pertinente y la que oficiosamente se considere útil, y luego de ello se tome la decisión que se amerite.

SEGUNDO:- En firme devolver el expediente a su juzgado de origen.

Contra esta providencia no procede recurso alguno.

NOTIFÍQUESE Y CÚMPLASE

EDWAR ENRIQUE MARTINEZ PEREZ
MAGISTRADO

6.4.12 Evaluating the intervention

The questionnaire was:

Respected sr/mrs:

Tanner   associated   Not associated

Tanner   on the riverbank   Out of the riverbank

Official employee

Other_____

Please answer to this short questionnaire on the conflict resolution process, which started in 2004:

Compared to the previous experiences in Villapinzón, how would you consider the process lead by Mrs Monica (1 being the lowest score and 5 the highest) with respect to:

   (a) Respect to tanner´s ideas and access to participation
   (b) How reliable was the process
   (c) How just was the process
Thank You!!!

Results of the Evaluation of the intervention process done by Evidalia Fernández with 50 tanners, one Car official and the Mayor.

Number of tanners: 50

	Out of riverbank	On the riverbank
Associated ACURTIR	41	3
Not associated	5	1

The results showed:

(a)	Out of riverbank	On the riverbank
Associated ACURTIR	95% 4-5 5% 3	100% 4-5
Not associated	100% 4-5	Score of 5

(b)	Out of riverbank	On the riverbank
Associated ACURTIR	100% 4-5	100% 4-5
Not associated	100% 4-5	Score of 5

(c)	Out of riverbank	On the riverbank
Associated ACURTIR	100% 4-5	100% 4-5
Not associated	100% 4-5	Score of 5

It is interesting to highlight that while the CAR official scored 4 to all three items, the Mayor put a score of 3 to (a), 2 (b), and 3 to (c). This scanned copy is enclosed.

Another scanned copy to a tanner is enclosed

Respetado señor/a:

CURTIDOR   afiliado   no afiliado

CURTIDOR   de la ronda   fuera de la ronda

EMPLEADO PÚBLICO   X

OTRO-----------------------

Favor responder a esta pequeña evaluación del proceso que
empezó desde Junio de 2004 en resolución de conflictos:

1. Comparado a los años anteriores a 2004 califique el proceso
   vivido con la dra Mónica del 1 al 5 siendo el 5 la mejor
   calificación respecto a :

Respeto a las ideas y participación de los curtidores:
_____3_____

Transparencia del proceso:
_____2_____

Búsqueda de proceso
JUSTO___3_____

GRACIAS ¡!!!!!!!!!!!!

Respetado señor/a:

CURTIDOR   afiliado   (no afiliado)

CURTIDOR   (de la ronda)   fuera de la ronda

EMPLEADO PÚBLICO

OTRO----------------------

Favor responder a esta pequeña evaluación del proceso que empezó desde Junio de 2004 en resolución de conflictos:

1. Comparado a los años anteriores a 2004 califique el proceso vivido con la dra Mónica del 1 al 5 siendo el 5 la mejor calificación respecto a :

Respeto a las ideas y participación de los curtidores: 4
_____

Transparencia del proceso:
_____   4

Búsqueda de proceso
JUSTO_____   5

GRACIAS ¡!!!!!!!!!!!!

Mesa de trabajo Villapinzón, Junio 12 de 2008-06-12

Puntos para desbloquear e impulsar proceso en Villapinzón

1. **Reactivación de la reapertura temporal de industrias basadas en recirculación de cromo y sulfuro**
   - Se evita clandestinidad
   - Se estimulan a los que quieren cumplir
   - Se empieza a limpiar el río
   - Tiempo para la definición de ronda hidráulica a la CAR

2. **Solicitud a la CAR para seguir apoyando la Producción Más Limpia pero desde la coordinación con el Comando y Control.**
   - Plan de choque para la oficina de Almeidas por el evidente desorden y confusión en los expedientes
   - Ajustar parámetros del 1594. El 08 no es legal.

3. **Programa integral para el desmonte de la ronda una vez se haya coordinado y adecuado el lugar de relocalización**
   - Trabajo conjunto con Municipio, Gobernación y Presidencia de la República

4. **Firma del Plan Especial de curtiembres**
   - Solicitud a la CAR para que se le acepte al municipio los prediseños del alcantarillado pues los definitivos son sujeto de estudio (hace dos años recursos SINA II que se perdieron)
   - Contratación para los diseños definitivos con experto propuesto con anterioridad.
   - Permitir un tratamiento especial para los relocalizados de la ronda de menos de 2 hect.
   - sitio para valorización de residuos

5. **Invitación a participar de este proceso al ministerio de Comercio, Industria y Turismo**
   - Propuesta de fomento específica a Villapinzón como fomento a la Mipyme con líneas de crédito especiales de productividad y no de consumo.

Congreso de la República de Colombia

Bogotá, 11 de Junio 2008

Doctor
ANDRES BARRETO ROZO
Secretario Hábitat y Recursos Mineros
Gobernación de Cundinamarca
E.   S.   D.

Respetado Doctor Barreto:

De antemano agradezco su amable invitación y celebro que el gobierno departamental asuma con gran liderazgo la problemática que existe alrededor del tema de los curtiembres en la provincia de Almeidas y especialmente la jurisdicción de los municipios de Villapinzón, Chocontá y Cogua.

Desde hace un tiempo atrás hemos venido trabajando con la comunidad afectada y las distintas autoridades del nivel municipal, departamental, regional y nacional, tratando de concertar propuestas y soluciones.

Lamentablemente, en el Congreso de la República nos encontramos terminando legislatura, lo cual me impide acompañarlos en tan importante evento. Reitero mi compromiso de trabajo en este tema en beneficio de los cundinamarqueses.

Con sentimiento de consideración y aprecio,

NANCY PATRICIA GUTIERREZ CASTAÑEDA
Presidenta del Senado de la República

6.4.14 Reopening tanneries

Corporación Autónoma Regional de Cundinamarca - CAR
Oficina Provincial de Almeidas y el Municipio de Guatavita
República de Colombia

AUTO OPAG No. 1 5 7 DE 0 2 MAYO 2008

(Por el cual se registra un vertimiento y se adoptan otras disposiciones)

Que el día 24 de enero de 2005 la Secretaria de Gobierno y Oficina Asuntos Policivos de Villapinzón realiza la diligencia de CIERRE TEMPORAL de la industria del curtido del cuero ubicada en el predio Carboneles, vereda Chigualá, jurisdicción del municipio de Villapinzón, propiedad del señor Manuel de Jesús Torres Segura, en la cual se concede un plazo de 28 días para que se termine el proceso del curtido. (Folio 110).

Que el día 1°. de marzo de 2005 la Secretaria de Gobierno y oficina Asuntos Policivos de Villapinzón, realiza la diligencia de verificación de CIERRE TEMPORAL de la industria del curtido del cuero La Pradera ubicada en el predio Carboneles, vereda Chiguala, jurisdicción del municipio de Villapinzón. (Folio 123).

Que mediante Resolución 2460 del 01 de diciembre de 2005 la Corporación establece el Plan de Manejo Ambiental presentado por el señor Manuel de Jesús Torres Segura para la industria de curtido de pieles de propiedad. (Folios 146 y 147).

Que mediante Resolución 1653 del 08 de junio de 2006, la Corporación suspende la ejecución de la sanción de CIERRE TEMPORAL impuesta mediante Resolución 903 del 15 de septiembre de 2004, al señor Manuel de Jesús Torres Segura en calidad de propietario de la industria de curtido de pieles ubicada en el predio Carboneles, vereda Chigualá, jurisdicción del municipio de Villapinzón, por un término de seis (6) meses, tiempo dentro del cual deberá construir el sistema de tratamiento primario. (Folios 166 a 168).

Q ue mediante Resolución 3096 del 08 de noviembre de 2006 la Corporación rechaza el recurso de reposición presentado por el señor Manuel de Jesús Torres Segura contra la Resolución 1653 del 08 de junio de 2006. (Folios 172 a 174).

Que el día 04 de enero de 2007 la Secretaria de Gobierno y Oficina Asuntos Policivos de Villapinzón realiza la diligencia de suspensión de ejecución de la sanción de CIERRE TEMPORAL de la industria del curtido de pieles La Pradera ubicada en el predio Carboneles, vereda Chiguala, jurisdicción del municipio de Villapinzón, propiedad del señor Manuel de Jesús Torres Segura. (Folio 190).

Que mediante Auto 994 del 13 de julio de 2005, la corporación inicia un trámite de permiso de vertimientos y se realiza el cobro por concepto del servicio de evaluación ambiental a nombre de l señor Manuel de Jesús Torres Segura, para beneficio del predio de su propiedad denominado Carbonelá, vereda Chigualá, jurisdicción del municipio de Villapinzón. (Folios 2 y 3 cuaderno2).

Seguimiento al trámite de permiso de vertimientos

Que mediante Auto 994 del 13 de julio de 2005, la Corporación inicia un trámite de permiso de vertimientos a nombre del señor Manuel de Jesús Torres Segura y realiza el cobro por concepto del servicio de evaluación ambiental. (folios 2-3 carpeta2)

6.4.15 Responsibilities in SWITCH project

Id	Nombre de tarea	Comienzo	Fin	Predec	Código producto	Responsable
1	Plan operativo aprobado por CAR para desarrollo del proyecto	mié 27/02/08	mié 27/02/08		CAR-OR- 01	Tania Santos
3	Diseñar lista de chequeo para curtiembres	jue 06/03/08	jue 06/03/08			Tania Santos
4	Definición de criterios de selección de empresas con la CAR Bogotá y Chocontá	mar 18/03/08	mar 18/03/08			Tania Santos
5	Visita a curtiembres preseleccionadas por CAR de Villapinzón y Chocontá	jue 06/03/08	jue 20/03/08			Tania Santos
6	Seleccionar "2 industrias piloto para pruebas de PML	vie 21/03/08	vie 21/03/08	5		Tania Santos
7	Firma de acuerdo de compromiso participación en el proyecto	vie 18/04/08	vie 18/04/08	6	CAR- 01	Monica Sanz
9	Conocimientos sobre asociatividad	lun 21/04/08	lun 21/04/08			Monica Sanz
10	Encuesta de conocimientos de PML a curtidores	lun 21/04/08	lun 21/04/08			Monica Sanz
11	Sistema de Costos por industria	lun 21/04/08	lun 21/04/08			Jose Stalin Rojas
12	Muestreo de calidad de agua de efluentes(pelambre, curtido, teñido, lavado y todos los efluentes	lun 21/04/08	lun 21/04/08			amilla y Sofia Duarte
13	Balance de materia y energía por empresa	lun 21/04/08	lun 21/04/08			camilla, Sofia Duarte
14	Indicadores	lun 21/04/08	lun 21/04/08			Javier Burgos
16	Balance de materia y energía 12 industrias(incluir revisión SWITCH)	lun 21/04/08	jue 22/05/08	7		rte, Carlos Escamilla
17	Indicadores de línea base social, económico, ambiental	lun 21/04/08	jue 22/05/08	7		Javier Burgos
18	Muestreos de efluente(diseño de toma de muestras)	lun 21/04/08	jue 22/05/08	7		rte, Carlos Escamilla
19	Sistema de costos de cada empresa	lun 21/04/08	jue 22/05/08	7		Jose Stalin Rojas
20	Encuesta de conocimiento de los curtidores sobre PML, Gestión Empresarial, asociatividad	lun 21/04/08	jue 22/05/08	7		to y Carlos Escamilla
22	Firma de acuerdo de implementación de PML, asociatividad	vie 23/05/08	vie 23/05/08		CAR-03	Monica Sanz
25	Apreciación del material por CAR Bogotá, CAR Chocontá y representante de los curtidores	vie 30/05/08	vie 30/05/08			Laura Osorio
26	Diseño de material para medición y seguimiento de productividad, competitividad y asociatividad	lun 21/04/08	vie 30/05/08		CAR-05	Laura Osorio
28	200 Afiches, 100 carpetas con instructivos, procedimientos y fichas de seguimiento	lun 02/06/08	vie 20/06/08		CAR-06	Laura Osorio
30	Programación de pruebas de implementación de PML. Invitación a 7 empresas y uno de sus o	lun 05/05/08	vie 23/05/08			Tania Santos
32	Pruebas de laboratorio de efluentes	lun 25/05/08	mar 26/08/08			Jose Stalin Rojas
33	Análisis de laboratorio del cuero	lun 25/05/08	mar 26/08/08			Jose Stalin Rojas
34	Sistema de costos	lun 25/05/08	mar 26/08/08			Sofia y Carlos E
35	Capacitación con los formatos para realizar seguimiento	lun 25/05/08	mar 26/08/08		COL-08	Carolina Y Tatiana
36	Firma de acta de participación	mar 26/08/08	mar 26/08/08			Jose Stalin Rojas
38	Medición de calidad de productos	lun 25/05/08	lun 25/05/08			Jose Stalin Rojas
39	Análisis precio de venta, punto de equilibrio	lun 25/05/08	lun 25/05/08		COL 11	Jose Stalin Rojas
40	Programa de mejoramiento de la productividad en cada curtiembre: aumento portafolio de produc	mar 26/08/08	mar 26/08/08			Jose Stalin Rojas
41	Documento sobre condiciones para obtener sello verde de calidad para el producto	vie 31/10/08	vie 31/10/08		COL 10	Jose Stalin Rojas
42	Capacitación gerencial a 12 curtidores	vie 29/08/08	vie 29/08/08			Jose Stalin Rojas
44	Indicadores económicos sociales, ambientales	jue 28/05/08	vie 05/09/08			Jose Stalin Rojas
45	Comparación con indicadores de línea base	jue 28/05/08	vie 05/09/08			Jose Stalin Rojas
47	Visita a las 12 curtiembres seleccionadas y a 30 curtidores	lun 03/11/08	vie 19/12/08	31	CAR-09	rte, Carlos Escamilla
49	Implementación de PML y SGA en las 6 industrias	mié 01/10/08	vie 31/10/08		COL-07	amilla y Sofia Duarte
51	Tesis: Eficiencia de la herramienta para ejecución de sal en curtiembres	lun 02/06/08	vie 02/01/09		COL 18	Javier Burgos
52	Compostaje de los residuos obtenidos en el pelambre ecológico	mar 01/04/08	vie 30/05/08		COL 16	alson Cuervo y Prof.
53	Recirculación de efluentes e impacto en el manejo del agua y calidad de los cueros	vie 27/02/09	vie 27/02/09		COL-12(17)	Profesor UNAL
56	Reuniones de aprestamiento individuales: ACURTIR, proveedores, compradores, entidades de a	mar 27/05/08	mar 27/05/08			Monica Sanz
57	Reuniones de aprestamiento con proveedores de insumos químicos	mié 28/05/08	mié 28/05/08	56		Sofia Duarte
58	Reunión de concertación entre las entidades e instituciones de apoyo a la cadena	vie 27/06/08	vie 27/06/08			Monica Sanz
59	Reunión para compra de cueros: ANDI, CEINNOVA, CC3	vie 27/06/08	vie 27/06/08			Jose Stalin Rojas
60	Reunión posibles clientes Unche	lun 30/06/08	lun 30/06/08			Jose Stalin Rojas
61	Reunión de aprestamiento Venta de productos, proexport	mar 30/09/08	mié 28/05/08	56		Jose Stalin Rojas
62	Firma de acuerdo de agentes de la cadena	mar 30/09/08	mar 30/09/08	56		Monica Sanz
63	Realizar capacitaciones a los curtidores sobre asociatividad.	vie 29/08/08	vie 29/08/08			Monica Sanz
64	Reuniones de evaluación de acuerdos	vie 28/11/08	vie 28/11/08	62		Monica Sanz
65	Presentación de resultados de la Tesis. Aplicabilidad de la asociatividad como herramienta par	lun 01/12/08	lun 01/12/08	64	COL 01	Jose Stalin Rojas
70	Definición de una comercializadora apoyada por la asociación	lun 01/12/08	lun 01/12/08	64		Monica Sanz
71	Indagar en el mercado de las grasas los requisitos de calidad de cada posible cliente a nivel naci	vie 30/05/08	vie 30/05/08			Jose Stalin Rojas
72	Caracterización del producto final para analizar opciones de comercialización.	vie 29/08/08	vie 29/08/08			Sofia Duarte
73	Comparar las propiedades físicas y químicas obtenidas del producto final con los requerimientos	lun 01/09/08	lun 01/09/08	71		Sofia Duarte
74	Mediante un estudio de costos dar un valor comercial al producto y compararlo con la actual com	mar 02/09/08	mar 02/09/08	72		Jose Stalin Rojas
75	Realizar alianzas comerciales para la valorización de este residuo.	mié 27/02/08	mié 27/02/08			
76	Indagar las condiciones para obtener un sello de calidad o sello verde para el producto.	vie 27/06/08	vie 27/06/08			
77	Realizar un análisis general de la cadena productiva y ver como se puede incorporar este nuevo	vie 29/08/08	vie 29/08/08			

6.4.16 Land planning barriers to CP

Villapinzón, Cundinamarca. 15 de Enero de 2009.

Doctor
Edgar Bejarano
Director
Corporación Autónoma de Cundinamarca CAR

Respetado doctor:

De la manera más respetuosa queremos poner en su conocimiento la situación que nos agobia. La Corporación CAR ha expedido unos Autos y Resoluciones con copia a la Fiscalía a curtidores que han presentado Plan de Manejo Ambiental aprobado técnicamente por la Corporación, que se han movido de la zona de los 30 metros caso "Arturo Enrique Gómez Melo" y/o que están invirtiendo en soluciones en Producción Limpia dentro del marco del proyecto SWITCH apoyado por la Corporación CAR, caso "Paulino Monroy" y/o que llevan cinco años sin trabajar, habiendo desmontado sus equipos caso "Jose Teodoro Segura Cufiño", obligado a demoler.

Como presidenta de ACURTIR solicito que así como el 2 de diciembre del año 2008, en reunión con curtidores en la Corporación CAR, el director Edgar Bejarano, planteó que la definición de la ronda hidráulica así como la legalización de los terrenos industriales no se daría antes de finales del mes de mayo de 2009, se de un compás de espera a los curtidores siempre y cuando implementen producción limpia y hayan realizado esfuerzos en solucionar su problema.

Los curtidores no pueden plantear soluciones definitivas mientras los temas de Ordenamiento Territorial que llevan 3 años dilatados y los límites de vertimientos no sean adaptados a la realidad de las condiciones locales y al Acuerdo 043 sobre la calidad del agua de río Bogotá.

Invocamos a que se haga justicia siempre y cuando las soluciones del estado sean una realidad.

Agradecemos su amable apoyo

Evidalia Fernández
Presidenta ACURTIR
C.C.: 21.101.705 Villapinzón

c.c.: **Alejandro Rueda.** Procurador Delegado para Asuntos Ambientales y Agrarios
Andrés González Díaz. Gobernador Cundinamarca
Henry Polania. Delegado Sector Curtiembres. Gobernación de Cundinamarca
Juan Lozano. Ministro Ambiente Vivienda y Desarrollo Territorial.
José Severo González. Viceministerio de Agua y Saneamiento Básico.

6.4.17 The ministry's support to judicial order switching penal fines

Si logramos constituir esa junta y darle el mandato a ese gerente y encargarlo como secretario técnico de las mesas de avance, de las mesas de verificación de los compromisos, creo señor Contralor que haríamos un gran aporte. Tenemos como base el documento CONPES que se aprobó en el años 2004, pero como los documentos CONPES en sí mismos corren el riesgo de quedarse en letra muerta sino se les hace respiración en la nuca y seguimiento diario de todos los compromisos, pensamos que con ese instrumento de la gerencia, con esa junta directiva, con ese ánimo de concertación, de convocatoria de identificación de propósitos avanzaremos en esa línea que nos hemos propuesto de generar conciencia y generar compromisos.

Con respecto al segundo tema que es la problemática puntual del Municipio de Villapinzón debo ser muy sincero: cuando uno llega aquí, lo primero que le ocurre es pedirle a los compañeros de los medios de comunicación que nos dejen contarle a Cundinamarca y al país de la bondad de la gente de Villapinzón, de la belleza que hay aquí, pero además del infinito servicio ambiental que se le presta a la ciudad capital como quiera que está aquí muy cerca nuestro Guacheneque, nuestra laguna, y como quiera que está aquí muy cerca el Alto de la Calavera que es el verdadero nacimiento del río Bogotá. Es por eso que tiene, señor Contralor, tanto significado que hayamos hecho esta reunión aquí y que desde Villapinzón levantemos una voz para que podamos trabajar en algo que venimos planteando desde el ministerio y es el reconocimiento por servicios ambientales a los municipios que tienen las fábricas de agua. Claro que a Villapinzón que alberga la laguna, que alberga el nacimiento del río y al conjunto de municipios de la cuenca.

La comunidad toda le debe gratitud a Villapinzón y no solo con palabras y aplausos y con palmaditas en la espalda sino con plata, con billete sobre la mesa, para que se pueda compensar el esfuerzo que hace el municipio en este proceso. Y lo digo no solamente en un empeño de generar una voz abstracta. Esto ha sido concreto: el municipio toma dinero legítimamente de sus arcas para comprar los predios que son protectores del agua y yo quiero señalar y reconocer el esfuerzo de la CAR para ayudarles a los municipios que tienen nacederos de agua en la compra de los predios para proteger nuestros nacederos, eso hay que hacerlo, hay que continuarlo, el Plan de Desarrollo que se aprobó recientemente en el Congreso de la República, reproduce la norma que autoriza la destinación de los recursos de los municipios en el 1% por ciento de sus ingresos corrientes para la compra de estos predios.

El tercer tema puntual de Villapinzón tiene que ver con la curtiembre. Lo ha dicho el señor Contralor con toda claridad, nosotros necesitamos garantizar el respeto a la normatividad ambiental dentro de un marco en el que sean posibles las actividades productivas, las actividades productivas honradas de los colombianos en todo el país. Hemos tenido una reunión con los distinguidos miembros del Concejo Municipal, con la presencia del señor alcalde y el director de Planeación, con representantes de la Contraloría y con la presencia del señor Director encargado de la CAR y hemos acordado, doctor Turbay, lo siguiente: no se puede seguir demorando la solución, hay una concertación hecha en la que han participado los representantes de las curtiembre, de los municipios y los representantes del ministerio. Hoy les digo que para efectos de todos los acuerdos futuros está en firme ese entendimiento, está en firme esa decisión.

Había una discusión en relación con el área mínima de las curtiembres en la oficina de Planeación del municipio que ha dado su visto bueno para que se trabaje sobre la base de lo que indica la norma y tenemos el compromiso del señor director de la CAR para que el proceso de contratación del estudio técnico, pueda concluir dentro del término de los siguientes dos meses para que una solución integral nos permita dejar en firme el plan especial de manejo de curtiembres de Villapinzón, responsabilidad ambiental y oportunidades para el trabajo, ese es el espíritu con el que debemos trabajar a la mayor velocidad.

Pasando al cuarto tema de la problemática que vive Villapinzón, por coincidencias de la vida, mi compañero de la facultad de derecho de la Universidad de los Andes, Misael Garzón es el apoderado de un alto número de personas que está viviendo una situación dramática por que han sido condenadas en un proceso penal y hasta les han embargado sus bienes. Estas, fueron personas que no recibieron en su momento la información, la ilustración, la guía, el apoyo, el acompañamiento para poder cumplir bien sus actividades y hoy tenemos familias que están apunto de perder la totalidad de su patrimonio. Inclusive hay niños que se pueden quedar sin techo donde vivir, si no les encontramos una solución.

El Dr. Misael Garzón viene avanzando en unas formulas judiciales que nos permitan buscar mecanismos de compensación en las sanciones para que las personas sancionadas sobre la base de asumir la responsabilidad ambiental futura, de cambiar los procesos de producción como lo señalaba el señor Contralor, de manera que no se cause dañe a los recursos naturales.

Queremos dentro del marco de la buena voluntad y la ley contar con el apoyo de los integrantes del Consejo Superior de la Judicatura, como quiera que haya situaciones jurídicas que se han consolidado para que les podamos ayudar a estas buenas familias de Villapinzón que hoy tienen un enorme sufrimiento.

Con respecto al quinto tema que es el de la aproximación integral a la problemática del río, debemos tener muy presente que se requiere el reconocimiento de la problemática particular de cada uno de los tramos en la cuenca, por eso nosotros con la CAR, tenemos que seguir apoyando los procesos de reforestación en la cuenca alta, tenemos que adelantar la tarea que ya tiene el aval del Gobierno nacional, aval quiere decir la firma para pagar en el momento que no lo haga la Corporación, pero el compromiso es que mediante un crédito de 50 millones de dólares la CAR y todos y cada uno de los municipios de la cuenca puedan tener su planta de tratamiento de aguas residuales.

De manera que a la reforestación y a la adopción de mecanismos de producción limpia en la cuenca alta le podamos agregar la instalación de estas plantas en los distintos municipios. Tenemos con nuestros compañeros de gobierno, con la Superintendencia de Servicios Públicos, la tarea conjunta de resolver una dificultad que hoy se ha planteado en virtud de una interpretación que tiene origen en el Consejo de Estado, para que los municipios puedan avanzar en sus plantas de tratamiento de aguas residuales.

Hemos logrado un acuerdo yo diría histórico en Bogotá y yo quiero hacerle hoy un reconocimiento en su ausencia al señor alcalde Luis Eduardo Garzón, porque con su apoyo logramos que finalmente se cambiara una doctrina que estaba generando una dificultad al interior de la CAR. Con el apoyo enorme del doctor Edgar Ruiz, de la entonces directora de la CAR, Gloria Álvarez, se pudo firmar el entendimiento para que los recursos que estaban a disposición para completar las obras de descontaminación en Bogotá se pudieran aplicar.

Hay que decirlo con franqueza: hemos votado billones de pesos rebuilendo aguas en Bogotá en una planta inconclusa en el Salitre. La planta de tratamiento de aguas residuales del Salitre que no se terminó, les ha costado a los bogotanos y a los colombianos una suma astronómica con mínimo efecto porque no se terminaron las obras y no se adelantaron las obras necesarias para generar el impacto integral en la ciudad de Bogotá.

Este acuerdo busca terminar la planta de tratamiento de aguas residuales del Salitre. Hacer la segunda fase del Salitre, contempla los esfuerzos que viene adelantando el Distrito Capital frente a los interceptores del Fucha, del Salitre y del Tunjuelo. Contempla las estaciones elevadoras que no se habían presupuestado inicialmente. Tenemos que hacer el cierre financiero de la PTAR Canoas. Tenemos unos instrumentos en el plan de desarrollo, requerimos del concurso del Distrito Capital y hay una voluntad del Alcalde y del señor Edgar Ruiz. Hay una decisión de la Nación de hacer la planta de tratamiento de Canoas y tenemos un instrumento que se ha desarrollado en el plan de desarrollo que son los planes departamentales de agua potable y saneamiento básico, que nos van a permitir, señor Contralor, orientar recursos para este propósito.

Yo quiero aquí recoger sus palabras, Señor Contralor, porque todo este modelo tiene que ser sobre la base de que los colombianos podamos ver la ejecución de las obras y que no se roben la plata. Tiene usted razón en que no ha existido en el pasado una visión integral del manejo de la cuenca, muy importante fue que en

LA ASOCIACIÓN DE CURTIDORES DE VILLAPINZÓN " ACURTIR"
NIT 83.203.853-8
CERTIFICA

Que gracias a la intervención, desde el año 2004, de la Doctora Mónica Sanz identificada con cédula de ciudadanía número 39.683.812 de Bogotá, en el marco de la realización de su tesis doctoral, aprendimos en resolución de conflictos incorporando estos principios a la realización de nuestras actividades diarias; tanto en la relación con nuestros vecinos y colegas, como con la instituciones gubernamentales entre ellas: la Alcaldía Municipal, la Gobernación y la Corporación Autónoma Regional de Cundinamarca CAR (autoridad ambiental), en trabajo conjunto con instituciones académicas (DELF IHE de Holanda, Universidad Nacional de Colombia y Universidad de Los Andes – Colombia)

Durante estos años la gestión realizada por la Doctora Mónica fue de vital importancia, y se obtuvo como resultados:

1. La consolidación de nuestra Asociación ACURTIR, con estatutos y personeria jurídica.
2. Participación en Consejos Comunales con el presidente de la Republica de Colombia, así como reuniones con delegados del presidente y el congreso de la republica para establecer acuerdos en la problemática ambiental y juridica de los curtidores de Villapinzon con la CAR.
3. Establecer confianza para el trabajo conjunto con las autoridades ambientales y gubernamentales para obtener el Plan de Ordenamiento Territorial, que define el uso del suelo y da las directrices para el desarrollo de la actividad industrial,
4. La participación activa de ACURTIR en la Mesa Nacional del Cuero dirigida por el Ministerio del Medio Ambiente, en donde se establecen las directrices normativas en materia ambiental que rigen al sector curtidor.
5. Se observo una celeridad en los tramites que presentaban los curtidores ante la autoridad ambiental CAR.
6. La obtención de 7 concesiones de uso de agua para las curtiembres,
7. Se realizaron 64 planes de manejo ambiental con el apoyo de la Cámara de Comercio de Bogotá, lo que permitió legalizar la actividad comercial de estos curtidores y como consecuencia detener los procesos sancionatorios y penales que se venian adelantando por parte de la CAR y la Fiscalía.
8. Obtener recursos de entidades nacionales e internacionales como La Cámara de Comercio de Bogotá, La Comunidad Económica Europea, COLCIENCIAS y la CAR, para realizar proyectos de capacitación, producción limpia, gestión de residuos, que permite a los curtidores realizar mejor su actividad comercial sin dañar el entorno.
9. Genero un cambio cultural en los empresarios para minimizar sus vertimientos con mejores prácticas operativas y la construcción de aproximadamente 40 plantas de tratamiento en las diferentes industrias que cumplen con los requerimientos del uso del suelo.
10. Se gestiono el reconocimiento de los curtidores de ronda de rio como propietarios para evitar su desalojo y establecer acuerdo y/o programas para su relocalización mediante la compra de sus predios
11. Se lograron acuerdo asociativos entre los curtidores, como es la compra de un terreno para el manejo comunitario de lo residuos generados por las curtiembres para minimizar costos de disposición con terceros y/o daño al medio ambiente por su mala gestión.

12. La obtención por parte de la presidencia de ACURTIR (EVIDALIA FERNÁNDEZ) del reconocimiento entregado por la revista Semana a los mejores lideres del país en el año 2011.

Todos estos logros, y la transformación que ha tenido la zona al igual que el cambio de imagen del Sector de curtiembres de Villapinzón y Choconta ante el país y en especial ante las autoridad gubernamentales, así como el reconocimiento internacional y el nacimiento de nuevos lideres dentro del sector industrial; es gracias a la entrega, compromiso, calidad humana, liderazgo, apoyo, capacidad de escuchar, acompañamiento y gestión del Doctora Mónica.

No nos queda mas que agradecimiento por todos los conocimientos transmitidos para gestionar y resolver los conflictos de una manera ordenada, pacifica y asociativa. Pedirle a Dios que la siga bendiciendo a ella y todos los suyos; igualmente que Dios permita que nos siga acompañando para contar con su apoyo y dirección para dar continuidad en el trabajo articulado con las instituciones académicas y gubernamentales para lograr la sostenibilidad y trabajo con responsabilidad social de todas las curtiembres de Villapinzon y Choconta en armonía con nuestro Rio Bogotá.

En constancia se firma a los 5 días del mes de junio de 2012

JUNTA DIRECTIVA DE ACURTIR

EVIDALIA FERNÁNDEZ
Presidenta

JUAN CARLOS CALDERÓN
vicepresidente

PAULINO MONROY

ROSENDO LÓPEZ

ANANÍAS LÓPEZ CASTRO

ANTONIO CONTRERAS

Líderes
AMBIENTALES

Cueros que no contaminan

Al sector de las curtiembres le ha tocado cargar con la "mala fama" de ser uno de los más contaminantes. Lo que pocos saben es que hay personas que vienen desarrollando un trabajo importante para que los empresarios mejoren sus prácticas de producción y de esta manera impacten menos a la madre naturaleza.

Una de ellas es Evidalia Fernández, presidenta de la Asociación de Curtidores de Villapinzón y Chocontá, quien con el apoyo de profesionales como la bióloga Mónica Sáenz, trabaja con 150 microempresarios de estos municipios. La idea es enseñar a estas personas técnicas para una producción más limpia, algo que también ha mejorado las relaciones interpersonales de la comunidad.

Con el tiempo 'el enano' se creció al punto de lograr el apoyo de la Unión Europea bajo el proyecto Switch, que busca encontrar solución a los problemas causados en el agua del Río Bogotá por los vertimientos de esta industria. Los resultados dan cuenta de una reducción de 95% de los residuos, 60% de la reconversión de aguas y el aprovechamiento de grasas para abonos y creación de nuevos productos.

Ingenio carioca

Alessandro Carlucci es un brasilero que ha pasado los últimos 20 años de su vida trabajando para la multinacional Natura, empresa dedicada a la fabricación y comercialización de cosméticos naturales y que hace poco entró a Colombia clasificada como una pyme. Carlucci es un ingeniero comercial especializado en Kellogg School of Management y es el responsable de que en las últimas dos décadas Natura genere 6.000 empleos directos en Brasil y Latinoamérica, y según cuentas del mismo Carlucci, factura al año 2.000 millones de dólares.

Esta empresa de origen brasilero se caracteriza por la calidad de sus productos y el cuidado que rige en su cadena de producción. Su línea Ekos, por ejemplo, se produce con base en activos de frutos naturales que se encuentran en la Amazonia, y su publicidad siempre usa frases que invita al individuo a conectarse consigo mismo y con la naturaleza.

Sus empaques también tienen componentes reciclados y una tabla ambiental en donde se aprecia los productos biodiversos que usa. En Colombia ya obtuvo el Premio de Responsabilidad Ambiental Fundación Siembra Colombia. Su objetivo es transformar retos socio ambientales, en oportunidades de negocio.

ISSN 1794 - 7456

Colombia ▪ Febrero - Marzo 2010 / año 6 ▪ www.misionpyme.com

Edición 32 ▪ Precio $7.500

REVISTA

EN LA MIRA
Diseños amigos del planeta

RANKING
Tecnología

+MisiônPyme

SU HERRAMIENTA DE GESTIÓN EMPRESARIAL

25
Líderes Pyme

Hilda Bernal
CB-Group

Alfredo Gómez
Dismet S.A.

Evidalia Fernández
Acurtir

Héroes
con empresa

Un café con
Harry Sasson

CARA A CARA
Academia vs. Sena

Pymes
en la era tecnológica

ESPECIAL
Pensiones y cesantías 'estrenan' look

LOS MEJORES **LÍDERES DE COLOMBIA**

Evidalia Fernández | Líder comunitaria

◄ EL PODER DE LA PALABRA

Al principio, Evidalia Fernández no hablaba mucho. Pero en 2004, cuando los curtidores artesanales de cueros de Villapinzón, un pequeño municipio en las afueras de Bogotá, estaban en el cénit de un conflicto con las autoridades ambientales, que estaban a punto de cerrar sus negocios y dejar sin sustento a cientos de familias, esta mujer empezó a mostrar sus dotes de líder. Se propuso tocar las puertas de las autoridades y la comunidad internacional, para explicar que su gremio estaba dispuesto a usar métodos limpios para garantizar un menor impacto ambiental. Disminuyeron en 90 por ciento el uso de sustancias químicas como el sulfuro y el cromo, aprendieron a usar el pelo en la fabricación de abonos y a verter cada vez menos desechos, lo que ha evitado que 70 por ciento de estos lleguen al río y ha permitido ahorrar 70 por ciento de agua. Hoy 12 familias, lideradas por Evidalia, impulsan un plan piloto de producción limpia de cueros que tiene el apoyo de las entidades ambientales que años atrás querían cerrar sus negocios.

Foto: Juan Carlos Sierra

Científico | **Alberto Ospina Taborda**

EL AVANZADO DE LA CIENCIA ►

La suya ha sido una vida dedicada al avance y la innovación en ciencia y tecnología en el país, a través del estímulo a la educación y la investigación. De origen humilde, aprovechó su paso por la Armada para especializarse en Ingeniería Eléctrica, Electrónica y de Telecomunicaciones. En 1958 trajo a Colombia nuevas estrategias para la enseñanza de la Física. Diez años más tarde creó el Fondo Colombiano de Investigaciones Científicas y Proyectos Especiales Francisco José de Caldas (Colciencias) y fue su primer director; dirigió la Fundación para el Fomento Educativo y el Avance de la Ciencia y la Tecnología en Colombia y aprovechó su vinculación con universidades de Estados Unidos para establecer el fondo de becas Harvard Colombia, que apoya a colombianos de escasos recursos para que cursen estudios de posgrado en esa universidad.

Foto: Daniel Reina

Luis Humberto Soriano | Maestro

◄ EL JUGLAR DE LOS LIBROS

Una idea sencilla le dio al tema de la lectura una relevancia que muy pocas veces había tenido en el país. Luis Humberto Soriano, maestro del corregimiento de La Gloria, en Nueva Granada, Magdalena, y licenciado en Lengua Castellana, revivió la tradición de los juglares vallenatos. Con un burro y setenta libros de su colección personal, comenzó a recorrer los pueblos y veredas del departamento para compartir la lectura con quienes no tienen acceso a ella. Fue tan llamativo y novedoso su esfuerzo que en 2004 fue declarado colombiano ejemplar y diversos medios colombianos, así como *The New York Times* y la cadena CNN destacaron su labor. El proyecto tuvo tanto impacto que se ha convertido en el símbolo del Plan Nacional de Lecturas y Bibliotecas del Ministerio de Cultura. Su iniciativa ya ha sido replicada en la Sierra Nevada de Santa Marta y es uno de los protagonistas de la última edición de la revista *Colours*, de Benetton, dedicada a los superhéroes latinoamericanos.

Foto: Juan Carlos Sierra

trayectoria. Porque así como un líder es el cuerpo prestado del ímpetu de una causa superior, la sociedad también se ve encarnada en su lucha y de esa manera le da vida y lo hace brillar.

Las sociedades necesitan símbolos y referentes para romper el *status quo* y ejercer su poder transformador. Las naciones, para avanzar, deben elevar sus causas, volverlas visibles para que puedan ser apropiadas por la gente. Sea la causa que sea: la protección de las víctimas, la recuperación de la memoria, la defensa de los derechos de la mujer o la lucha contra el crimen o la corrupción, de tal forma que todos la podamos criticar o defender, y resolvamos nuestras diferencias con la razón en un debate democrático y no con el maniqueísmo y la violencia, como ha ocurrido en Colombia durante tantos años.

Por eso es esencial resaltar a nuestros mejores líderes públicos: porque son

un ejemplo y queremos que ese ejemplo ayude a iluminar a los demás colombianos. Para que los inspire, los sensibilice y los motive. Para que ayuden a despejar esa niebla de nihilismo y desconfianza que se ha venido apoderando de la esfera pública colombiana. Los escándalos de corrupción, la cultura del dinero fácil, la falta de visión y de grandeza, la indiferencia de la sociedad o la sensación de parálisis del Estado han venido colonizando el imaginario de la Nación, donde en la mente de muchos colombianos se ha gestado una especie de Leviatán, y ahí el hombre es un lobo para el hombre.

Pero los finalistas y ganadores de este premio son la demostración de que Colombia se parece más a *El contrato Social* de Rousseau que al *Leviatán* de Tomás Hobbes. Más a los esfuerzos de una sociedad por salir adelante que a los intereses, ambiciones y odios que la

dividen y la anulan. Este premio es la muestra más palpable de la modernización del país. Los ganadores ejercen un liderazgo que está vinculado a las ideas y a que son capaces de mover a la sociedad con una visión universal. Muy distinto, dentro de las categorías de liderazgo de Max Weber, al liderazgo tradicional o al liderazgo carismático, más propios, según él, de las sociedades premodernas.

Con estos finalistas uno ve hombres y mujeres que mueven a la sociedad hacia un mundo mejor. Es el retrato de una sociedad que quiere modernizarse. Los diez ganadores tienen un común denominador: el conocimiento. Los hacen líderes sus ideas, no el poder de sus cargos. No son líderes por su poder, sino que son influyentes por su ejemplo, su conocimiento y su autoridad moral.

La historia ya no la hacen los llaneros solitarios. En un mundo

globaliz
son cole
liderazg
pregunt
son sol
o una ti

L
natural
y socie
avasall
como u
esta lo

Telefónica

globalizado y de redes sociales las acciones son colectivas. Y estos líderes muestran un liderazgo diverso y un país plural. La gran pregunta es si estos hombres y mujeres son solo una fotografía de la coyuntura o una tendencia de largo plazo.

Lo importante, en el fondo, es la naturaleza de la dialéctica entre líderes y sociedad. Que el líder no termine avasallando a la sociedad y guiándola como un rebaño, sino permitiendo que esta lo supere.

Carlos Marx, el gran crítico del capitalismo, escribió: "*Los filósofos no se han cansado de interpretar el mundo de diferentes maneras, pero de lo que se trata es de transformarlo*". Y esto es lo que están haciendo estos líderes, cada uno desde su espacio: ayudando a transformar a Colombia. ▪

Vea video de los ganadores en
Semana.com

Telefónica

MUNDO: LIBIA SIN GADAFI

Semana

INFORMACIÓN DE COLOMBIA Y DEL MUNDO · 30 DE AGOSTO AL 5 DE SEPTIEMBRE · EDICIÓN n.º 1530 · semana.com

Carmen Palencia
Líder del movimiento de víctimas

Óscar Naranjo
Director de la Policía Nacional

Viviane Morales
Fiscal general de la Nación

Alejandro Reyes
Gestor de restitución de tierras

LOS MEJORES LÍDERES DE COLOMBIA

Estos son los ganadores del primer gran premio de liderazgo realizado en el país.

Francisco de Roux
Gestor de paz

Moisés Wasserman
Educador

Mónica Roa
Activista

Daniel Coronell
Periodista

José Alejandro Cortés
Empresario

Juanita León
Periodista

ACTA DE REUNIÓN
SECTOR DE CURTIEMBRES VILLAPINZÓN – CHOCONTA

FECHA DE REUNIÓN: **04 de noviembre de 2004**

PARTICIPANTES:

Mónica Sanz	UNESCO-IHE
Evidalia Fernández	Asociación de Curtidores de Villapinzón-Choconta
Nancy Patricia Gutiérrez	Congreso de la Republica
Edgar Erazo	CAR
Mauricio Bayona Pulido	CAR
Marlene Salazar	CAR
Luis F. Martínez	Cámara de Comercio
Julián Martínez Sánchez	Alcaldía de Villapinzón
Alejandro Gutierrez	Alcaldía de Choconta
Gloria Lucia Álvarez	CAR
Camilo Ospina	Presidencia de la Republica
Liliana Gaitán	MAVDT
Diana Moreno	MAVDT

ANTECEDENTES: Dado el compromiso obtenido por el señor presidente en el concejo comunal, en relación con las curtiembres de Villapinzon y Choconta se adelanta la segunda reunión para determinar compromisos por parte de cada una de las entidades relacionadas con el tema para la solución de la problemática ambiental y social que enfrenta el sector de curtiembres de la región de Villapinzon y Chocontá, habiéndose radicado el proyecto el 19 de octubre y obtenido respuesta por parte de la CAR el 3 de noviembre. La fase I se acortó, de 6 meses a mes y medio.

DESARROLLO DE LA REUNIÓN: Se mostró un panorama sobre la localización de las curtiembres de Villapinzón y Choconta, en donde se mencionó que 60 de dichas curtiembres se encuentran en la ronda del río, para los cuales se solicitó por parte de los industriales una alternativa de compra de terrenos para reubicar dichas empresas, estudiando para lo anterior la información de 50 escrituras entregada por doña Evidalia Fernández, representante de las curtiembres de Villapinzón y Choconta. Por otra parte se manifestó la necesidad del Municipio de Villapinzón de tener la colaboración del IGAC en lo relacionado con la digilitalización de la cartografía para definir la situación de ubicación de las curtiembres en la Jurisdicción de Villapinzón.

En relación con el proyecto de producción más limpia, se solicitó por parte de la Cámara de Comercio información en lo relacionado con 65 estudios que reposan en los archivos de la CAR para el desarrollo el proyecto de producción más limpia en el sector de curtiembres en Villapinzón.

Así mismo, se mencionó que las actividades del proyecto estarían enfocadas a dar soluciones ambientales, empresariales y de asociatividad para el sector, en este sentido se realizaría la capacitación en producción más limpia y en temas empresariales a 100 curtiembres, dado la imposibilidad de avanzar en este tema con los industriales ubicados en la ronda y con 60 industriales que se encuentran fuera de la ronda se trabajaría en un diagnóstico de dichas empresas.

En esta oportunidad el Ministerio de Ambiente Vivienda y Desarrollo Territorial, ofreció apoyo económico para el desarrollo de la segunda etapa del proyecto de Producción Más Limpia, en donde dichos recursos son provenientes del Programa de Apoyo al Sistema Nacional Ambiental –SINA II, en donde para acceder a estos recursos se participa en una convocatoria que realiza el Ministerio, este proceso debe ser acompañado por la Corporación Autónoma Regional. Ante este ofrecimiento la CAR mencionó que ellos tenían priorizado otros temas como el de reforestación pero que en la segunda convocatoria podría ser apoyado.

Sobre las medidas de cierre definitivo y temporal, aclaró la CAR que estas medidas seguirían el proceso administrativo sin parar hasta tanto no se tuvieran resultados concretos, sin embargo y dado que los procesos administrativos de las entidades de gobierno son más demorados que los del sector privado, esto daría tiempo para acometer el proyecto.

COMPROMISOS: Una vez discutidos los aspectos antes mencionados, se establecieron los siguientes compromisos:

1. La Gobernación de Cundinamarca se comprometio a analizar la posibilidad de compra los terrenos de la ronda del río, para permitir la posibilidad de reubicación de las curtiembres afectadas.
2. El IGAC se comprometio a entregar el 31 de diciembre de 2004 la digilitalización de la cartografía para que el Municipio pudiera adelantar su trabajo en la definición de la zona de desarrollo de la industria del sector de curtido.
3. Compromiso de la CAR de facilitar cualquier información que requiera la Cámara de Comercio para el desarrollo del proyecto.

ACTA DE REUNIÓN
SECTOR DE CURTIEMBRES VILLAPINZÓN – CHOCONTA

FECHA DE REUNIÓN: **14 de Octubre 2004**

PARTICIPANTES:

Mónica Sanz	UNESCO-IHE
Evidalia Fernández	Asociación de Curtidores de Villapinzón-Choconta
Nelly Yolanda Villamizar	Tribunal administrativo Bogotá y Cundinamarca
Edgar Erazo	CAR
Marlene Salazar	CAR
Adriana Guillén	Procuraduría Ambiental y Agraria
	Alcalde de Chocontá
Jorge Garzón	Alcaldía de Villapinzón
Martha Alonso	Procuraduría Ambiental y Agraria
Liza Paola Grueso	Gobernación de Cundinamarca
	Alcalde de Cogua
I	
Camilo Ospina	Presidencia de la República
Liliana Gaitán	MAVDT

Como resultado del concejo comunal del 2 de Octubre se llevó a cabo la primera reunión sobre el conflicto de Villapinzón y Chocontá el día 14 de Octubre.

La CAR planteó que los términos legales estaban vencidos y que la comunidad había sido renuente a implementar sus cambios de descontaminación.
UNESCO-IHE planteó por su parte que se encontraba en un proceso constructivo de resolución de conflictos desde marzo de 2004 y que ofrecía un proyecto que podría dar solución a toda la región. Manifestó haber radicado una propuesta en junio y que en la actualidad el proyecto contaba con la financiación de la Cámara de Comercio de Bogotá y que se basaba en el modelo del Cerrito en el Valle del Cauca.
Las entidades de control plantearon la necesidad de dar mayor celeridad al proyecto. Plantearon que la fase I debía hacerse en menos tiempo.

Compromisos

La CCB estudiaría los tiempos y cronogramas y radicaría nueva versión de dicho proyecto.

La Corporación respondería ante de la siguiente reunión definida para el 4 de noviembre.

ACTA DE REUNIÓN
SECTOR DE CURTIEMBRES VILLAPINZÓN – CHOCONTA

FECHA DE REUNIÓN: **25 de enero de 2005**

PARTICIPANTES:

Mónica Sanz	UNESCO-IHE
Evidalia Fernández	Asociación de Curtidores de Villapinzón-Choconta
Nancy Patricia Gutiérrez	Congreso de la República
Edgar Erazo	CAR
Marlene Salazar	CAR
Luis.F Martínez	Cámara de Comercio de Bogotá
Mauricio Molina	Cámara de Comercio de Bogotá
Jorge Garzón	Alcaldía de Villapinzón
Martha Alonso	Procuraduría Ambiental y Agraria
Liza Paola Grueso	Gobernación de Cundinamarca
Lucia Rueda Villamizar	Fiscalia
Itala Rodríguez	Presidencia de la República
Camilo Ospina	Presidencia de la República
Liliana Gaitán	MAVDT
Diana Moreno	MAVDT

No pones al resto de curtidores, ni al defensor del pueblo?

ANTECEDENTES: En la reunión del pasado 04 de noviembre de 2004 se adquirieron compromisos por parte de las diferentes entidades involucradas en el proceso de las curtiembres de Villapinzón en tal sentido se dió la oportunidad de conocer sobre los avances de dichos compromisos y aclarar el panorama de requerimientos por parte de la fiscalia, procuraduría y CAR, dada la condición de cierre de la industria de curtido en el Municipio de Villapinzón que se ha presentado en los últimos días

DESARROLLO DE LA REUNIÓN: Se hizo un recuento de la gestión realizada desde la primera reunión, en donde se mencionaron temas como: la fechas de inicio del proyecto, las reuniones realizadas en Villapinzón para dar información de los cierres de las curtiembres, sobre la entrega de informes del proyecto.

Se resaltó la participación decidida de la comunidad, al pagar las caracterizaciones de la primera fase y al haber trabajado en la definición de la concesión de aguas.

En esta reunión se preciso por parte de la Dra. Marta Alonso delegada de la Procuraduría, sobre la no respuesta por parte de los industriales del cuero de Villapinzón ante los requerimientos de la CAR. Por otra parte la Fiscalia menciona la existencia de algunas contradicciones en las resoluciones ya elaboradas por la CAR que de alguna manera recargan el trabajo de esta entidad. este punto es importante pero no copie con exactitud.

Se menciono igualmente que la sanción económica más alta asciende al valor de $ 1.500.000 pesos. Diana no entiendo esto puesto que ellos tienen 100 salarios mínimos.

La CAR, hablo sobre el tema de los cierres temporales y definitivos, ante lo cual se menciono que los cierres temporales hasta tanto no lleguen a la CAR resultados de las Medidas solicitadas por la Corporación al Industrial, como permiso de concesión este ya se hizo, implementación de tratamiento primario, permisos de vertimiento, y/o Planes de Manejo Ambiental; estas medidas no podrían ser levantadas; en cuanto a los cierres definitivos, necesariamente se debería dar la reubicación de las curtiembres que se encuentran en la ronda del río.
Al respecto la Dra Nancy Patricia Gutiérrez planteó abrir el espacio para una solución política dónde se levante el cierre ante la aprobación de la fase II para una buena ejecución del proyecto y dónde los la ronda puedan trabajar en procesos limpios hasta que la Gobernación y el municipio determinen las acciones y terrenos apropiados para su reubicación.

En este espacio también se habló y se dio claridad sobre el contenido del proyecto de producción más limpia fase II que se esta desarrollando en Villapinzón, en el cual se solicitarán los permisos de vertimientos y elaboraran los Planes de Manejo Ambiental de las curtiembres que no lo han presentado y la modificación de los Planes existentes si así se requiere, de tal forma que algunos industriales podrán tramitar sus créditos para ir avanzando igualmente en algunas obras de tratamiento primario.

La Cámara de Comercio aclara que este proyecto contara con recursos y personal definido, se recalca además que para el adecuado desarrollo de este proyecto se necesitaba claridad en relación con las exigencias de la CAR para el sector. En este sentido la CAR menciona que los requerimientos establecidos se encuentran en cada uno de los actos administrativos emitidos.

Se pregunta a la CAR la posibilidad de suspender durante cuatro meses la medida de cierre al presentar el plan de trabajo del proyecto de producción más limpia que se desarrolla en Villapinzón por parte de la Cámara de Comercio, con el fin de poder desarrollar dicho proyecto en las curtiembres. Ante esto, la CAR y la Procuraduría no se encuentran de acuerdo, puesto que las medidas tomadas solo pueden ser levantadas por decisión entre procuraduría y CAR o con los resultados de los requerimientos exigidos por la autoridad ambiental.

En relación con el avalúo de los terrenos de la ronda, la Gobernación manifiesta que la comisión topográfica iniciaría trabajos el primero (1) de febrero, dado que el dinero se obtendría del 1% de los ingresos de los departamentos y municipios, en relación con la adquisición de áreas de interés para Acueductos Municipales, de acuerdo a los estipulado en la Ley 99 de 1993, art. 111, ante lo cual surgió un interrogante que no permitiría la compra de los terrenos de la ronda del río, puesto que de acuerdo a un comunicado del Ministerio de Ambiente, Vivienda y Desarrollo Territorial este 1% no podría ser utilizado para el caso de Villapinzón y Choconta, esta inquietud se planteo ser aclarada en este mismo día, con la Ministra de ambiente en su despacho.

Se aclaro igualmente que el IGAC cumplió el compromiso de entregar el 31 de diciembre la digitalización de la cartografía al Municipio de Villapinzon, pero que al detectar errores en las coordenadas estas fueron devueltas con el compromiso de que el IGAC devolviera las planchas ya corregidas el 31 de enero al municipio.

COMPROMISOS: Una vez discutidos los aspectos antes mencionados, se establecieron los siguientes compromisos:

1. La Gobernación de Cundinamarca se compromete a aclarar la inquietud sobre la destinación del 1% para el apoyo en la compra de los terrenos de la ronda del río.
2. Los funcionarios designados por la Corporación se comprometieron a consultar con la dirección y la jurídica la posibilidad de suspender temporalmente las medidas de cierre con la presentación del plan de trabajo del proyecto de producción más limpia segunda fase, por parte de la Cámara de Comercio, de forma tal que se pueda adelantar este proyecto.

3. La Cámara de Comercio se comprometio a radicar el plan de trabajo del proyecto de producción más limpia segunda fase el 31 de enero de 2005
4. La CAR se comprometio a tener el concepto de la evaluación de la propuesta presentada por la Cámara de Comercio el 7 de febrero.

T - #1010 - 101024 - C322 - 240/170/18 - PB - 9781138027695 - Gloss Lamination